T0257029

TAMING THE UNKNOWN

TAMING THE UNKNOWN

A History of Algebra from Antiquity
to the Early Twentieth Century

VICTOR J. KATZ
AND
KAREN HUNGER PARSHALL

PRINCETON UNIVERSITY PRESS
PRINCETON AND OXFORD

Copyright © 2014 by Princeton University Press
Published by Princeton University Press, 41 William Street,
Princeton, New Jersey 08540
In the United Kingdom: Princeton University Press, 6 Oxford Street,
Woodstock, Oxfordshire, OX20 1TW

Jacket images: Background images: Title page of *Discours de la method* by
René Descartes, from the Plimpton Collection of the Rare Book and Manuscript
Library, Columbia University. Problem on parabolas from *Elementa curvarum
linearum* by Jan de Witt, in the 1659 edition of Descartes's *Geometria*, from the David
Eugen Smith Collection of the Rare Book and Manuscript Library, Columbia
University. Mathematicians, left to right: Luca Pacioli, Pierre de Fermat, Évariste
Galois, James Joseph Sylvester, Georg Frobenius, Emmy Noether. Detail portrait of
the mathematician Luca Pacioli and an unknown young man by Jacopo de'Barbari.
Museo Nazionale di Capodimonte, Naples, Italy. © Scala/Ministero per I Beno e le
Attivitá culturali/Art Resources, NY. Image of Noether courtesy of Drs. Emiliana and
Monica Noether.

All Rights Reserved

Library of Congress Cataloging-in-Publication Data

Katz, Victor J.
Taming the unknown: history of algebra from antiquity to the
early twentieth century / Victor J. Katz and Karen Hunger Parshall.
pages cm
Includes bibliographical references and index.
ISBN 978-0-691-14905-9 (hardcover: alk. paper)
1. Algebra–History. I. Parshall, Karen Hunger, 1955- II. Title.
QA151.K38 2014
512–dc23
2013024074

British Library Cataloging-in-Publication Data is available

This book has been composed in ITC New Baskerville

press.princeton.edu

Typeset by S R Nova Pvt Ltd, Bangalore, India

2   4   6   8   10   9   7   5   3   1

We dedicate this book to our spouses,

PHYLLIS KATZ

and

BRIAN PARSHALL,

and to our parents,

BESS and MARTIN KATZ (in loving memory)

and

JEAN KAY "MIKE" and MAURICE HUNGER

Contents

Acknowledgments

The present book came about almost serendipitously. In 2006, the first author reviewed *Unknown Quantity: A Real and Imaginary History of Algebra*, a work on the history of algebra by John Derbyshire.[1] Although very well written, that book contained so many errors of fact and inadequacies of interpretation that it would have been a disservice to let it go unchallenged. In January of 2008, the first author heard the second author's invited hour lecture on the history of algebra at the Joint Mathematics Meetings in San Diego. Realizing that her ideas were similar to those he had expressed in his own invited lecture on the history of algebra at the International Congress of Mathematics Education (ICME) in Copenhagen in 2004, he asked if she would be interested in jointly writing a book on the history of algebra. Curiously, although we live only about two hours apart, we only finally met to outline our ideas and to agree to submit the project to the Princeton University Press when we both attended a meeting of the International Study Group on the Relations Between History and Pedagogy of Mathematics in Mexico City in July of 2008. The present book is the fruit of our ideas developed over long careers teaching and working on the history of mathematics and, in particular, on the history of algebra.

The first author would like to thank Clemency Montelle and John Hannah for inviting him to the University of Canterbury in Christchurch, New Zealand in the spring of 2011 to lecture on the history of algebra. He was able to develop some of his ideas during those lectures. In addition, both Clemency and John graciously read drafts of several of the chapters of the book and offered many suggestions for improvement. The first author is also grateful to Mogens Niss of Roskilde University, Denmark for inviting him to present a lecture at ICME 10 that gave him the chance further to organize his ideas on the history of algebra. His greatest debt of gratitude, however, is to his wife, Phyllis, for her encouragement over

---

[1] See *Science* 312 (2006), 1473–1474.

the years, for her many conversations at every hour of the day and night, and for everything else.

We embarked on this book project at the same moment that the second author began a three-year term as Associate Dean for the Social Sciences at the University of Virginia. She would like to thank both Meredith Woo, the Dean of the College and Graduate School of Arts and Sciences, for asking her to serve and all of her colleagues in the Dean's Office and in the Division of Social Sciences for their support as she wore the very distinct hats of administrator and researcher. Her special thanks go to Rick Myers and Nancy Bertram for their endless patience and sage advice as she learned the administrative ropes; to Cristina Della Coletta and Jim Galloway, the Associate Deans of Arts and Humanities and of Sciences, respectively, for their friendship and sense of common cause; and to Seth Matula for always reminding her when it was Thursday. Three years of Thursdays and a year-long sabbatical leave later, the book is finished.

The second author is also indebted to her friends and colleagues, Joe Kett in the Corcoran Department of History and Fred O'Bryant in the Brown Science and Engineering Library, who, although not mathematicians, seemingly never tired of listening to tales of the latest twists and turns in the research and writing that went into the book over what for her, at least, were restorative lunches; to Ivor Grattan-Guinness, Professor Emeritus at Middlesex University and doyen of the history of mathematics, who always challenged her with questions she had not thought to ask; and to her husband and colleague in the Department of Mathematics, Brian Parshall, who read and critiqued the book in manuscript with his algebraist's eye.

Both authors extend their heartiest thanks to Greg Hays, John Miller, and Tony Woodman in the Department of Classics and to Elizabeth Meyer in the Corcoran Department of History, all at the University of Virginia, for their help with questions involving both Latin and Greek; to Lew Purifoy and his coworkers in Interlibrary Loan at the University of Virginia for never failing to secure even the hardest-to-find sources; to Jennifer Lee of the Rare Book and Manuscript Division of the Columbia University Library for her assistance in finding most of the manuscript and book pages that serve as illustrations in the text; to Julie Riddleberger, the TEX guru of Virginia's Department of Mathematics, for cheerfully and masterfully transforming our often crude sketches into the myriad figures that now illuminate our text; and to the Press's readers, who gave us numerous valuable suggestions for improvement. Last, but certainly not least, we thank our editor, Vickie Kearn, and the staff at Princeton University Press: Quinn Fusting, Vickie's assistant; Dimitri Karetnikov, who worked on the figures; Carmina Alvarez, the designer for the internal book design and the jacket; Sara Lerner, the production editor; Alison Durham, the copyeditor; and Rob McCaleb, who drew the maps. All of them worked with us to see the book into hard covers.

Victor J. Katz
Silver Spring, MD

Karen Hunger Parshall
Charlottesville, VA

30 June 2013

TAMING THE UNKNOWN

1

Prelude: What Is Algebra?

What is algebra? It is a question to which a high school student will give one answer, a college student majoring in mathematics another, and a professor who teaches graduate courses and conducts algebraic research a third. The educated "layperson," on the other hand, might simply grimace while retorting, "Oh, I never did well in mathematics. Wasn't algebra all of that x and y stuff that I could never figure out?" This ostensibly simple question, then, apparently has a number of possible answers. What do the "experts" say?

On 18 April 2006, the National Mathematics Advisory Panel (NMAP) within the US Department of Education was established by executive order of then President George W. Bush to advise him, as well as the Secretary of Education, on means to "foster greater knowledge of and improved performance in mathematics among American students."[1] Among the panel's charges was to make recommendations on "the critical skills and skill progressions for students to acquire competence in algebra and readiness for higher levels of mathematics." Why should competence in algebra have been especially singled out?

When it issued its final report in March 2008, the panel stated that "a strong grounding in high school mathematics through Algebra II or higher correlates powerfully with access to college, graduation from college, and earning in the top quartile of income from employment."[2] Furthermore, it acknowledged that "although our students encounter difficulties with many aspects of mathematics, many observers of

---

[1] US Dept. of Education, 2008, p. 71. The next quotation is also found here.
[2] US Dept. of Education, 2008, p. xii. For the next two quotations, see pp. xiii and 16, respectively.

educational policy see Algebra as a central concern." The panel had thus sought to determine how best to prepare students for entry into algebra and, since algebra was of such concern, it had first to come to terms with the question, what is the essential content of school algebra? In answer, it identified the following as the major topics: symbols and expressions, linear equations, quadratic equations, functions, the algebra of polynomials, and combinatorics and finite probability. Of course, each of these topics encompasses several subtopics. For example, the "algebra of polynomials" includes complex numbers and operations, the fundamental theorem of algebra, and Pascal's triangle. Interestingly, the panel mentioned "logarithmic functions" and "trigonometric functions" under the topic of "functions" but made no explicit mention of analytic geometry except in the special case of graphs of quadratic functions. Although the details of the panel's list might prompt these and other quibbles, it nevertheless gives some idea of what high school students, in the United States at least, generally study—or should study—under the rubric of "algebra."

These topics, however, constitute "school algebra." What about algebra at the college level? Most courses entitled "college algebra" in the United States simply revisit the aforementioned topics, sometimes going into slightly greater depth than is expected in high school. Courses for mathematics majors—entitled "modern algebra" or "abstract algebra"—are quite another matter, however. They embrace totally different topics: groups, rings, fields, and, often, Galois theory. Sometimes such courses also include vectors, matrices, determinants, and algebras (where the latter is a technical term quite different from the broad subject under consideration here).

And then there is algebra at the graduate and research levels. Graduate students may take courses in commutative or noncommutative algebra, representation theory, or Lie theory, while research mathematicians styled "algebraists" may deal with topics like "homological functors on modules," "algebraic coding theory," "regular local rings," or any one of hundreds of topics listed in the American Mathematical Society's "Mathematics Subject Classification." How do all of these subjects at all of these levels of sophistication fit together to constitute something called "algebra"? Before addressing this question, we might first ask why we need *this* book about it?

WHY THIS BOOK?

To be sure, the historical literature already includes several more or less widely ranging books on the history of algebra that are targeted, like the present book, at those with a background equivalent to a college major in mathematics;[3] a recent "popular" book assumes even less in the way of mathematical prerequisites.[4] Most in the former group, however, are limited either in the eras covered or in geographical reach, while that in the latter has too many errors of fact and interpretation to stand unchallenged. *This* book thus grew out of a shared realization that the time was ripe for a history of algebra that told the broader story by incorporating new scholarship on the diverse regions within which algebraic thought developed and by tracing the major themes into the early twentieth century with the advent of the so-called "modern algebra."

We also believe that this is a story very much worth telling, since it is a history very much worth knowing. Using the history of algebra, teachers of the subject, either at the school or at the college level, can increase students' overall understanding of the material. The "logical" development so prevalent in our textbooks is often sterile because it explains neither why people were interested in a particular algebraic topic in the first place nor why our students should be interested in that topic today. History, on the other hand, often demonstrates the reasons for both. With an understanding of the historical development of algebra, moreover, teachers can better impart to their students an appreciation that algebra is not arbitrary, that it is not created "full-blown" by fiat. Rather, it develops at the hands of people who need to solve vital problems, problems the solutions of which merit understanding. Algebra has been and is being created in many areas of the world, with the same solution often appearing in disparate times and places.

And this is neither a story nor a history limited to school students and their teachers. College-level mathematics students and their

---

[3] In fact, the prerequisites for reading the first ten chapters are little more than a solid high school mathematics education. The more general histories of algebra include van der Waerden, 1985; Scholz, 1990; Bashmakova and Smirnova, 2000; and Cooke, 2008, while the more targeted include Nový, 1973; Sesiano, 1999 and 2009; Kleiner, 2007; and Stedall, 2011.

[4] Derbyshire, 2006.

professors should also know the roots of the algebra they study. With an understanding of the historical development of the field, professors can stimulate their students to master often complex notions by motivating the material through the historical questions that prompted its development. In absorbing the idea, moreover, that people struggled with many important mathematical ideas before finding their solutions, that they frequently could not solve problems entirely, and that they consciously left them for their successors to explore, students can better appreciate the mathematical endeavor and its shared purpose. To paraphrase the great seventeenth- and early eighteenth-century English mathematician and natural philosopher, Sir Isaac Newton, mathematicians have always seen farther by "standing on the shoulders" of those who came before them.

One of our goals in the present book is thus to show how—in often convoluted historical twists and turns—the deeper and deeper consideration of some of the earliest algebraic topics—those generally covered in schools—ultimately led mathematicians to discover or invent the ideas that constitute much of the algebra studied by advanced college-level students. And, although the prerequisites assumed of our readers limit our exploration of the development of the more advanced algebraic topics encountered at the graduate and research levels, we provide at least a glimpse of the origins of some of those more advanced topics in the book's final chapters.

SETTING AND EXAMINING THE HISTORICAL PARAMETERS

Nearly five decades before the National Mathematics Advisory Panel issued its report, historian of mathematics, Michael Mahoney, gave a more abstract definition of algebra, or, as he termed it, the "algebraic mode of thought":

> What should be understood as the "algebraic mode of thought"? It has three main characteristics: first, this mode of thought is characterized by the use of an operative symbolism, that is, a symbolism that not only abbreviates words but represents the workings of the combinatory operations, or, in other words, a symbolism with which one operates. Second, precisely because of the central role of combinatory operations, the algebraic mode of thought deals

with mathematical relations rather than objects. Third, the algebraic mode of thought is free of ontological commitment....In particular, this mode of thought is free of the intuitive ontology of the physical world. Concepts like "space," "dimension," and even "number" are understood in a purely mathematical sense, without reference to their physical interpretation.[5]

Interestingly, Mahoney's first characteristic of algebraic thought as an "operative symbolism"—as well as the discussion of symbolism—is the first of the topics mentioned in the NMAP report. If, however, we believed that an operative symbolism is a necessary characteristic of algebra, this book would not begin before the seventeenth century since, before that time, mathematics was generally carried out in words. Here, we shall argue that symbolism is *not* necessary for algebra, although it has certainly come to characterize it—and, indeed, all of mathematics—over the past three centuries. We shall also argue that, initially, algebra dealt with objects rather than relations and that the beginnings of algebra actually *required* physical interpretations.

The roots of algebra go back thousands of years, as we shall see in the next chapter, but the two earliest texts that serve to define a subject of algebra are the *Arithmetica* of Diophantus (third century CE) and *The Compendious Book on the Calculation of al-Jabr and al-Muqābala* of al-Khwārizmī (ninth century CE). Although neither of these books required physical interpretations of the problems they presented, they did deal with objects rather than relations and neither used any operative symbolism. However, as we shall see below, al-Khwārizmī's book in particular was on the cusp of the change from "physical interpretations" to "abstract number" in the development of algebra. And, although the term "algebra" is absent from the texts both of Diophantus and al-Khwārizmī, it is clear that their major goal was to find unknown numbers that were determined by their relationship to already known numbers, that is, in modern terminology, to solve equations. This is also one of the goals listed in the NMAP report, so it would be difficult to deny that these works exhibit "algebraic thought." Thus, in order to study algebra historically, we need a definition of it somewhat different from that of Mahoney, which applies only to the algebra of the past three centuries.

---

[5] Mahoney, 1971, pp. 1–2.

It is interesting that school algebra texts today do not even attempt to define their subject. In the eighteenth and nineteenth centuries, however, textbook writers had no such compunction. The standard definition, in fact, was one given by Leonhard Euler in his 1770 textbook, *Elements of Algebra*. Algebra, for Euler, was "the science which teaches how to determine unknown quantities by means of those that are known."[6] He thus articulated explicitly what most of his predecessors had implicitly taken as the meaning of their subject, and we follow his lead here in adopting his definition, at least in the initial stages of this book when we explore how "determining unknowns" was accomplished in different times and places.

Now, there is no denying that, taken literally, Euler's definition of algebra is vague. It is, for example, not immediately clear what constitutes the "quantity" to be determined. Certainly, a "number" is a quantity—however one may define "number"—but is a line segment a "quantity"? Is a vector? Euler was actually clear on this point. "In algebra," he wrote, "we consider only numbers, which represent quantities, without regarding the different kinds of quantity."[7] So, unless a line segment were somehow measured and thus represented by a number, Euler would not have considered it a legitimate unknown of an algebraic equation. Given, however, the close relationship between geometry and what was to evolve into algebra, we would be remiss here not to include line segments as possible unknowns in an equation, regardless of how they may be described, or line segments and areas as "knowns," even if they are not measured. By the time our story has progressed into the nineteenth century, moreover, we shall see that the broadening of the mathematical horizon will make it necessary also to consider vectors, matrices, and other types of mathematical objects as unknowns in an equation.

Besides being vague, Euler's definition, taken literally, is also quite broad. It encompasses what we generally think of as "arithmetic," since the sum of 18 and 43 can be thought of as an "unknown" that can be expressed by the modern equation $x = 18 + 43$. To separate arithmetic from algebra, then, our historical analysis will generally be restricted to efforts to find unknowns that are linked to knowns in a more complicated way than just via an operation. This still leaves room for debate, however,

---

[6] Euler, 1770/1984, p. 186.
[7] Euler, 1770/1984, p. 2.

as to what actually constitutes an "algebraic" problem. In particular, some of the earliest questions in which unknowns are sought involve what we term proportion problems, that is, problems solved through a version of the "rule of three," namely, if $\frac{a}{b} = \frac{x}{c}$, then $x = \frac{ac}{b}$. These appear in texts from ancient Egypt but also from Mesopotamia, India, China, Islam, and early modern Europe. Such problems are even found, in geometric guise, in classical Greek mathematics. However, al-Khwārizmī and his successors generally did not consider proportion problems in discussing their own science of *al-jabr* and *al-muqābala*. Rather, they preferred to treat them as part of "arithmetic," that is, as a very basic part of the foundation of mathematical learning. In addition, such problems generally arose from real-world situations, and their solutions thus answered real-world questions. It would seem that in ancient times, even the solution of what we would call a linear equation in one variable was part of proportion theory, since such equations were frequently solved using "false position," a method clearly based on proportions. Originally, then, such equations fell outside the concern of algebra, even though they are very much part of algebra now.

Given these historical vagaries, it is perhaps easiest to trace the development of algebra through the search for solutions to what we call quadratic equations. In the "West"—which, for us, will include the modern-day Middle East as far as India in light of what we currently know about the transmission of mathematical thought—a four-stage process can be identified in the history of this part of algebra. The first, *geometric stage* goes back some four millennia to Mesopotamia, where the earliest examples of quadratic equations are geometric in the sense that they ask for the unknown length of a side of a rectangle, for example, given certain relations involving the sides and the area. In general, problems were solved through manipulations of squares and rectangles and in purely geometric terms. Still, Mesopotamian mathematicians were flexible enough to treat quadratic problems not originally set in a geometric context by translating them into their geometric terminology. Mesopotamian methods for solving quadratic problems were also reflected in Greek geometric algebra, whether or not the Greeks were aware of the original context, as well as in some of the earliest Islamic algebraic texts.

Al-Khwārizmī's work, however, marked a definite shift to what may be called the static, equation-solving, *algorithmic stage* of algebra. Although al-Khwārizmī and other Islamic authors justified their methods through geometry—either through Mesopotamian cut-and-paste geometry or through formal Greek geometry—they were interested not in finding *sides* of squares or rectangles but in finding *numbers* that satisfied certain conditions, numbers, in other words, that were not tied to any geometric object. The procedure for solving a quadratic equation for a number is, of course, the same as that for finding the side of a square, but the origin of a more recognizable algebra can be seen as coinciding with this change from the geometric to the algorithmic state, that is, from the quest for finding a geometric object to the search for just an unknown "thing." The solution of cubic equations followed the same path as that of the quadratics, moving from an original geometric stage, as seen initially in the writing of Archimedes (third century BCE) and then later in the work of various medieval Islamic mathematicians, into an algorithmic stage by the sixteenth century.

Interestingly, in India, there is no evidence of an evolution from a geometric stage to an algorithmic one, although the ancient Indians knew how to solve certain problems through the manipulation of squares and rectangles. The earliest written Indian sources that we have containing quadratic equations teach their solution via a version of the quadratic formula. In China, on the other hand, there is no evidence of either geometric or algorithmic reasoning in the solution of quadratic equations. All equations, of whatever degree above the first, were solved through approximation techniques. Still, both Indian and Chinese mathematicians developed numerical algorithms to solve other types of equations, especially indeterminate ones. One of our goals in this book is thus to highlight how each of these civilizations approached what we now classify as algebraic reasoning.

With the introduction of a flexible and operative symbolism in the late sixteenth and seventeenth century by François Viète, Thomas Harriot, René Descartes, and others, algebra entered yet another new stage. It no longer reflected the quest to find merely a numerical solution to an equation but expanded to include complete curves as represented by equations in *two* variables. This stage—marked by the appearance of analytic geometry—may be thought of as the *dynamic stage*, since studying curves as solutions of equations—now termed differential equations—arose in problems about motion.

New symbolism for representing curves also made it possible to translate the complicated geometric descriptions of conic sections that Apollonius had formulated in the third century BCE into brief symbolic equations. In that form, mathematics became increasingly democratic, that is, accessible for mastery to greater numbers of people. This was even true of solving static equations. The verbal solutions of complicated problems, as exemplified in the work of authors like the ninth-century Egyptian Abū Kāmil and the thirteenth-century Italian Leonardo of Pisa, were extremely difficult to follow, especially given that copies of their manuscripts frequently contained errors. The introduction of symbolism, with its relatively simple rules of operation, made it possible for more people to understand mathematics and thus, ultimately, for more mathematics to be created. It also provided a common language that, once adopted, damped regional differences in approach.[8]

Moreover, spurred by Cardano's publication in 1545 of the algorithmic solutions of cubic and quartic equations, the new symbolism enabled mathematicians to pursue the solution of equations of degree higher than four. That quest ultimately redirected algebra from the relatively concrete goal of finding solutions to equations to a more *abstract stage*, in which the study of structures—that is, sets with well-defined axioms for combining two elements—ultimately became paramount. In this changed algebraic environment, groups were introduced in the nineteenth century to aid in the determination of which equations of higher degree were, in fact, solvable by radicals, while determinants, vectors, and matrices were developed to further the study of systems of linear equations, especially when those systems had infinitely many solutions.

Complex numbers also arose initially as a result of efforts to understand the algorithm for solving cubic equations, but subsequently took on a life of their own. Mathematicians first realized that the complex numbers possessed virtually the same properties as the real numbers, namely, the properties of what became known as a field. This prompted the search for other such systems. Given fields of various types, then, it

---

[8] This is not to say that indigenous techniques and traditions did not persist. Owing to political and cultural mores, for example, Japan and China can be said to have largely maintained indigenous mathematical traditions through the nineteenth century. However, see Hsia, 2009 and Jami, 2012 for information on the introduction of European mathematics into China beginning in the late sixteenth century.

was only natural to look at the analogues of integers in those fields, a step that led ultimately to the notions of rings, modules, and ideals. In yet a different vein, mathematicians realized that complex numbers provided a way of multiplying vectors in the plane. This recognition motivated the nineteenth-century Irish mathematician, William Rowan Hamilton, to seek an analogous generalization for three-dimensional space. Although that problem proved insoluble, Hamilton's pursuits resulted in a four-dimensional system of "generalized numbers," the quaternions, in which the associative law of multiplication held but not the commutative law. Pushing this idea further, Hamilton's successors over the next century developed the even more general notion of algebras, that is, n-dimensional spaces with a natural multiplication.

At the close of the nineteenth century, the major textbooks continued to deem the solution of equations the chief goal of algebra, that is, its main defining characteristic as a mathematical subject matter. The various structures that had been developed were thus viewed as a means to that end. In the opening decades of the twentieth century, however, the hierarchy flipped. The work of the German mathematician, Emmy Noether, as well as her students and mathematical fellow travelers fundamentally reoriented algebra from the more particular and, in some sense, applied solution of equations to the more general and abstract study of structures per se. The textbook, *Moderne Algebra* (1930–1931), by one of those students, Bartel van der Waerden, became the manifesto for this new definition of algebra that has persisted into the twenty-first century.

THE TASK AT HAND

Here, we shall trace the evolution of the algebraic ideas sketched above, delving into some of the many intricacies of the historical record. We shall consider the context in which algebraic ideas developed in various civilizations and speculate, where records do not exist, as to the original reasoning of the developers. We shall see that some of the same ideas appeared repeatedly over time and place and wonder if there were means of transmission from one civilization to another that are currently invisible in the historical record. We shall also observe how mathematicians, once they found solutions to concrete problems, frequently generalized to situations well beyond the original question. Inquiries into these and other issues will allow us to reveal not only the historicity but also the complexity of trying to answer the question, what is algebra?, a question, as we shall see, with different answers for different people in different times and places.

2

Egypt and Mesopotamia

The earliest civilizations to have left written mathematical records—the Egyptian and the Mesopotamian—date back thousands of years. From both, we have original documents detailing mathematical calculations and mathematical problems, mostly designed to further the administration of the countries. Both also fostered scribes of a mathematical bent who carried out mathematical ideas well beyond the immediate necessity of solving a given problem. If mathematics was thus similarly institutionalized in Egypt and Mesopotamia, it nevertheless took on dramatically different forms, being written in entirely distinct ways in the two different regions. This key difference aside, the beginnings of algebra, as we defined it in the preceding chapter, are evident in the solutions of problems that have come down to us from scribes active in both of these ancient civilizations.

PROPORTIONS IN EGYPT

Although Egyptian civilization dates back some 7000 years, Egypt was only united under the pharaohs around 3100 BCE. Thought of as intermediaries between mortals and gods, the pharaohs fostered the development of Egypt's monumental architecture, including the pyramids and the great temples at Luxor and Karnak, and generally encouraged the development of civil society. Writing, which began in Egypt around the third millennium BCE, was one manifestation of those broader societal developments, but since much of the earliest writing concerned accounting, written number systems also developed. Two different styles of Egyptian writing—hieroglyphic writing for monumental inscriptions and hieratic, or cursive, writing, done with a brush and ink on papyrus—had associated with them two written number systems. The former system was a base-ten grouping system with separate symbols for 1, 10, 100, 1000, and so on. The latter, also base ten, was a ciphered system with different symbols for the numbers 1 through 10, then symbols for every multiple of 10 through 100, and so on. Except for the special case of $\frac{2}{3}$, fractions were all expressed as sums of unit fractions, so, for example, instead of writing $\frac{5}{6}$ as we would, the Egyptians wrote $\frac{1}{2} + \frac{1}{3}$. In what follows, we shall translate both systems of Egyptian numbers using ordinary Arabic numbers, but we shall express unit fractions of the form $\frac{1}{n}$ as $\bar{n}$.

Mathematics developed in at least two key settings in the Egyptian context: architecture and government. Relative to the former, numerous archaeological remains—in particular, scale-model drawings used by builders—demonstrate that mathematical techniques, particularly of proportionality, were used in both design and construction. In the civil context, since scribes supported a pharaoh's rule by dealing with the collection and distribution of goods, the calculation of the calendar, the levying of taxes, and the payment of wages, they likely developed many of the Egyptian mathematical techniques. Evidence for this appears not only in various papyrus documents—among them actual letters—but also in illustrations found on the walls of tombs of highly placed officials, where scribes are often depicted working together, engaged in tasks such as accounting for cattle or produce. Similarly, extant three-dimensional models represent scenes like the filling of granaries, which always involved a scribe to record quantities. From all these records, we deduce that the scribes of Egypt were invariably male.

Some of the scribes' tasks are also described in problems written in the only two surviving papyri that contain collections of mathematical problems and their solutions: the *Rhind Mathematical Papyrus*, named for the Scotsman Alexander H. Rhind (1833–1863) who purchased it at Luxor in 1858, and the *Moscow Mathematical Papyrus*, purchased in 1893 by Vladimir S. Golenishchev (1856–1947) who later sold it to the Moscow Museum of Fine Arts. The former was copied about 1650 BCE by the scribe A'h-mose from an original about 200 years older, while the latter also dates from the nineteenth century BCE. Unfortunately, although many papyri have survived due to the generally dry Egyptian climate, only a few short fragments of other original Egyptian mathematical papyri are still extant.

Besides the display of arithmetic techniques, the most important mathematical concept apparent in the two mathematical papyri is

Figure 2.1. Egyptian scribe, from the Louvre, Paris, courtesy Jose Ignacio Soto / Shutterstock.

proportionality. As noted in the opening chapter, this idea generally underlies the earliest problems introduced in school mathematics that require the determination of an unknown quantity, so the fact that problems involving proportionality are present on the Egyptian papyri makes these perhaps the simplest and earliest algebra problems in existence. To get a feel for the types of proportion problems that engaged the Egyptians, consider several examples from their mathematical papyri.

Problem 75 of the *Rhind Papyrus* asks for the number of loaves of *pesu* 30 that can be made from the same amount of flour as 155 loaves of *pesu* 20, where a *pesu* is an Egyptian measure that can be expressed as the ratio of the number of loaves to the number of *hekats* of grain and where a *hekat* is a dry measure approximately equal to $\frac{1}{8}$ bushel. The problem is thus to solve the proportion $x : 30 = 155 : 20$, which the scribe accomplished by first dividing 155 by 20 and then multiplying the result by 30 to get $232\frac{1}{2}$. Similar problems occur elsewhere in both the *Rhind* and the *Moscow* papyri.

A slightly more sophisticated use of proportions may be found in problem 7 of the *Moscow Papyrus*, where we are given a triangle of area 20 and a ratio of height to base of $2\frac{1}{2}$ and asked to find the triangle's height and base. The scribe began by multiplying the area by 2 (to give 40 as the area of the corresponding rectangle). Realizing that the product of this area (height × base) and the given ratio gives the square of the desired height, he found that this square is 100 and therefore that the desired height is 10. Multiplying 10 by the reciprocal of $2\frac{1}{2}$ then gives 4 for the desired base.

Both of these problems are "practical," but ideas of proportionality are also used to solve "abstract" problems in the papyri. In the *Rhind Papyrus*, for example, several problems translate into simple linear equations. In problem 24, we are asked to find a quantity such that when it is added to $\frac{1}{7}$ of itself the result is 19. The scribe tackled the question this way: "Assume [the answer is] 7. Then 7 plus $\overline{7}$ of 7 is 8. Multiply 8 so as to get 19. The answer is 2 $\overline{4}$ $\overline{8}$. Multiply this value by 7. The answer is 16 $\overline{2}$ $\overline{8}$."[1] In modern notation, this was a matter of solving the linear equation $x + \frac{1}{7}x = 19$. The scribe opened by making the (false) assumption that the answer is 7, because $\frac{1}{7}$ of 7 is an integer. (Thus, the technique is often called the "method of false position.") He then noted, however, that $7 + \frac{1}{7} \cdot 7 = 8$, and not the desired quantity 19. Since, in modern terms, the value of the "linear function" $x + \frac{1}{7}x$ is proportional to x, the scribe then needed to find the multiple of 8 that produces 19. He did this in the next step and found that this value is $2\frac{3}{8}$. Thus, to find the correct answer to the original question, he had finally to multiply 7 by $2\frac{3}{8}$ to get $16\frac{5}{8}$.

Even in the only example extant in the Egyptian papyri of what we would style a quadratic equation, the scribe used false position, essentially based on the idea of proportions. The *Berlin Papyrus*, a small fragment dating from approximately the same era as the *Rhind* and *Moscow* papyri, contains a problem in which the scribe is to divide a square of area 100 square cubits into two other squares, where the ratio of the sides of the two squares is 1 to $\frac{3}{4}$. (The Egyptian cubit has length approximately 18 inches.) The scribe began by assuming that, in fact, the sides of the two needed squares are 1 and $\frac{3}{4}$ and then calculated the sum of the areas of

[1] Chace, 1979, p. 36.

these two squares to be $(1)^2 + (\frac{3}{4})^2 = 1\frac{9}{16}$. Since the desired sum of the areas is 100, the scribe realized that he could not compare the squares' areas directly but had to compare their sides. He thus took the square root of $1\frac{9}{16}$, namely, $1\frac{1}{4}$ and compared this to the square root of 100, namely, 10. Since 10 is 8 times as large as $1\frac{1}{4}$, the scribe concluded that the sides of the two other squares must be 8 times the original guesses, namely, 8 and 6 cubits, respectively.

A final problem, the twenty-third of the *Moscow Papyrus*, is what would now be called a "work" problem: "Method of calculating the work rate of a cobbler. If you are told the work rate of a cobbler: If he [only] cuts [leather for sandals] it is 10 [pairs] per day; if he [only] finishes [the sandals] it is 5 [pairs] per day. How much is [his work rate] per day if he cuts and finishes [the sandals]."[2] Today's students would solve this using algebra, but the scribe used a different method. He noted that the shoemaker cuts 10 pairs of sandals in one day and decorates 10 pairs of sandals in two days, so that it takes three days for him to both cut and decorate 10 pairs. The scribe then divided 10 by 3 to find that the shoemaker can cut and decorate $3\frac{1}{3}$ pairs in one day.

This algebra of proportions is the only "algebra" evident in the extant Egyptian mathematical papyri. Whether the Egyptians understood any other algebraic concepts is unknown. However, the *Moscow Papyrus* contains the famous calculation of the volume of a truncated pyramid, a pyramid "under construction" and missing its top. There has been much speculation as to how the scribes figured out this algorithm, assuming that they, in fact, had a procedure that worked in general and not just in the specific case noted in the papyrus. One suggestion is that they employed the algorithm for finding the volume of a pyramid by somehow subtracting the volume of the upper "missing" pyramid from the completed pyramid. If this were true, it would show that they used, first, proportions to determine the height of the completed pyramid and, second, some kind of algebraic calculation, probably couched in geometric terms, to find the algorithm for the truncated pyramid.[3] Interestingly, it was the Mesopotamian scribes who developed this "geometric" form of algebra in much detail.

---

[2] Imhausen, 2007, p. 39.
[3] Gillings, 1972, pp. 187–193 and the references listed there.

GEOMETRICAL ALGEBRA IN MESOPOTAMIA

The Egyptian and Mesopotamian civilizations are roughly the same age, with the latter having developed in the Tigris and Euphrates River valleys beginning sometime in the fifth millennium BCE. Many different governments ruled the Mesopotamian region over the centuries, some small city-states and others large empires. For example, from approximately 2350 to 2150 BCE, the region was unified under a dynasty from Akkad. Shortly thereafter, the Third Dynasty of Ur rapidly expanded until it controlled most of southern Mesopotamia. Under this dynasty's centralized bureaucracy, a large system of scribal schools developed to train future officials. Although the Ur Dynasty collapsed around 2000 BCE, the small city-states that succeeded it still demanded numerate scribes. By 1700 BCE, Hammurapi, the ruler of one of those city-states, Babylon, had expanded his rule to much of Mesopotamia and had instituted a legal and civil system to help regulate his empire.

Although the scribes in Mesopotamia, some of whom were female,[4] did not have quite the social status of their Egyptian counterparts, they nevertheless dealt with many issues important in running a government, including trade, building, managing the labor force, and surveying. And, although sources relative to Egyptian mathematics are rare, numerous original mathematical documents from ancient Mesopotamia survive, especially from the so-called Old Babylonian period of Hammurapi.

Mesopotamian scribes wrote with styluses on clay tablets, which dried in the sun and were sometimes baked for better preservation. These tablets were generally of a size that fit comfortably in the hand, but sometimes they could be as small as a postage stamp or as large as an encyclopedia volume. Thousands of these tablets have been unearthed, usually from the ruins of buildings, and hundreds of them are mathematical in nature. The scholarly study of these texts, especially in the last half century, has resulted in a very detailed picture not only of Mesopotamian mathematics itself but also of how mathematics fit into the broader culture.

The earliest mathematics in Mesopotamia was associated with accounting procedures involved in trade and the distribution of goods.

---

[4] Robson, 2008, pp. 122–123.

In this context, a sexagesimal place-value system developed—not only for integers but also for fractions—in the course of the third millennium BCE. This system was generally used in "pure" calculation, while various nonsexagesimal systems evolved for use in measurement from which and into which the scribes had to "translate" their sexagesimal system. The scribes recorded numbers between 1 and 60 in terms of a grouping system (with two different symbols, one for 1 and another for 10), but we will translate these numbers using the now-standard notation where $a, b, c; d, e$, for example, stands for $a \times 60^2 + b \times 60 + c + d \times 60^{-1} + e \times 60^{-2}$. And, although the Mesopotamians did not use any symbol to designate an empty place value in a given number (they did sometimes leave a slightly larger space between digits), we will do so for clarity using our 0.

In addition to that of accounting, Mesopotamian mathematics also developed in the context of surveying. To compare the areas of fields, for example, surveyors evidently conceived of them as divided into various squares and rectangles that they could then mentally rearrange. Out of this practice evolved so-called "geometrical algebra," a method of manipulating areas ("cut and paste") to determine unknown lengths and widths. In general, as we shall see, the problems in geometrical algebra that show up on the clay tablets are artificial, in the sense that they do not reflect actual problems that the surveyors (or the scribes) needed to solve. Clearly, there were Mesopotamian "mathematicians" who were able to abstract new methods and patterns out of the real-world practices of the surveyors and who carried the problems far beyond any immediate necessity.

Since geometrical algebra comes from manipulating two-dimensional regions, its methods involve what we think of as "quadratic" problems. Before considering these, however, we note that Mesopotamian scribes could also handle linear equations, and, like their Egyptian counterparts, they solved them using methods based on an understanding of proportions. The simplest such problems, equations that we would write as $ax = b$, were solved by multiplying each side by the reciprocal[5] of a. (Such equations often occurred in the process of solving a more complicated problem.) This method, of course, is based on understanding that $ax : a = x : 1$. In trickier situations, such as those involving systems of

---

[5] Extensive tables of reciprocals were developed in Mesopotamia for exactly this purpose.

two linear equations, the Mesopotamians, like the Egyptians, used the method of false position.

Consider this example from the Old Babylonian text VAT 8389:[6] One of two fields yields 4 *gur* per *bur*, while the second yields 3 *gur* per *bur*. Here, a "*gur*" is a unit of capacity equal to 300 *sila*, the basic unit of capacity, and a "*bur*" equals 1800 *sar*, the standard area measure. Rewriting these conditions, the first field yields 0;40 ($=\frac{2}{3}$) *sila* per *sar*, while the second yields 0;30 ($=\frac{1}{2}$) *sila* per *sar*. We are then told that the first field yielded 500 *sila* more than the second, while the total area of the two fields is 1800 *sar*. The question, then, is how large is each field?[7] We might translate this problem into a system of two equations

$$\frac{2}{3}x - \frac{1}{2}y = 500,$$

$$x + y = 1800,$$

with x and y representing the unknown areas. Solving the second equation for x, we would then substitute the result into the first to get a value for y that we could then use to determine a value for x. The scribe, however, made the initial assumption that x and y were both equal to 900. Thus, the grain collected from the first field would amount to 600 *sila*, while that from the second would amount to 450 *sila*, for a difference of 150. Since the desired difference is 500, the scribe needed to figure out how to adjust the areas to increase the difference by 350. He realized that each transfer of 1 *sar* from the second field to the first would increase the difference of the yield by $0;40 + 0;30 = 1;10$ (or $\frac{2}{3} + \frac{1}{2} = 1\frac{1}{6}$). Thus, he had to divide 350 by $1;10$ ($= 1\frac{1}{6}$), that is, determine what multiple of $1\frac{1}{6}$ was equal to 350. Since the answer was 300, the solution to the problem was that the first field had area $900 + 300 = 1200$ *sar*, while the second had area $900 - 300 = 600$ *sar*.

Presumably, the Mesopotamians also solved more complicated, single linear equations by false position, although the few such problems known

---

[6] The initials associated with the tablet generally indicate the collection in which the tablet is held. In this chapter, we refer to Yale University (YBC), the British Museum (BM), the Vorderasiatischer Museum, Berlin (VAT), the Schøyen Collection, Norway (MS), the Département des Antiquités Orientales, Louvre, Paris (AO), and the collection first published by Evert M. Bruins and Marguerite Rutten as *Textes mathématiques de Suse* (TMS).

[7] For a literal translation of this problem and its solution, see Høyrup, 2002, pp. 78–79.

do not reveal their method. Consider, for example, this problem from tablet YBC 4652: "I found a stone. I did not weigh it. A seventh I added. An eleventh [of the total] I added. I weighed it: 1 *mina* [= 60 *gin*]. What was the original [weight of the stone?]"[8] We can translate this into the modern equation $(x + \frac{1}{7}x) + \frac{1}{11}(x + \frac{1}{7}x) = 60$. On the tablet, the scribe just presented the answer, namely, $x = 48; 7, 30 (= 48\frac{1}{8}$ *gin*). If it had been solved by false position, the first guess would have been that $y = x + \frac{1}{7}x = 11$. Since, then, $y + \frac{1}{11}y = 12$ instead of 60, the guess must be increased by the factor $\frac{60}{12} = 5$ to the value 55. Then, to solve $x + \frac{1}{7}x = 55$, the next guess could have been $x = 7$. This value would have produced $7 + \frac{1}{7}7 = 8$ instead of 55. So the last step would have been to multiply the guess of 7 by the factor $\frac{55}{8}$ to get $\frac{385}{8} = 48\frac{1}{8}$, the correct answer.

While tablets containing explicit linear problems are limited, quadratic problems, which we certainly recognize as "algebraic," abound. To solve them, moreover, the scribes made full use of the "cut and paste" geometrical algebra developed by the surveyors. Standard problems such as finding the length and width of a rectangle given the area and the sum or difference of the length and width were thus solved, quite literally, by "completing the square." Such geometric manipulations allowed the Mesopotamians to move beyond the solution of basic proportion problems. This is why we call this earliest stage of algebra the "geometric" stage.

Consider, for example, problem 8 from tablet YBC 4663 where, as part of a larger problem, we are given that the length of a rectangle exceeds the width by $3; 30 (= 3\frac{1}{2})$ rods, while the area is $7; 30 (= 7\frac{1}{2})$ *sar*. (A rod is equal to 12 cubits, while a *sar* is a square rod.) The solution procedure translates literally as "Break off $\frac{1}{2}$ of that by which the length exceeds the width so that it gives you $1; 45 [= 1\frac{3}{4}]$. Combine $1; 45$ so that it gives you $3; 03, 45 [= 3\frac{1}{16}]$. Add $7; 30$ to $3; 03, 45$, so that it gives [you] $10; 33, 45 [= 10\frac{9}{16}]$. Take its square-side, so that it gives you $3; 15 [= 3\frac{1}{4}]$. Put down $3; 15$ twice. Add $1; 45$ to 1 [copy of $3; 15$], take away $1; 45$ from 1

---

[8] Melville, 2002, p. 2. Note here that the two fractions are not regular in the sexagesimal system, that is, 7 and 11 do not have finite sexagesimal reciprocals. In fact, that was part of the point of the problem. Since the reciprocals of these numbers could not actually be used, it was necessary to make guesses and only afterward to multiply by regular fractions, such as $\frac{1}{12}$ and $\frac{1}{8}$. The word that we have translated as "a seventh" is literally "the reciprocal of 7."

Figure 2.2. Babylonian tablet YBC 4663, courtesy of the Yale Babylonian Collection.

[copy of 3; 15], so that it gives you length and width. The length is 5 rods; the width $1\frac{1}{2}$ rods."[9] The wording shows that the Mesopotamian scribe considered this a geometric problem and not merely a problem

---

[9] Robson, 2008, p. 89. Note that $\frac{1}{2}$ appears in two places rather than 0; 30. In both of these situations, the scribe used a special symbol for that fraction, rather than, as in the previous problem, an expression denoting "the reciprocal of."

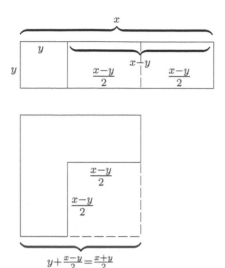

Figure 2.3.

in manipulating numbers. The scribe began by drawing a rectangle of unknown length x and width y to represent the area and next marked off the difference between length and width ($x - y = 3; 30 = 3\frac{1}{2}$) along the length (figure 2.3). He or she then cut off half of the rectangle determined by y and $x - y$ and moved the cut-off piece so that it formed a symmetrical L shape with the remainder; the total area is still 7; 30. The missing square enclosed by the L then has area 1; 45 × 1; 45 = 3; 03, 45 ($= \left(\frac{x-y}{2}\right)^2$). Thus, to "complete the square," the scribe had to add the areas of the missing square and the L-shaped region. The total area of the completed square is then 7; 30 + 3; 03, 45 = 10; 33, 45 ($= xy + \left(\frac{x-y}{2}\right)^2$). The side of this large square ($= y + \frac{x-y}{2} = \frac{x+y}{2}$) must be 3; 15. Thus, if we subtract from this one side of the small square, we find that the original width y is 3; 15 − 1; 45 = 1; 30, while if we add the side of the small square to that of the large one we get the original length $x = 3; 15 + 1; 45 = 5$ ($\frac{x+y}{2} - \frac{x-y}{2} = y$; $\frac{x+y}{2} + \frac{x-y}{2} = x$). Figure 2.3 makes the procedure clear, and translating it into a modern formula to solve the generic system $x - y = b$, $xy = c$, the solution is

$$x = \sqrt{c + \left(\frac{b}{2}\right)^2} + \frac{b}{2}, \qquad y = \sqrt{c + \left(\frac{b}{2}\right)^2} - \frac{b}{2}.$$

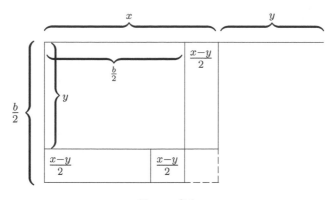

Figure 2.4.

Similarly, problem 25 from tablet BM 85200 + VAT 6599[10] generates the solution of the generic system $x + y = b$, $xy = c$. We are told that the area of a rectangle is 8; 20 ($= 8\frac{1}{3}$), while the sum of the length and width is 5; 50 ($= 5\frac{5}{6}$). The scribe wrote, "$\frac{1}{2}$ of 5; 50 break [giving 2; 55 ($= 2\frac{11}{12}$)]; Combine [2; 55 to itself, that is, square it]: it gives you 8; 30, 25 [$= 8\frac{73}{144}$]. From the inside, tear out 8; 20; it gives you 0; 10, 25 [$= \frac{25}{144}$]. The square-side is 0; 25 [$= \frac{5}{12}$]. To 2; 55 append and tear out. 3; 20 [$= 3\frac{1}{3}$] is the length; 2; 30 [$= 2\frac{1}{2}$] is the width."[11] The scribe's procedure arose from considering a figure like figure 2.4 and realizing that the square on half the sum of the length and width exceeds the original rectangle by the square of half the difference, that is,

$$\left(\frac{x+y}{2}\right)^2 - xy = \left(\frac{x-y}{2}\right)^2.$$

It follows that since the two values on the left-hand side are known, one can find the value of the right-hand side and, therefore, the difference of the length and width. Adding this value to and subtracting it from half the sum then gives the length and width. Again, if we put the scribe's procedure into modern terms to solve the generic system $x + y = b$,

---

[10] This particular tablet is in two pieces, one in London and the other in Berlin.
[11] Høyrup, 2002, pp. 146–147 (slightly modified for clarity).

$xy = c$, we could write the answer as

$$x = \frac{b}{2} + \sqrt{\left(\frac{b}{2}\right)^2 - c}, \qquad y = \frac{b}{2} - \sqrt{\left(\frac{b}{2}\right)^2 - c}.$$

Interestingly, tablet BM 85200 + VAT 6599 is one of the few from the Old Babylonian period with a known author. According to Assyriologist Eleanor Robson, this tablet and ten other extant ones were all written by Iskur-mansum, son of Sin-iqisam, a teacher in the town of Sippar in the 1630s BCE.[12]

Geometry also underlies the Mesopotamian solution to what we would interpret as a single quadratic equation. We take as our example problem 1 from tablet MS 5112, where the translation again reflects the geometric flavor of the problem: "The field and 2 'equal-sides' heaped [added together] give 2, 00 [= 120]. The field and the sameside are what?"[13] (In modern notation, this asks us to find x satisfying $x^2 + 2x = 120$.) The scribe solved the problem this way: "To 2, 00 [= 120] of the heap, 1 the extension add, giving 2, 01 [= 121]. Its equal-side resolve, giving 11. From 11 the extension 1 tear off, then 10 is the equal-side." Although it may not seem "geometrical" to add the area (field) and the two sides, the scribe meant that the square field is extended by a unit in two directions as in figure 2.5, and that extended field has total area 120. The scribe then "completed the square" by adding the small square of area 1 (the square

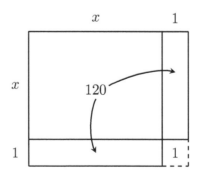

Figure 2.5.

[12] Robson, 2008, pp. 94–97.
[13] Friberg, 2007, p. 309.

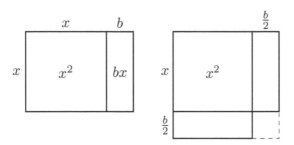

Figure 2.6.

of the extension). Thus, the total area of the enlarged square is 121, so that its side is 11. Then, when one "tears off" the extension of 1, the result is that the side x of the original square is 10. Again, the Mesopotamian rule exemplified by this problem translates easily into a modern formula for solving $x^2 + bx = c$, namely,

$$x = \sqrt{\left(\frac{b}{2}\right)^2 + c} - \frac{b}{2},$$

which we recognize as a version of the quadratic formula. Figure 2.6 shows the geometric meaning of the procedure in the generic case, where we start with a square of side x adjoined by a rectangle of width x and length b. The procedure then amounts to cutting half of the rectangle off from one side of the square and moving it to the bottom. Adding a square of side $\frac{b}{2}$ completes the square. It is then evident that the unknown length x is equal to the difference between the side of the new square and $\frac{b}{2}$, exactly as the formula implies.

For the analogous problem $x^2 - bx = c$, the Mesopotamian geometric procedure (figure 2.7) is equivalent to the formula

$$x = \sqrt{\left(\frac{b}{2}\right)^2 + c} + \frac{b}{2}.$$

To see this, consider problem 2 from tablet BM 13901, a problem we translate as $x^2 - x = 870$: "I took away my square-side from inside the area and it was 14, 30 [= 870]. You put down 1, the projection. You break off half of 1. You combine 0; 30 [= $\frac{1}{2}$] and 0; 30. You add 0; 15 [= $\frac{1}{4}$] to 14, 30. 14, 30; 15 [= $870\frac{1}{4}$] squares 29; 30 [= $29\frac{1}{2}$].

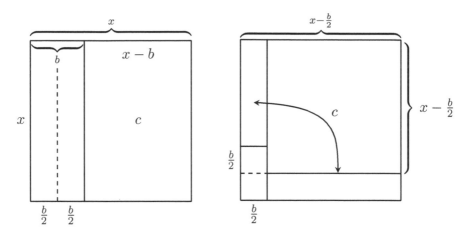

Figure 2.7.

You add $0;30$ which you combined to $29;30$ so that the square-side is 30."[14]

It is important to keep in mind here that the "quadratic formulas" given above are modern interpretations of the Mesopotamian proce-dures.[15] The scribes did not have "formulas," but they did have algo-rithms for the two types of equations that we express as $x^2 + bx = c$ and $x^2 - bx = c$. For the Mesopotamians, problems represented by these two equations were different, because their geometric meanings differed. To a modern mathematician, however, they are the same because the coefficient of x can be taken as positive or negative. We should also note that the modern quadratic formula in these two cases gives a positive and a negative solution for each equation. Since negative solutions make no geometrical sense, however, they were completely ignored by the Mesopotamians.

Note, too, that in both of these quadratic problems, the coefficient of the x^2 term is 1. How did the Mesopotamians treat quadratic equations of the form $ax^2 \pm bx = c$ when $a \neq 1$? There are problems on BM 13901 showing that the scribes usually scaled up the unknown to reduce the problem to the case $a = 1$.[16] For example, problem 7 on that tablet

[14] Robson, 2007, p. 104.

[15] Otto Neugebauer first interpreted many of these problems as quadratic equations in Neugebauer, 1951, but his interpretation of the problems as "pure algebra" has been superseded by the interpretation in terms of geometrical algebra presented here.

[16] Robson, 2007, p. 104.

can be translated into the modern equation $11x^2 + 7x = 6\frac{1}{4}$. The scribe multiplied by 11 to turn the equation into a quadratic in $11x$, namely, $(11x)^2 + 7(11x) = 68\frac{3}{4}$. He or she then solved

$$11x = \sqrt{\left(\frac{7}{2}\right)^2 + 68\frac{3}{4}} - \frac{7}{2} = \sqrt{81} - \frac{7}{2} = 9 - 3\frac{1}{2} = 5\frac{1}{2}.$$

To find x, the scribe would normally have multiplied by the reciprocal of 11 but noted, in this case, that the reciprocal of 11 "cannot be solved." Nevertheless, because the problem was probably manufactured to give a simple answer, it is easy to see that the unknown side x is equal to $\frac{1}{2}$.

This idea of "scaling" also enabled the scribes to solve quadratic-type equations that did not directly involve squares but that, instead, involved figures of different shapes. For example, consider this problem from TMS 20:[17] The sum of the area and side of the convex square is $\frac{11}{18}$. Find the side. In this case, the "convex square" is a figure bounded by four quarter-circle sides (figure 2.8). The Mesopotamians had found that, with appropriate approximations, the area A of such a "square" with side s is given by $A = \frac{4}{9}s^2$. Thus, we can translate this problem into the equation $A + s = \frac{11}{18}$, or, on multiplying by the scaling factor $\frac{4}{9}$, into $\frac{4}{9}A + \frac{4}{9}s = \frac{22}{81}$. But, given the relationship between A and s, the scribe was able to think

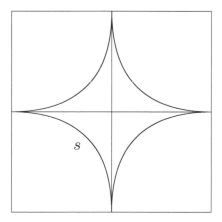

Figure 2.8.

[17] Melville, 2002.

of this in the form of an equation for $\frac{4}{9}s$ in this way:

$$\left(\frac{4}{9}s\right)^2 + \frac{4}{9}s = \frac{22}{81}.$$

He then solved this as usual to get $\frac{4}{9}s = \frac{2}{9}$ and concluded by multiplying by the reciprocal $\frac{9}{4}$ to find the answer $s = \frac{1}{2}$.

Although the methods described above were standard when solving quadratic equations, the scribes did occasionally employ other techniques in particular settings. Consider, for example, problem 23 on BM 13901.[18] There, we are told that the sum of four sides and the (square) surface is $\frac{25}{36}$. Although this problem is of the type $x^2 + bx = c$, in this case the b is 4, the number of sides of the square, which is more "natural" than the coefficients we saw earlier. Interestingly, this same problem turns up frequently in much later manifestations of this early tradition in both Islamic and medieval European mathematics. The scribe's method here depended explicitly on the "4." In the first step of the solution, $\frac{1}{4}$ of the $\frac{25}{36}$ was taken to get $\frac{25}{144}$. To this, 1 was added to give $\frac{169}{144}$. The square root of this value is $\frac{13}{12}$. Subtracting 1 gives $\frac{1}{12}$. Thus, the length of the side is twice that value, namely, $\frac{1}{6}$. This new procedure is best illustrated by figure 2.9. What the scribe intended is that the four "sides" are really projections

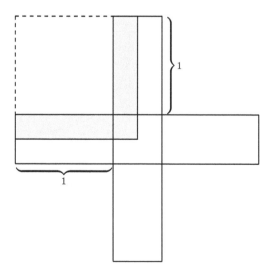

Figure 2.9.

[18] Robson, 2007, p. 107.

of the actual sides of the square into rectangles of length 1. Taking $\frac{1}{4}$ of the entire sum means considering only the shaded gnomon which is one-fourth of the original figure. When we add a square of side 1 to that figure, we get a square the side of which we can then find. Subtracting 1 from the side gives half the original side of the square.

Other problems on BM 13901 treat various situations involving squares and sides, with each of the solution procedures having a geometric interpretation. As a final example from this text,[19] consider problem 9: $x^2 + y^2 = \frac{13}{36}$, $x - y = \frac{1}{6}$. The solution to this system, which we generalize into the system $x^2 + y^2 = c$, $x - y = b$, was found by a procedure describable in terms of the modern formula

$$x = \sqrt{\frac{c}{2} - \left(\frac{b}{2}\right)^2} + \frac{b}{2}, \qquad y = \sqrt{\frac{c}{2} - \left(\frac{b}{2}\right)^2} - \frac{b}{2}.$$

It appears that the Mesopotamians developed the solution using the geometric idea expressed in figure 2.10. This figure shows that

$$x^2 + y^2 = 2\left(\frac{x+y}{2}\right)^2 + 2\left(\frac{x-y}{2}\right)^2.$$

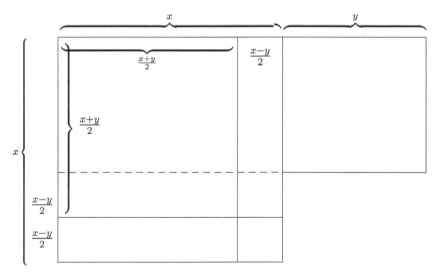

Figure 2.10.

From this, we see that

$$c = 2\left(\frac{x+y}{2}\right)^2 + 2\left(\frac{b}{2}\right)^2$$

and, therefore, that

$$\frac{x+y}{2} = \sqrt{\frac{c}{2} - \left(\frac{b}{2}\right)^2}.$$

Because $x = \frac{x+y}{2} + \frac{x-y}{2}$ and $y = \frac{x+y}{2} - \frac{x-y}{2}$, the result follows.

However, not every "quadratic" problem came from a geometric situation, even though the solution procedure itself was always geometric. For instance, we find the following problem on YBC 6967: "A reciprocal exceeds its reciprocal by 7. What are the reciprocal and its reciprocal?"[20] Since, in Mesopotamia, reciprocals were pairs of numbers the product of which was 1, 60, or any power of 60, we can translate the problem into our system as

$$x - y = 7, \ xy = 60.$$

The scribe clearly conceptualized this problem, as in the earlier problem from YBC 4663, as finding the sides of a rectangle of area 60, for he or she wrote, "You break in half the 7 by which the reciprocal exceeds its reciprocal; $3;30 \ [= 3\frac{1}{2}]$ will come up. Combine $3;30$ and $3;30$ [that is, square $3;30$]. $12;15 \ [= 12\frac{1}{4}]$ will come up. Append 60 to the $12;15$ and $72;15 \ [= 72\frac{1}{4}]$ will come up. The square side of $72;15$ is $8;30 \ [= 8\frac{1}{2}]$. Put $8;30$ down twice. Tear out $3;30$ from one and append $3;30$ to the other. One is 12, the other is 5." Thus, the geometrical algebra procedure is the same as that displayed in figure 2.3.

An even more interesting quadratic equation arises from the sum of an arithmetic progression, where it is evident that the scribe understood that

---

[20] Robson, 2008, p. 113.

the solution came from an algorithm originally motivated geometrically, but that could simply be applied arithmetically. Problem 5 from AO 6770 asks us to find the length of a reed if it is applied an unknown number of times to measure a length of 4 cubits, where for each application of the reed, a piece of length 1 finger falls off.[21] (Here 30 fingers equal 1 cubit.) If n represents the number of times the reed is used before it disappears, then the formula for the sum of the numbers from 1 to n, which was known to the Mesopotamians, allows us to translate this problem into the equation $\frac{n(n+1)}{2} = 120$. The scribe considered this equation in the form that we would write as $n^2 + n = 240$ and solved it just as he or she solved similar equations, namely, by completing the square on the left-hand side by adding $\left(\frac{1}{2}\right)^2 = \frac{1}{4}$ to get $\left(n + \frac{1}{2}\right)^2 = 240\frac{1}{4}$. Then $n + \frac{1}{2} = 15\frac{1}{2}$, so $n = 15$, and the length of the original reed was 15 fingers or $\frac{1}{2}$ cubit.

Although there are numerous Mesopotamian tablets dealing with algebra problems, mostly quadratic in form, none of the tablets discovered to date explains the reasoning behind the algorithms employed. Yet, somehow, the scribes had reasons for doing what they did, and this information was likely passed down from one generation of scribes to the next. In our discussions of the examples, we have elucidated how most contemporary scholars understand the scribes' reasoning, but, interestingly, there is no commonly held explanation as to why the Mesopotamian quadratic problems were developed and explored in the first place. Various scholars have speculated about these origins, however. Jens Høyrup, for example, believes that they were originally developed as riddles to demonstrate the professional competence of surveyors and scribes, while Eleanor Robson holds that the development of techniques for accurately calculating lines and areas reflected a concern with justice, part of the ideology of the Mesopotamian states around 2000 BCE. In her interpretation, the Mesopotamians' concern with finding the correct values in measuring land yielded "numerate justice" in imposing taxes.[22] Regardless of why the Mesopotamians originally developed their geometric version of algebra, the scribes were fascinated with the possibilities it opened up and were eager to demonstrate their cleverness in solving new problems.

[21] Friberg, 2007, p. 247.
[22] See Høyrup, 2002, pp. 362–385 and Robson, 2008, pp. 123–124 for more details.

****

The extant papyri and tablets containing Egyptian and Mesopotamian mathematics were generally teaching documents used to transmit knowledge from one scribe to another and, frequently, from a senior scribe to an apprentice. In the Mesopotamian case, in particular, long lists of quadratic problems were given as "real-world" problems, although they are, in fact, just as contrived as problems found in most current algebra texts. As in our modern texts, however, the ancient authors used such problems as teaching tools. To learn a technique requires practice and to learn a mathematical technique requires practice using systematic variation in the parameters of the problems. Thus, the mathematical texts often contained problems that grew in complexity. Their function was to provide trainee scribes with a set of example types, problems the solutions of which could be applied in other situations. Learning mathematics for these trainees meant first learning how to select and perhaps modify an appropriate algorithm, and then mastering the arithmetic techniques necessary to carry out the algorithm in order to solve any new problem. Thus, several different problems in the *Rhind Papyrus* deal with solving linear equations, even though these types of equations were not used to solve real-world problems in Egypt. Similarly, in the Mesopotamian context, it was not really that important to solve quadratic equations per se because these could rarely be applied to the actual situations that a scribe would encounter. What was important for the students who read these documents, however, was the development of general problem-solving skills.

Even though the reasoning behind the algorithms displayed in the Egyptian and Mesopotamian mathematical documents was not made explicit, the algorithms themselves were apparently transmitted. The technique of false position for solving linear equations shows up in many later civilizations, and was even used in texts in the early years of the United States. As we shall see in later chapters, there is at least some evidence that geometrical techniques for solving quadratic equations were transmitted to Greece, and even more evidence that they were eventually transmitted to the Islamic world, where the development of algebra accelerated considerably.

3

The Ancient Greek World

Mathematics began to develop in Greece around 600 BCE and differed greatly from what had emerged in the more ancient civilizations of Egypt and Mesopotamia. Owing perhaps to the fact that large-scale agriculture was impossible in a topography characterized by mountains and islands, Greece developed not a central government but rather the basic political organization of the *polis* or city-state with a governmental unit that, in general, controlled populations of only a few thousand. Regardless of whether these governments were democratic or monarchical, they were each ruled by law and therefore fostered a climate of argument and debate. It was perhaps from this ethos that the necessity for proof in mathematics—that is, for argument aimed at convincing others of a particular truth—ensued.[1]

We have only fragmentary documentation of the actual beginnings of mathematics in Greece. Eudemus of Rhodes (ca. 350–290 BCE), for example, evidently wrote a formal history of Greek mathematics around 320 BCE. No longer extant, it was summarized in the fifth century CE by Proclus (411–485 CE). From this and other scattered sources, it appears that Thales (ca. 624–547 BCE) was the first to state and prove certain elementary geometric theorems, although it is not known exactly what the concept of "proof" may have meant to him. Other Greek mathematicians of the fifth and fourth centuries BCE—Pythagoras (ca. 572–497 BCE), Hippocrates of Chios (ca. 470–410 BCE), Archytas (fifth century BCE), Theaetetus (417–369 BCE), Eudoxus (406–355 BCE), and Menaechmus (fourth century BCE)—reportedly stated and proved the basic triangle congruence theorems, the Pythagorean theorem, various theorems about circles and inscribed polygons, results on ratio, proportion, and similarity,

---

[1] See Lloyd, 1996 for more discussion as to why the notion of deductive proof developed in Greece.

and even facts about conic sections. Work of Autolycus (fourth century BCE) on spherical geometry in connection with astronomy, however, actually survives and employs a basic theorem–proof structure. Most of these early developments took place in Athens, one of the richest of the Greek states at the time and one where public life was especially lively and discussion particularly vibrant.[2]

A key figure in this culture of rational debate, Aristotle (384–322 BCE) codified the principles of logical argument and insisted that the development of any area of knowledge must begin with the definition of terms and with the articulation of axioms, that is, statements the truth of which is assumed. Aristotle, however, also emphasized the use of syllogisms as the building blocks of logical argument, whereas later Greek mathematicians, especially Chrysippus (280–206 BCE) and others in the third century BCE, preferred to employ propositional logic. The earliest treatise on geometry to come down to us in this style—that is, based on axioms and using logical argument—is the *Elements* of Euclid, written sometime around 300 BCE.

GEOMETRICAL ALGEBRA IN EUCLID'S *ELEMENTS* AND *DATA*

Although we know essentially nothing about the life of Euclid, it is generally assumed that he worked in Alexandria, the city founded by Alexander the Great that served, beginning in 332 BCE, as the capital of Egypt under the dynasty of the Ptolemies and then soon fell under Roman rule after the death of Cleopatra in 30 BCE. Whoever Euclid was, he evidently based his *Elements* on Aristotle's principles, and his text soon became the model for how mathematics should be written—with definitions, axioms, and logical proofs—a model that has persisted to this day. The *Elements* contains not only the basic results of plane geometry that still form part of the secondary curriculum around the world, but also material on elementary number theory and solid geometry. Whether the *Elements* contains algebra, however, has been hotly contested since the late nineteenth century, with the debate centering around the meaning and purpose of Book II. In some sense, the dispute has dealt with two different issues: the meaning of Book II to Euclid, that is, the actual *history*

---

[2] For more on the context of early Greek mathematics, see Cuomo, 2001.

of why and how it was written, and the *heritage* of Book II up to our own day, that is, how it has been used and interpreted over time.[3]

Book II deals with the relationships between various rectangles and squares and has no obvious goal. In fact, its propositions are only infrequently used elsewhere in the *Elements*. One interpretation of this book, dating from the late nineteenth century but still common today, is that it, together with a few propositions in Books I and VI, can best be interpreted as "geometrical algebra," the representation of algebraic concepts through geometric figures. In other words, the square constructed on a line AB can be thought of as a geometric representation of $\overline{AB}^2$, and, if we assign a numerical length a to AB, then we can think of a^2 as representing this square. Similarly, a rectangle with perpendicular sides AB and BC of lengths a and b, respectively, could be interpreted as either the product $\overline{AB} \cdot \overline{BC}$ or, more simply, as ab. From this point of view, relationships among such objects can be interpreted as equations.

The majority of scholars today believe that, in Book II, Euclid only intended to display a relatively coherent body of *geometric* knowledge that could be used in the proof of other geometric theorems, if not in the *Elements* themselves then in more advanced Greek mathematics.[4] That Euclid was thinking geometrically seems clear even from his opening definition: "Any rectangle is said to be *contained* by the two straight lines forming the right angle."[5] This does not state that the area of a rectangle is the product of the length by the width. In fact, Euclid never multiplied two lines (or their associated lengths) together because he had no way of defining such a process for arbitrary lengths. At various places, he did multiply lines by numbers (that is, positive integers), but he otherwise wrote only of rectangles contained by two lines.

Consider, for example, the following.

Proposition II.4:   *If a straight line be cut at random, the square on the whole is equal to the squares on the segments and twice the rectangle contained by the segments.*

---

[3] See Grattan-Guinness, 2004a and 2004b for more discussion of the distinction between history and heritage.

[4] See, for example, Unguru, 1975.

[5] All quotations from Euclid's *Elements* are taken from the Heath translation, as reissued in Euclid, 2002.

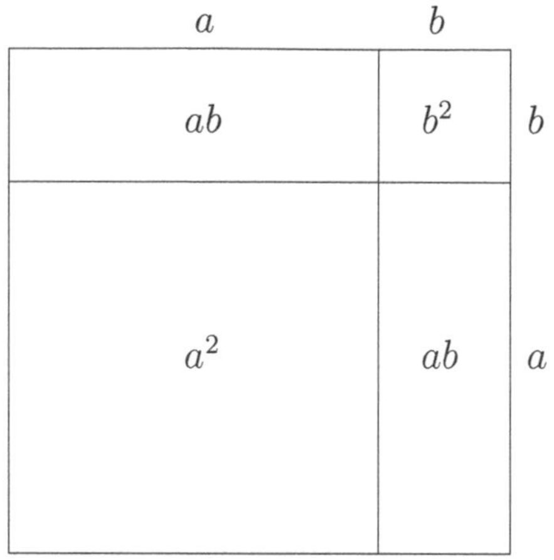

a	b	
ab	b^2	b
a^2	ab	a

Figure 3.1.

For Euclid, this was a geometric result proved by drawing and comparing the appropriate squares and rectangles (figure 3.1). Here, and elsewhere in Book II, Euclid seems to be proving a result about "invisible" figures—namely, figures referred to in the theorem with respect to an initial line and its segments—by using "visible" figures—namely, the actual squares and rectangles drawn. By assigning lengths a and b to the two segments of the given line in the proposition, we can easily translate the result into the rule for squaring a binomial: $(a + b)^2 = a^2 + b^2 + 2ab$, a decidedly "algebraic" result. As we shall see later, medieval Islamic mathematicians interpreted some of Euclid's results as "algebra" in very much this way and applied them to the solution of quadratic equations. Book II thus has both a geometric history and an algebraic heritage.

If, however, we think of "algebra" as meaning the search for unknown quantities given certain relationships between those and known quantities (even if the "quantities" are geometric and are not assigned numerical values), then there certainly is algebra in Book II as well as in other books of the *Elements*. For example, we can consider the following problem from Book II as "algebraic" in this sense:

Proposition II.11: *To cut a given straight line so that the rectangle contained by the whole and one of the segments is equal to the square on the remaining segment.*

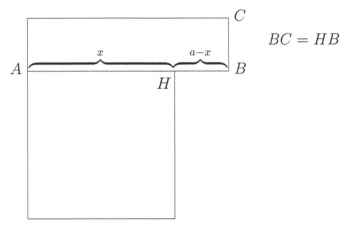

Figure 3.2.

Although this may appear to be geometric in that it deals with rectangles and squares, it meets the above definition of an algebra problem, since it involves the phrase "is equal to" and since, to Euclid, that phrase means "is equal in area."

In any case, the goal of proposition II.11 is to find a point H on the line AB—that is, to determine an unknown line AH—so that the rectangle contained by AB and HB is equal to the square on side AH (figure 3.2). (Note that finding the unknown line AH also determines the line HB). Now, even if we consider this an "algebra" problem, it does not ask us to find a "number" as the problems from Egypt and Mesopotamia do. Rather, it asks that we determine a certain unknown line segment that is part of a given line segment. With twenty-twenty hindsight, this problem translates easily into an algebraic equation with literal coefficients by setting the length of the line AB to be a and that of AH to be x. Then, $HB = a - x$, and the problem amounts to solving the equation[6]

$$a(a - x) = x^2 \text{ or } x^2 + ax = a^2.$$

[6] Keep in mind that the symbolism here and elsewhere in this section is not found in the *Elements*. We are using a modern interpretation in order to represent the heritage of the *Elements*, not their history.

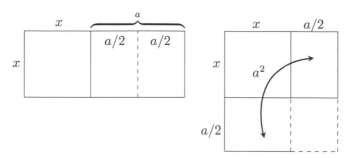

Figure 3.3.

As we have seen, the Mesopotamians could solve this problem in terms of a geometric solution that we can represent by the formula

$$x = \sqrt{\left(\frac{a}{2}\right)^2 + a^2} - \frac{a}{2}.$$

How did Euclid tackle it?

Typical of his style of exposition, Euclid presented a solution with no indication as to how he had arrived at it. To gain some insight into what his problem-solving technique may have been, let us try an analysis of the problem, that is, let us assume that the appropriate line segment has been found and see what this implies. The Mesopotamian method would have us represent $x^2 + ax$ by a square and an attached rectangle (figure 3.3) and then cut the rectangle in half, move one half to the bottom of the square, and "complete the square" by adding a square of side $\frac{a}{2}$. The line of length $x + \frac{a}{2}$ would then be the side of a square of area $a^2 + \left(\frac{a}{2}\right)^2$. Thus, to determine x, we would need first to construct a square out of the sum of two squares, an easy thing to do in light of the Pythagorean theorem by taking the hypotenuse of a right triangle whose legs are the sides of the two original squares.

This is, in fact, exactly what Euclid did. He drew the square on AB and then bisected AC at E (figure 3.4). It follows that EB is the desired hypotenuse. To subtract the line of length $\frac{a}{2}$ from EB, he drew EF equal to EB and subtracted off AE to get AF; this is the needed value x. Since he wanted the line segment marked off on AB, he simply chose H so that $AH = AF$.

With his construction completed, however, Euclid still needed to prove the result. Rather than draw numerous other squares and rectangles to

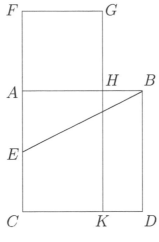

Figure 3.4.

show that the square on AH is equal to the rectangle contained by HB and $BD = AB$, he used an important result that he had proved earlier, in fact, a result probably designed precisely to prove proposition II.11:

Proposition II.6: *If a straight line is bisected and a straight line is added to it, the rectangle contained by the whole with the added straight line and the added straight line together with the square on the half is equal to the square on the straight line made up of the half and the added straight line.*

In this case, the line AC has been bisected and a straight line AF added to it. Therefore, according to this proposition, the rectangle on FC and AF plus the square on AE equals the square on FE. But, the square on FE equals the square on EB, which, in turn, is the sum of the squares on AE and AB. It follows that the rectangle on FC and AF (equal to the rectangle on FC and FG) equals the square on AB. Subtraction of the common rectangle contained by AH and HK yields the result that the square on AH equals the rectangle on HB and BD, as desired.

This now reduces the problem to proving proposition II.6 explicitly. To do that, Euclid did draw all the relevant squares and rectangles (figure 3.5). Here, AB is the original straight line, to which the segment BD is added. Thus, the goal of the proposition is to prove that the rectangle contained by AD and $BD = DM$, together with the square on CB (equal to the square on LH), is equal to the square on CD. This reduces to showing that the rectangle AL (equal to the rectangle CH) is

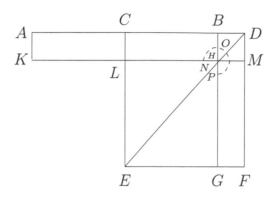

Figure 3.5.

equal to the rectangle HF, and Euclid accomplished this by using the proposition that the complements of a parallelogram about the diagonal are equal.

It is important to notice that this diagram, where we "complete the square" (on the gnomon NOP) by adding in the square on LH, is remarkably similar to figure 2.3. That was the diagram central to the Mesopotamian solution of both the simultaneous equations $y - x = b$, $xy = c$ and the equivalent quadratic equation $x(x + b) = c$ (or $x^2 + bx = c$). The relevance of this proposition and accompanying diagram for solving those equations was noticed by Islamic mathematicians in the ninth century; this proposition thus formed the basis of their proof of what amounts to the Mesopotamian procedure for the solutions. If we label AB as b and BD as x, the proposition shows that

$$(b + x)x + \left(\frac{b}{2}\right)^2 = \left(\frac{b}{2} + x\right)^2$$

and, therefore, that

$$x + \frac{b}{2} = \sqrt{\left(\frac{b}{2}\right)^2 + c} \quad \text{or} \quad x = \sqrt{\left(\frac{b}{2}\right)^2 + c} - \frac{b}{2}.$$

As we have just seen, Euclid, too, justified his solution of what amounts to a quadratic equation of this type by quoting proposition II.6.

Islamic mathematicians also noticed that proposition II.5 could be used to justify the analogous solution of the simultaneous

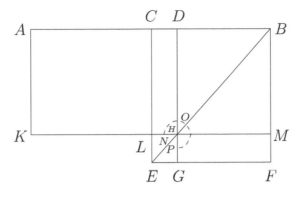

Figure 3.6.

equations $x + y = b$, $xy = c$ or the quadratic equation $x(b - x) = c$ (or $x^2 + c = bx$).

> **Proposition II.5:** *If a straight line is cut into equal and unequal segments, the rectangle contained by the unequal segments of the whole together with the square on the straight line between the points of section is equal to the square on the half.* (See figure 3.6; note its similarity to figure 2.4).

Euclid had used this same proposition to justify his own solution of the slightly different "equation" that follows.

> **Proposition II.14:** *To construct a square equal to a given rectilinear figure.*

We can think of this latter construction as an algebraic problem because we are asked to find an unknown side of a square satisfying certain conditions. Since Euclid had already shown that a rectangle could be constructed equal to any rectilinear figure, this problem translates into the modern equation $x^2 = cd$, where c, d are the lengths of the sides of the rectangle constructed. Again, an analysis may prove helpful.

Assume that $c \neq d$ (for otherwise the problem would be solved). We know from proposition II.5 that

$$cd = \left(\frac{c+d}{2}\right)^2 - \left(\frac{c-d}{2}\right)^2.$$

Since we want the right-hand side of this equation to be a square, we simply need to find the side of a square that is a difference of two squares. As before, by the Pythagorean theorem, this is one leg of a right triangle,

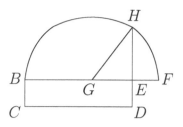

Figure 3.7.

when the two given squares are the squares of the hypotenuse and the other leg. Thus, we need to create a right triangle with one leg equal to $\frac{c-d}{2}$ and hypotenuse equal to $\frac{c+d}{2}$. This is exactly what Euclid did. He placed the sides of the rectangle $BE = c$, $ED = EF = d$ in a straight line and bisected BF at G (figure 3.7) to get $GE = \frac{c-d}{2}$, thereby obtaining one side of the desired right triangle. To get a line equal to GF as the hypotenuse, he constructed the semicircle BHF of radius GF, where H is the intersection of that semicircle with the perpendicular to BF at E. The desired hypotenuse is then GH, and HE is the side of the square that solves the problem.

Note, here, that we by no means claim that Euclid's intention in stating propositions II.5 and II.6 was to solve quadratic equations; in fact, he used them to justify *geometric* solutions to two very specific literal equations—articulated in propositions II.11 and II.14—asking for unknown line segments. For Euclid, the latter "equations" were probably just interesting problems to solve, and the identities II.5 and II.6 just happened to be of service in solving them. Interestingly, though, he treated both II.11 and II.14 a second time in the *Elements*, in Book VI, employing similarity techniques. While some of the other identities in Book II also proved useful in solving particular problems—especially those concerning conic sections, a topic not treated in the *Elements*—Euclid did not apply them in the later books of his treatise.

Euclid came closer actually to "solving a quadratic equation" in Book VI using the notion of "application of areas" that he had introduced in the following.

Proposition I.44: *To a given straight line to apply, in a given rectilinear angle, a parallelogram equal to a given triangle.*

This construction involves finding a parallelogram of given area with one angle given and one side equal to a given line segment; the parallelogram is to be "applied" to the given line segment. If the angle is a right angle, we can interpret the problem algebraically. Taking the area of the triangle to be cd and the given line segment to have length a, the goal is to find a line segment of length x such that the rectangle with length a and width x has area cd, that is, to solve the equation $ax = cd$. Euclid treated this problem geometrically, of course, but within the context of Book I, he employed a fairly complicated procedure involving several areas.

In Book VI, however, he expanded the notion of "application of areas" to applications that are "deficient" or "exceeding." The importance of these notions will be apparent in the discussion of conic sections below, but for now we note that, in modern terms, Euclid gave geometric solutions to two types of quadratic equations as follows.

> **Proposition VI.28:** *To a given straight line to apply a parallelogram equal to a given rectilinear figure and deficient by a parallelogram similar to a given one; thus the given rectilinear figure must not be greater than the parallelogram described on the half of the straight line and similar to the defect.*

> **Proposition VI.29:** *To a given straight line to apply a parallelogram equal to a given rectilinear figure and exceeding by a parallelogram figure similar to a given one.*

In the first case, Euclid proposed to construct a parallelogram of given area, the base of which is less than the given line segment AB. The parallelogram on the deficiency—that is, the line segment SB—is to be similar to a given parallelogram (figure 3.8). In the second case, the constructed parallelogram of given area has base greater than the given line segment AB, while the parallelogram on the excess—that is, the line segment BS—is again to be similar to one given (figure 3.9).

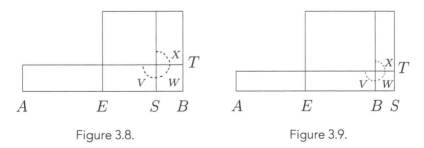

Figure 3.8.   Figure 3.9.

Figure 3.10. Euclid's *Elements* VI-28 (although labeled VI-26) from f. 58 of the Plimpton MS 165 (ca. 1294), courtesy of the Rare Book and Manuscript Library, Columbia University. Note that the parallelogram is drawn as a rectangle.

To simplify matters, and to show why we can think of Euclid's constructions as solving quadratic equations, assume, as in the figures, that the given parallelogram in each case is a square, that is, that the defect or the excess will itself be a square. This implies that the constructed parallelograms are rectangles. (Note that this is, in fact, the assumption of the medieval scribe who prepared the pictured manuscript of proposition VI.28 of the *Elements*. See figure 3.10.). Designate *AB* in each case by b and the area of the given rectilinear figure by c. The problems reduce to finding a point S on AB (proposition VI.28) or on AB extended (proposition VI.29) so that $x = BS$ satisfies $x(b-x) = c$ in the first case and $x(b+x) = c$ in the second. That is, the goals of the problems are, in modern terms, to solve the quadratic equations $bx - x^2 = c$ and $bx + x^2 = c$, respectively.

In each case, Euclid found the midpoint E of AB and constructed the square on BE with area $\left(\frac{b}{2}\right)^2$. In the first, he chose S so that ES is the side of a square of area $\left(\frac{b}{2}\right)^2 - c$ (hence the condition stated in the proposition that, in effect, c cannot be greater than $\left(\frac{b}{2}\right)^2$). This choice for ES implies that

$$x = BS = BE - ES = \frac{b}{2} - \sqrt{\left(\frac{b}{2}\right)^2 - c}.$$

In the second, he chose S so that ES is the side of a square of area $\left(\frac{b}{2}\right)^2 + c$, and thus

$$x = BS = ES - BE = \sqrt{\left(\frac{b}{2}\right)^2 + c} - \frac{b}{2}.$$

In both cases, one can prove that the choice is correct by showing that the desired rectangle equals the gnomon XWV and that the gnomon is, in turn, equal to the given area c. Algebraically, that amounts to showing, in the first case, that

$$x(b-x) = \left(\frac{b}{2}\right)^2 - \left[\left(\frac{b}{2}\right)^2 - c\right] = c$$

and, in the second,

$$x(b+x) = \left[\left(\frac{b}{2}\right)^2 + c\right] - \left(\frac{b}{2}\right)^2 = c.$$

Given that these two theorems are about parallelograms—rather than rectangles and squares—Euclid's proof involved constructing various new parallelograms and demonstrating their relationships. It is easy enough to see from the diagrams, however, that his constructions generalize the proofs of propositions II.5 and II.6. If he had stated these propositions explicitly for the special case that the given parallelogram is a square, he could simply have referred to II.5 and II.6 for the proofs.

This idea is made even clearer if we turn to one of Euclid's other works, the *Data*, effectively a supplement to Books I–VI of the *Elements*. Each proposition in the *Data* takes certain parts of a geometric configuration as given or known and shows that, therefore, certain other parts are determined. ("Data" means "given" in Latin). Generally, Euclid did this by showing *exactly how* to determine them. The *Data*, in essence, transformed the synthetic purity of the *Elements* into a manual appropriate to one of the goals of Greek mathematics: the solution of new problems.

To get the flavor of this, consider two propositions from the *Data*—propositions 84 and 59—closely related to proposition VI.29 in the *Elements*.

Proposition 84: *If two straight lines contain a given area in a given angle, and one of them be greater than the other by a given straight line, each of them will be given, too.*[7]

If, as in the discussion of proposition VI.29, it is assumed that the given angle is a right angle—and, as above, the diagram for this proposition in the surviving medieval manuscripts shows such an angle—the problem is closely related to one of the standard Mesopotamian problems, namely, to find x, y if the product and difference are given, that is, to solve the system

$$xy = c, \qquad y - x = b.$$

Euclid began by setting up the rectangle contained by the two straight lines TS, AS (figure 3.11). He then chose point B on AS so that $BS = TS$. Thus, $AB = b$ was the given straight line. He now had a given area—the rectangle ($AT = c$) applied to a given line b—exceeding by a square figure, to which he could apply

Proposition 59: *If a given area be applied to a given straight line, exceeding by a figure given in form [that is, up to similarity], the length and width of the excess are given.*

---

[7] All quotations from the *Data* are taken from Taisbak, 2003.

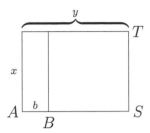

Figure 3.11.

It is in this context that Euclid really solved the problem of proposition 84, using a diagram similar to that of proposition VI.29 in the *Elements* (figure 3.12). As there, he bisected the line $AB = b$ at E, constructed the square on $BE = \frac{b}{2}$, noted that the sum of that square and the original area (the rectangle $AT = c$) is equal to the square on $SE = x + \frac{b}{2}$ (or $y - \frac{b}{2}$), and thereby showed how either of those quantities would be determined as the side of that square. Algebraically, this amounts to the standard Mesopotamian formulas

$$y = \sqrt{\left(\frac{b}{2}\right)^2 + c} + \frac{b}{2}$$

and

$$x = \sqrt{\left(\frac{b}{2}\right)^2 + c} - \frac{b}{2}.$$

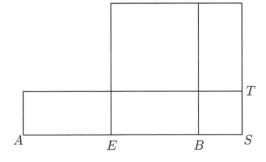

Figure 3.12.

As before, Euclid dealt only with geometric figures and never actually wrote out rules like these. However, while his formulation of the problem—in terms of finding two lengths satisfying certain conditions—was nearly identical to the Mesopotamian formulation, it nevertheless enabled him to generalize the Mesopotamian problem from rectangles to parallelograms. He followed this same pattern in propositions 85 and 58, solving the geometric equivalent of the system

$$xy = c, \quad x + y = b$$

(analogous to proposition VI.28 in the *Elements*).

Since the notion of application of areas is important in the definition of the conic sections and since Euclid is credited with a book on that subject, it is not surprising that he explored these ideas in the *Elements*. Because his version of the conics is no longer extant, however, we look instead at the classic work that Apollonius (fl. 200 BCE) wrote on the topic.

GEOMETRICAL ALGEBRA IN APOLLONIUS'S *CONICS*

Apollonius hailed from the city of Perga on the southern coast of the Anatolian Peninsula (modern-day Turkey) across the Mediterranean Sea and almost due north of the city of Alexandria. There, he reportedly studied under the followers of Euclid. It is perhaps not surprising, then, that Apollonius became interested in the kinds of mathematical questions—such as conic sections—that Euclid had explored. There are, however, various indications that the Greeks were studying conics in detail well before Euclid, but there is only speculation as to the theory's exact origins. It may have originated in the context of the classic problem of doubling the cube, that is, of finding a cube whose volume is double that of a given cube or, in algebraic terms, of finding a side x so that $x^3 = 2a^3$, where a is the side of the given cube. The Greeks evidently reduced this problem (what we can characterize as a pure cubic equation) to the problem of finding two mean proportionals between the side of the original cube a and its double, that is, of determining x, y such that $a : x = x : y = y : 2a$. In modern terms, this is equivalent to solving simultaneously any two of the three equations $x^2 = ay$, $y^2 = 2ax$, and

$xy = 2a^2$, equations that represent parabolas in the first two instances and a hyperbola in the third.

While the Greeks did not write equations like these, by some time in the fourth century BCE, there were certainly scholars who recognized that if the points on a curve had their abscissa and ordinate in the relationships expressed by them, then the resultant curves could be realized as sections of a cone. The theory of conic sections was thus part of Greek geometry. In the seventeenth century, however, conic sections were central to the development of analytic geometry—the application of algebra to geometry—and the creators of that subject were convinced that the Greeks had developed conic sections using a form of algebra (see chapter 10). To understand why, consider Apollonius's treatment of the hyperbola in Books I and II of his magnum opus, the *Conics*.

Apollonius defined the hyperbola initially by cutting the cone with a plane through its axis. The intersection of this plane with the base circle is a diameter BC, and the resulting triangle ABC is called the axial triangle. The hyperbola is then a section of the cone determined by a plane that cuts the plane of the base circle in the straight line ST perpendicular to BC so that the line EG—the intersection of the cutting plane with the axial triangle—intersects one side of the axial triangle and the other side produced beyond A (figure 3.13).[8] In fact, in this situation, there are two branches of the curve. Given this definition, Apollonius derived the "symptom" of the curve, the characteristic relationship between the ordinate and abscissa of an arbitrary point on the curve. Apollonius, of course, presented his results in geometric language, but we will modify this somewhat for ease of reading, replacing, in particular, "the square on a line AB" by "$\overline{AB}^2$," "the rectangle contained by AB and BC" by "$AB \cdot BC$," and "the ratio of AB to CD" by "$\frac{AB}{CD}$."

Apollonius began by picking an arbitrary point L on the section and passing a plane through L parallel to the base circle. The section of the cone produced by the plane is a circle with diameter PR. Letting M be the intersection of this plane with the line EG, LM is perpendicular to PR, and therefore $\overline{LM}^2 = PM \cdot MR$. Now let D be the intersection of EG with the second side of the axial triangle produced, and let EH (the so-called

---

[8] Apollonius, 1998, pp. 21–23 (prop. I-12) for the definition and symptom of the hyperbola.

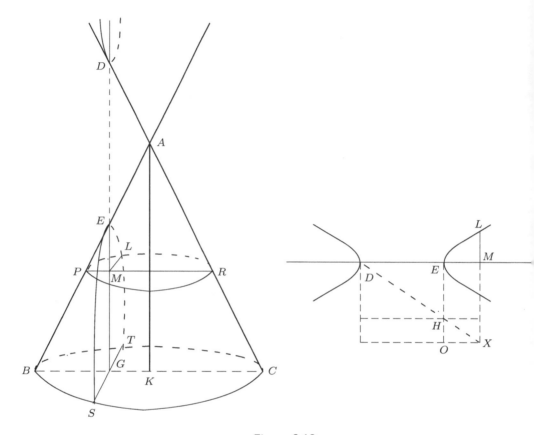

Figure 3.13.

"parameter" of the curve) be chosen so that

$$\frac{DE}{EH} = \frac{\overline{AK}^2}{BK \cdot KC} = \frac{AK}{BK} \cdot \frac{AK}{KC},$$

where AK is parallel to DE. (EH is drawn perpendicular to DE). Apollonius then proved that the square on LM is equal to a rectangle applied to the line EH with width equal to EM and exceeding (*yperboli*) by a rectangle similar to the one contained by DE and EH, whence the name "hyperbola" for the curve.

Now, by similarity,

$$\frac{AK}{BK} = \frac{EG}{BG} = \frac{EM}{MP} \quad \text{and} \quad \frac{AK}{KC} = \frac{DG}{GC} = \frac{DM}{MR}.$$

Therefore,

$$\frac{DE}{EH} = \frac{EM \cdot DM}{MP \cdot MR};$$

but also

$$\frac{DE}{EH} = \frac{DM}{MX} = \frac{DM}{EO} = \frac{EM \cdot DM}{EM \cdot EO}.$$

It follows that $MP \cdot MR = EM \cdot EO$ and, therefore, that $\overline{LM}^2 = EM \cdot EO = EM \cdot (EH + HO)$. Because the rectangle contained by EM ($= OX$) and HO is similar to the one contained by DE and EH, Apollonius has therefore proved his result.

It is easy enough for us to translate Apollonius's symptom into modern terms since we can conceive of the axis of the curve as our x-axis, with the ordinates measured perpendicularly. First, note that because $\frac{EM}{HO} = \frac{DE}{EH}$, we have $HO = \frac{EM \cdot EH}{DE}$. Therefore, if we set $LM = y$, $EM = x$, $EH = p$, and $DE = 2a$, Apollonius's symptom becomes the modern equation

$$y^2 = x\left(p + \frac{p}{2a}x\right)$$

for the hyperbola, where the parameter p depends only on the cutting plane determining the curve. By similar arguments, Apollonius showed that, for the ellipse, the square on the ordinate is equal to a rectangle applied to a line—the parameter—with width the abscissa and deficient (*ellipsis*) by a rectangle similar to the rectangle contained by the diameter of the ellipse and the parameter, while for the parabola, the square on the ordinate is equal to the rectangle applied (*paraboli*) to the parameter with width equal to the abscissa.[9] In modern terms, with the same conventions as before, the equations of the ellipse and parabola are

$$y^2 = x\left(p - \frac{p}{2a}x\right) \qquad \text{and} \qquad y^2 = px,$$

respectively.

[9] For these arguments, see Apollonius, 1998, pp. 24–26 (prop. I-13) and pp. 19–21 (prop. I-11), respectively.

Having derived the symptoms of the three curves from their definitions as sections of a cone, Apollonius showed conversely that given a parameter and a vertex (or a pair of vertices) at the end(s) of a given line (line segment), a cone and a cutting plane can be found, the section of which is a parabola (ellipse or hyperbola) with the given vertex (vertices), axis, and parameter.[10] Therefore, an ancient or medieval author could assert the "construction" of a conic section with given vertices, axes, and parameter in the same manner as the construction of a circle with given center and radius. Furthermore, after showing how to construct the asymptotes of a hyperbola, Apollonius demonstrated how to construct a hyperbola given a point on the hyperbola and its asymptotes, thus providing an additional construction possibility.[11] Finally, he showed that the symptoms of a hyperbola could be expressed in terms of its asymptotes instead of its parameter and axis. In particular, he proved

Proposition II.12: *If two straight lines at chance angles are drawn to the asymptotes from some point on the hyperbola, and parallels are drawn to the two straight lines from some other point on the section, then the rectangle contained by the parallels will be equal to that contained by those straight lines to which they were drawn parallel.*[12]

In other words, in figure 3.14, D and G are arbitrary points on the hyperbola, the asymptotes of which are AB and BC. Drawing DF and DE so that they meet the asymptotes and drawing GH parallel to DE and GK parallel to DF, the theorem then claims that $DE \cdot DF = GH \cdot GK$. If we translate this result into modern terms in the special case where the asymptotes are the coordinate axes, then we get a new defining equation for the hyperbola, namely, $xy = k$. Geometrically, this result means that, given a hyperbola with perpendicular asymptotes, every point on it defines a rectangle with the same area.

In deriving the properties of the conics, Apollonius generally used the symptoms of the curves—rather than the original definition—just as in modern practice these properties are derived from the equation. Still, his proofs were all geometrical and frequently very complex. It was this complexity that led seventeenth-century mathematicians to claim that

---

[10] See Apollonius, 1998, pp. 96–115 (props. I-52 through I-60) for this argument.
[11] Apollonius effects this construction in Apollonius, 1998, pp. 120–121 (prop. II-4).
[12] Apollonius, 1998, p. 127.

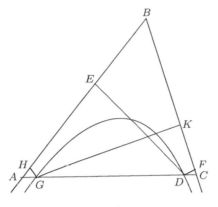

Figure 3.14.

he must have used some form of algebra to derive the properties; it was too difficult for them to imagine that such properties could have been derived purely geometrically. Although it is now generally held that Apollonius did not use algebra, it is frequently, as we have just seen, quite straightforward to translate some of his properties as well as his derivations and proofs into algebraic forms. Following the invention of analytic geometry in the seventeenth century, mathematicians generally read the *Conics* as "algebraic geometry" and translated Apollonius's arguments and results into algebraic language. It is, therefore, fair to say that, although Apollonius was *historically* a great geometer, his *heritage* is an algebraic form of geometry.

ARCHIMEDES AND THE SOLUTION OF A CUBIC EQUATION

Archimedes (287–212 BCE), arguably the greatest of the ancient mathematicians, lived in Syracuse on the island of Sicily between the times of Euclid and Apollonius, and he, like Apollonius, most probably journeyed to Alexandria to study under Euclid's successors. There, he would not only have met those Alexandrian scholars to whom he would later personally address some of his scientific works but also, like Apollonius, have acquired the knowledge of conic sections of which he would make such frequent use in his geometrical works. Here, we consider just one proposition from *The Sphere and the Cylinder II*:

Proposition 4: *To cut a given sphere so that the segments of the sphere have to each other the same ratio as a given [ratio].*[13]

In effecting the solution of this geometrical problem,[14] Archimedes not only displayed his considerable mathematical talents but also employed conics in such a way that eventually led to a general method of solving cubic equations.

Using earlier results, Archimedes reduced the problem to a special case of a general scenario about lines (figure 3.15): To cut a given straight line AB at a point E such that $AE : AG = \Delta : \overline{BE}^2$, where AG is a given line segment (here drawn perpendicular to AB) and Δ is a given area. Setting $AB = a$, $BE = x$, $AG = d$, and $\Delta = c^2$, we can translate this into the problem of finding x such that $(a - x) : d = c^2 : x^2$, that is, into solving the cubic equation $x^2(a - x) = c^2 d$. Aside from the pure cubic equations noted above, this is the first known instance in Greek mathematics translatable into modern terms as a cubic equation. Naturally, Archimedes did not consider this problem in this way, but he did solve it geometrically by means of conic sections.

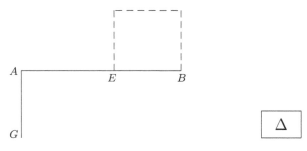

Figure 3.15.

He first produced an analysis of the problem. Assuming the construction completed, he drew GE, continued it to Z, drew GH parallel and equal to AB, and completed rectangle $GTZH$. Further, he drew KEL parallel to ZH and continued GH to M such that the rectangle on GH and HM is equal to the given area Δ (figure 3.16). With this,

$$\Delta : \overline{BE}^2 = AE : AG = GH : HZ \text{ (by similarity)} = \overline{GH}^2 : GH \cdot HZ.$$

[13] Netz, 2004b, p. 202 (slightly modified for clarity).
[14] See Netz, 2004a for a detailed discussion of this problem and its history.

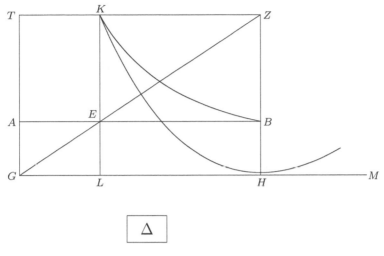

Figure 3.16.

Because $\overline{BE}^2 = \overline{KZ}^2$, we have $\overline{GH}^2 : GH \cdot HZ = \Delta : \overline{KZ}^2$, or $GH \cdot HZ : \overline{KZ}^2 = \overline{GH}^2 : \Delta$. But $\Delta = GH \cdot HM$. Therefore,

$$\overline{GH}^2 : \Delta = \overline{GH}^2 : GH \cdot HM = GH : HM = GH \cdot HZ : HM \cdot HZ.$$

It follows that $HM \cdot HZ = \overline{KZ}^2$. For Archimedes, this result meant that K was on the parabola through H with axis HZ and parameter HM. (If we use a coordinate system with origin H, and set $HM = \ell$, then the parabola has equation $x^2 = \ell y$). Furthermore, rectangles $GTKL$ and $AGHB$ are equal, or $TK \cdot KL = AB \cdot BH$. Again, this equality of areas shows that the hyperbola through B with asymptotes TG and GH passes through K. (This hyperbola has equation $(a - x)y = ad$). Archimedes thus concluded the analysis by noting that K was on the intersection of the hyperbola and parabola. Because hyperbolas and parabolas could be constructed, as noted earlier, Archimedes, in his synthesis of the problem, constructed them, determined K as their intersection, and then found E as in the diagram to solve the original problem.

The synthesis, however, is not complete unless we know in advance that the two curves intersect, so Archimedes considered this issue in detail. First, he noted that the original proportion implies that the desired solid with base the square on BE and height AE is equal to the given solid with base of area Δ and height AG. Because AE and BE are segments of line AB, however, the volume of the desired solid cannot be arbitrarily large. In fact, as Archimedes showed, a maximum volume occurs when BE is twice AE, and the problem cannot be solved unless the given solid has volume less than this maximum. He also showed that if this condition is met, then there is one solution if the volume of the given solid equals the maximum and two if it is less than the maximum. Because our equation $x^2(a - x) = c^2d$ represents the relationship between the solids, Archimedes had therefore shown that the maximum of $x^2(a - x)$ occurs when $x = \frac{2}{3}a$ and that the equation can be solved if $c^2d \leq \frac{4}{9}a^2 \cdot \frac{1}{3}a = \frac{4}{27}a^3$. Finally, he found that the original sphere problem reduced to a special case of the general problem, where Δ is the square on $\frac{2}{3}$ of the line AB and AG is less than $\frac{1}{3}$ of line AB. In that case, it is clear that the inequality condition is met, and the problem is solvable. Archimedes could also easily determine which of the two solutions to the general problem actually split the diameter of the sphere so that the segments were in the given ratio.

It is often stated, in connection with this problem, that "Archimedes solved a cubic equation,"[15] that is, for a particular cubic equation with literal coefficients, he demonstrated that it could be solved through the intersection of a parabola and a hyperbola. Furthermore, he found the limits of solution, demonstrating the conditions under which the problem is solvable. In terms of our earlier definition, then, Archimedes solved an algebraic problem. Still, we should keep in mind that he did not set out to "solve an equation." Rather, he sought to construct the solution to an interesting geometric problem. All of the expressions used in the solution are geometric, including the conic sections. Archimedes also made no attempt to generalize his solution to any other type of cubic equation, the natural inclination of someone trying to solve equations. It is thus fair to say that the *history* of this problem in the time of Archimedes is geometric, but, given that Islamic mathematicians studied this solution and eventually generalized it to cover all possible cubic equations, its *heritage* is algebraic.

****

As we have seen, arguably the three most important Greek mathematicians in the last centuries before the common era all contributed to the

---

[15] See Archimedes, 1953, pp. cxxv ff. for one discussion of this notion.

enterprise of geometrical algebra. That is, although they wrote geometric treatises and solved geometric problems, aspects of their work came to be thought of as "algebraic," that is, as mathematics dealing with the solution of equations. From Euclid, later mathematicians derived the solution of quadratic equations; from Apollonius, they derived the equations and properties of conic sections; and from Archimedes, they derived a particular method for solving cubic equations. This transformation from pure geometry to geometrical algebra in the first and third cases took place during Islamic times, as we shall see more clearly in chapter 7. The transformation of Apollonius's work, on the other hand, did not occur until the seventeenth century when conic sections were required to deal with various physical questions (see chapter 10). In the next chapter, however, we consider a *purely algebraic* work from Hellenistic Alexandria, one that had apparently shed all aspects of classic Greek geometry.

4

Later Alexandrian Developments

Euclid and, as we saw, most probably Apollonius and Archimedes participated in the vibrant research and cultural complex that began to coalesce around the Museum and Library in Alexandria around 300 BCE.[1] Conceived by Alexander the Great (356–323 BCE) and built on a strip of land just to the west of the Nile delta with the Mediterranean on its northern and Lake Mareotis on its southern shore, the city of Alexandria had rapidly taken shape in the decades after Alexander's death in the hands of his boyhood friend and senior general, Ptolemy I. As students of Aristotle, both Alexander and Ptolemy valued education, and Ptolemy sought to promote it in the new city through the establishment of both the Museum where scholars met and worked and the Library where they consulted texts on topics as diverse as theater, medicine, rhetoric, mathematics, and astronomy. By the time Egypt officially came under Roman rule in 27 BCE, Alexandria's scholarly institutions had already hosted and nurtured such noted classical thinkers as our mathematicians Euclid, Apollonius, and Archimedes, as well as the astronomer and geographer Eratosthenes (ca. 276–195 BCE) and the medical theorists Herophilus (ca. 335–280 BCE) and Erasistratus (ca. 300–250 BCE).

Although one (unintended) consequence of the Roman takeover of the city was the burning of the Library and the loss of 400,000 to 500,000 of the perhaps 750,000 papyrus scrolls stored there, Alexandria continued to function as an intellectual hub well into the common era, albeit one with a character different from that of Ptolemaic Alexandria. Whereas the Museum and Library under the Ptolemies had functioned as high-level research facilities for a select group of scholars who enjoyed the patronage of the state, those same facilities under Roman domination in

[1] The brief sketch of the Alexandrian context that follows draws from Pollard and Reid, 2006.

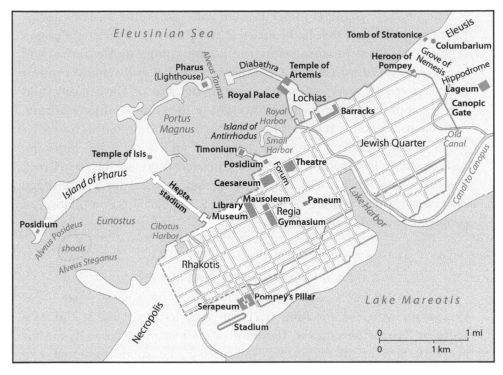

Figure 4.1. Map of ancient Alexandria.

the opening centuries of the common era came almost immediately to serve more practical purposes. The mathematician and engineer Hero of Alexandria (ca. 10–ca. 70 CE), for example, not only explored some of the theoretical aspects of mechanics but also engaged in applied engineering, amazing the rulers and citizenry of Alexandria with the pneumatic and hydraulic automata that he devised at the Museum. Others of his colleagues there embraced the more purely instructional goal of educating young nobles. If the latter scholars thus tended to focus more on gathering and organizing information and less on the creation of new ideas, they and their successors nevertheless continued to define an intellectual focal point within the Roman Empire. The noted physician Galen of Pergamum (ca. 130–200 CE) studied in Alexandria, and the astronomer Claudius Ptolemy (ca. 100–170 CE) worked and made observations there between 127 and 141 CE that informed the epicyclic mathematical models for the planetary orbits that he presented in his *Almagest*.

The city, too, changed under Roman rule. An ethnic mix of Greeks, Egyptians, and Jews that had coexisted and practiced a variety of pagan religions as well as Judaism before the Roman conquest, Alexandria had become by the third century CE a hotbed of civil and religious unrest characterized by waves of anti-Semitism, the rise of Christianity, and the development of Neoplatonic thought in the context of persevering, indigenous, Egyptian religious rites and practices. Despite this unsettled social climate, scholars continued to gather and work at the Museum. Among them, Diophantus (fl. 250 CE) produced one of the first Greek texts to come down to us that can be said to have both a heritage *and* a history that may be termed algebraic, namely, his compilation of problems entitled *Arithmetica*.

DIOPHANTINE PRELIMINARIES

Consonant with the teaching tradition that had evolved at the Museum in Roman Alexandria, Diophantus's *Arithmetica* was, as its author acknowledged in his dedication, intended to be an instructional tool for advanced students "anxious to learn how to investigate problems in numbers."[2] As Diophantus explained, he "tried, beginning from the foundations on which the science is built up, to set forth... the nature and power subsisting in numbers," but he also allowed that "perhaps the subject will appear rather difficult, inasmuch as it is not yet familiar." The material that Diophantus presented in the *Arithmetica* was not the geometry that Euclid had expounded in his *Elements*, material with which his readers would presumably have been familiar. Rather, he was providing a systematic, workbook-type treatment of the kinds of relations that can obtain between rational numbers—and especially their squares and cubes—and of how a priori to find rational numbers satisfying particular relations.

Of the thirteen books that originally comprised the *Arithmetica* only ten have come down to us—the first seven and then three more containing some of what was in the last six books. Of these, the first three—Books I through III—and the final three were part of a Greek manuscript tradition extending back to the thirteenth century, whereas Books IV

---

[2] Heath, 1964, p. 129. The quotations that follow in this paragraph are also from this page.

through VII only came to light in the early 1970s in an Arabic translation attributed to the ninth-century Islamic mathematician, Quṣṭā ibn Lūqā. Thus, until the 1970s, it was thought that the six Greek books of the *Arithmetica* were all that survived and that they were the work's first six consecutive books.[3] (See the conclusion of this chapter for more on the work's transmission from the third to the ninth century CE).

Although by no means a deductive, axiomatic treatment in the style of the *Elements*, the *Arithmetica* nevertheless opened with a series of definitions of what Diophantus termed "species" of numbers as well as with abbreviations that stand in for each one in the text. For Diophantus, our x^2 was called a "power" ($\delta \dot{\upsilon} \nu \alpha \mu \iota \varsigma$ or "dynamis") and denoted Δ^Y, and our x^3 was called "cube" ($\kappa \dot{\upsilon} \beta o \varsigma$ or "kubos") and denoted K^Y.[4] Indicative of the fact that Diophantus was engaged in what we would call algebra and not geometry, he broke the physical barrier of three dimensions by continuing up the scale and defining terms and notation additively to the sixth degree,[5] where x^4 was called "square-square" and denoted $\Delta^Y \Delta$, x^5 was called "square-cube" and denoted ΔK^Y, and x^6 was called "cube-cube" and denoted $K^Y K$. "But the number which has none of these characteristics," he continued, "but merely has in it an indeterminate multitude of units [that is, x] is called $\dot{\alpha} \rho \iota \theta \mu \dot{o} \varsigma$ ['arithmos'], 'number,' and its sign is ς." Finally, he specified that $\overset{\circ}{M}$ stand for "unit" ($\mu o \nu \dot{\alpha} \varsigma$ or "monas") or "that which is invariable in determinate numbers."

With these conventions in place, Diophantus proceeded to define the corresponding reciprocals of the powers of x and their notation. Thus, $\frac{1}{x}$ was termed $\dot{\alpha} \rho \iota \theta \mu o \sigma \tau \dot{o} \nu$ (or "arithmoston") and denoted ς^χ, $\frac{1}{x^2}$ was designated $\delta \upsilon \nu \alpha \mu o \sigma \tau \dot{o} \nu$ (or "dynamoston") and denoted $\Delta^{Y\chi}$, and so on. Notationally, too, Diophantus indicated addition by juxtaposition,

---

[3] Heath, 1964 is an edition of this sort, while Sesiano, 1982 provides the first English translation and critical edition of the Arabic manuscript of the original Books IV through VII. In what follows, we will use this ordering and notation for the books: I, II, and III for the first three books of Heath's translation; IV, V, VI, and VII for the next four books, that is, from the books translated and edited in Sesiano, 1982; and G-IV, G-V, and G-VI for the remaining Greek books that comprise part of what had been the original books VIII–XIII.

[4] See Heath, 1964, pp. 129–130 for this and the quotations that follow in this paragraph.

[5] Later on, in the fourth book, the sequence is extended to include x^8 and x^9, although not x^7. Since this book exists only in the ninth-century Arabic manuscript, we can only infer that the Greek denotations for these powers of the unknown might have been $\Delta K^Y K$ and $K^Y K^Y K$, respectively. Compare Sesiano, 1982, p. 177.

subtraction by ⅄ (which may have been an abbreviation associated with the Greek word for negation, that is, λεῖψις or "wanting") followed by a juxtaposition of all the terms to be subtracted, and "equals" by ι^σ, an abbreviation of the corresponding Greek word ἴσος. With these conventions, Diophantus would have written, for example, our $x^3 - 5x^2 + 8x - 1$ as

$$K^Y \bar{\alpha} s \bar{\eta} ⅄ \Delta^Y \bar{\epsilon} \overset{\circ}{M} \bar{\alpha},$$

where α is 1, β is 2, γ is 3, δ is 4, ϵ is 5, the ancient digamma ς is 6, ζ is 7, η is 8, and θ is 9, and where a bar over the letters distinguishes them as representing numbers.[6] The so-called "syncopated" style of mathematical exposition that resulted would prove very suggestive to European mathematicians like Raphael Bombelli and François Viète in the sixteenth century (see chapter 9).

Diophantus brought his introduction to Book I to a close with a discussion of the appropriate ways to manipulate an expression in order to find a desired solution. Like the good high school algebra teacher of today, he explained that "if a problem leads to an equation in which certain terms are equal to terms of the same species but with different coefficients, it will be necessary to subtract like from like on both sides until one term is found equal to one term"; in other words, he counseled his readers to combine like terms.[7] "If by chance," he continued, "there are on either side or on both sides any negative terms, it will be necessary to add the negative terms on both sides, until the terms on both sides are positive, and then again to subtract like from like until one term only is left on each side." These two steps are what Diophantus's later medieval Islamic readers would term the procedures of "al-jabr" and "al-muqābala," respectively, that is, "completion" and "balancing" (see chapter 7). Ideally, for Diophantus, all manipulations should result in what we would write as an expression of the form $ax^m = b$, but he promised his readers that "I will show you later how, in the case also

---

[6] For the example, see Heath, 1964, p. 42. Continuing through the Greek alphabet in sequence, ι is 10, κ is 20, ..., π is 80, the ancient Greek koppa ϙ is 90, ρ is 100, σ is 200, ..., ω is 800, and the ancient Greek sampi ϡ is 900. Numbers are then written positionally, so that, for example, 258 is σνη.

[7] For this and the quotations that follow in this paragraph, see Heath, 1964, p. 131.

where two terms are left equal to one term, such a problem is solved."[8]
With these preliminaries out of the way, Diophantus was ready to begin
his instruction in problem solving.

A SAMPLING FROM THE *ARITHMETICA*:
THE FIRST THREE GREEK BOOKS

Like Euclid in the *Elements*, Diophantus organized the material in the
Arithmetica from the simpler to the more complicated. In this spirit, Book
I primarily treated the case of determinate, that is, explicitly solvable,
equations either of or reducible to the first degree. The book opened in
problem I.1 asking the reader "to divide a given number into two having
a given difference."[9] Although Diophantus stated the problem generally,
he illustrated its solution method in terms of a particular example, taking
the given number to be 100 and the given difference to be 40, thereby
rendering the problem determinate. Letting x be the lesser of the two
numbers to be determined, the equation became $2x + 40 = 100$. Since 40
and 100 are "of the same species," it was necessary to subtract 40 from
both sides of the equation. Thus, $x = 30$, and the two numbers with the
desired difference were 30 and 70. This simple linear equation allowed
Diophantus to illustrate cleanly the fundamental manipulation technique
of combining like terms, a ploy that his readers would use over and over
again as they proceeded to master the problem-solving techniques that
he set out for them.

Diophantus followed this problem with others involving increasingly
complex scenarios for "dividing a given number," for example, "into two
having a given ratio" (I.2), into two "such that given fractions (not the
same) of each number when added together produce a given number"
(I.5), "twice into two numbers such that the first of the first pair may
have to the first of the second pair a given ratio, and also the second
of the second pair to the second of the first pair another given ratio"

---

[8] As Sesiano noted, a systematic exposition in fulfillment of this promise is unrealized in the
books that have come down to us, so it may have been the case that such a treatment comprised
all or part of the still missing books. See Sesiano, 1982, p. 78. Diophantus did, however, solve
isolated cases of such equations in the *Arithmetica*. See, for example, the discussion of problem
G-VI.6 below.

[9] Heath, 1964, p. 131.

(I.12), and "into three numbers such that the sum of each extreme and the mean has to the other extreme a given ratio" (I.20).[10] Consider, for example, Diophantus's approach to I.5. He took 100 as the initial number to be divided, $\frac{1}{3}$ and $\frac{1}{5}$ as the given fractions, and 30 as the given sum. By taking as the "second part $5x$," he was able to avoid fractions, since then one-fifth of the second part was simply x, and, moreover, since 30 equals one-third of the "first part" plus one-fifth of the "second part," the "first part" equals $3(30 - x)$. Thus, $(90 - 3x) + 5x = 100$, so $x = 5$, and the two numbers sought are 75 and 25. In treating I.5, however, Diophantus also, for the first of many times in the *Arithmetica*, needed to explain that the problem could not be solved unless a particular condition, which he articulated, was satisfied. As he noted, the second given number "must be such that it lies between the numbers arising when the given fractions respectively are taken of the first given number." To see what he meant, consider I.5 in modern notation.

We are asked to solve simultaneous linear equations of the form

$$x + y = a,$$

$$\frac{1}{m}x + \frac{1}{n}y = b,$$

where $m < n$. Diophantus's necessary condition required that $\frac{1}{n}a < b < \frac{1}{m}a$. Solving the two equations simultaneously yields the solution $x = \frac{m(nb-a)}{n-m}$. Since, for Diophantus, solutions must be positive, we must have $m(nb - a) > 0$ or $\frac{a}{n} < b$. To see the other inequality, note that the second equation yields $x + \frac{m}{n}y = mb$, but since $m < n$, we have $\frac{m}{n} < 1$, so the left-hand side of the equation, and hence mb, must be less than a.

Although most of the problems in Book I boil down to determinate systems of linear equations, three—I.27, I.28, and I.30—result in indeterminate systems—that is, systems for which there are more unknowns than equations and hence an infinite number of solutions a priori—reducible to quadratic equations. In I.27, for example, Diophantus asked his readers "to find two numbers such that their sum and product are given numbers."[11] In order to make the problem determinate, he took the sum to be 20 and the product to be 96, noting that a necessary

---

[10] Heath, 1964, pp. 131–137. For the quotations that follow in this paragraph, see p. 132.
[11] Heath, 1964, p. 140. The quotes that follow in this paragraph are also on this page.

condition must be satisfied, namely, "the square of half the sum must exceed the product by a square number" (here $\left(\frac{20}{2}\right)^2 = 96 + 4$). He then laid out the following steps to the solution. Let $2x$ be "the difference of the required numbers. Therefore the numbers are $10 + x$ and $10 - x$. Hence $100 - x^2 = 96$. Therefore $x = 2$, and the required numbers are 12, 8."

In modern notation, then, letting $\mathbb{X}$ and $\mathbb{Y}$ be the two numbers sought, Diophantus needed to solve the equations

$$\mathbb{X} + \mathbb{Y} = 20 \tag{4.1}$$

and

$$\mathbb{X}\mathbb{Y} = 96 \tag{4.2}$$

simultaneously. He introduced his unknown x by requiring the difference of the two desired numbers to be $2x$, that is, he required

$$\mathbb{X} - \mathbb{Y} = 2x, \tag{4.3}$$

where, in order to maintain positivity, he tacitly assumed that $\mathbb{X} > \mathbb{Y}$. Adding (4.1) and (4.3) and solving for $\mathbb{X}$ yielded $\mathbb{X} = 10 + x$, while subtracting (4.1) from (4.3) and solving for $\mathbb{Y}$ gave $\mathbb{Y} = 10 - x$. Substituting these values into (4.2) resulted in

$$96 = \mathbb{X}\mathbb{Y} = (10 + x)(10 - x) = 100 - x^2$$

or $x = 2$. Thus, the two desired numbers are 12 and 8.

We have, in fact, already encountered twice before the pair of simultaneous equations involved in I.27 in Diophantus's *Arithmetica*: in chapter 2 in our discussion of problem 25 on the Mesopotamian tablet BM 85200 + VAT 6599 and in chapter 3 in analyzing II.5 and II.14 in Euclid's *Elements*. In the first instance, the scribe provided an algorithm for determining the length and width of a rectangle given its area and the sum of its length and width, that is, an algorithm for effecting a geometric construction under certain conditions, while in the second, Euclid used the identity in II.5 to construct a square equal to a given rectangle. Each of these examples arose in an explicitly geometric context whereas Diophantus's did not. His problem was number-theoretic and, in fact, algebraic in the sense that it sought two unknown numbers, independently of any sort of geometrical construction. His method for finding them, however, was perfectly analogous to the Mesopotamian algorithm.

From the determinate—and for the most part linear—problems in Book I, Diophantus proceeded in Books II and III to introduce indeterminate equations of the second degree, the next level of sophistication. In the Greek edition that has come down to us, there are strong reasons to suspect that the first seven problems in Book II were interpolated later into the original Diophantine manuscript.[12] If that is, indeed, the case, then Diophantus's version of that book began with II.8, namely, "to divide a given square number into two squares."[13]

Diophantus opened his attack on the problem by specifying 16 as the given square. Translating into modern notation, he thus asked his readers to solve $x^2 + y^2 = 16$, one quadratic equation in two unknowns and hence an *indeterminate* quadratic. He next introduced the first of his five principal techniques for tackling problems involving squares; the others are laid out in II.9, II.10, II.11, and II.19.[14] Taking x^2 to be the other square, the problem reduces to solving for x such that $16 - x^2$ is also a square. To do this, Diophantus "form[ed] the square from any number of ἀριθμοί ["arithmoi" or "numbers"] minus as many units as there are in the side of 16"; in other words, he took "a square of the form $(mx - 4)^2$, m being any integer and 4 [being] the number which is the square root of 16." Because he required just one particular solution, he chose $m = 2$ to get $(2x - 4)^2$ and equated that to $16 - x^2$. Squaring and then combining like terms as in I.1 yielded $5x^2 = 16x$ or $x = \frac{16}{5}$, and the two squares sought were $\frac{256}{25}$ and $16 - \frac{256}{25} = \frac{144}{25}$. Notice that although Diophantus generated only one solution here, the method inherent in his problem II.8 produces as many different solutions as might be desired for an expression of the form $x^2 + y^2 = n^2$. For, taking m to be any (positive)

---

[12] See Heath, 1964, pp. 143–144 and Sesiano, 1982, p. 464.

[13] Heath, 1964, p. 144. For the quotes that follow in the next paragraph, see p. 145.

[14] In what follows, and merely by way of example, we look explicitly at the techniques laid out in II.8 and II.11 only. The other three techniques are in (II.9) "To divide a given number which is the sum of two squares into two other squares," (II.10) "To find two square numbers having a given difference," and (II.19) "To find three squares such that the difference between the greatest and the middle has to the difference between the middle and the least a given ratio." See Heath, 1964, pp. 145–151 for the quotations.

value and setting $y = mx - n$, we have $n^2 - x^2 = m^2 x^2 - 2mnx + n^2$ or $2mnx = (m^2 + 1)x^2$, that is, $x = \frac{2mn}{m^2+1}$.[15]

That Diophantus was fully aware that the techniques he presented in II.8, II.9, and II.10 had the power to produce more than the single solutions he provided is evidenced by a remark he made in the context of solving III.12, namely, "to find three numbers such that the product of any two added to the third gives a square."[16] Some preliminary spadework allowed him to reduce this to "find[ing] two squares differing by 48," and he noted that "this is easy and can be done in an infinite number of ways." This was precisely the sort of problem handled by the general technique he had presented to his readers in II.10, and one which, like those in II.8 and II.9, involved at one point the arbitrary selection of a positive value m. For every m, therefore, a different solution resulted.

Another of his general techniques, the so-called "double-equation" presented in II.11, also merits highlighting. In this case, the problem was "to add the same (required) number to two given numbers so as to make each of them a square," and, in particular, given the two numbers 2 and 3, to find x such that $x + 2$ and $x + 3$ are both squares.[17] Once again, the problem was indeterminate, this time involving two equations in three unknowns. "To solve it," Diophantus continued, "take the difference between the two expressions [to be made square, that is, $x + 2$ and $x + 3$] and resolve it into two factors; in this case let us say 4, $\frac{1}{4}$. Then take either (a) the square of half the difference between these factors and equate it to the lesser expression, or (b) the square of half the sum and equate it to the greater." Diophantus thus instructed his readers to consider, in modern notation, the difference between two squares $\mathbb{X}^2 - \mathbb{Y}^2$ and to write it as $(\mathbb{X} + \mathbb{Y})(\mathbb{X} - \mathbb{Y})$, where he took the first of these factors to be 4 and the second to be $\frac{1}{4}$. Now, the sum of the two factors is $2\mathbb{X} = \frac{17}{4}$, while their difference is $2\mathbb{Y} = \frac{15}{4}$, and at this point, there is a choice: either set $x + 2 = \left(\frac{2\mathbb{Y}}{2}\right)^2$ to solve for the answer or use $x + 3 = \left(\frac{2\mathbb{X}}{2}\right)^2$ to complete

---

[15] Compare Sesiano, 1982, pp. 6–7. As noted repeatedly above and in what follows, we present here an algebraic interpretation in modern notation of Diophantus's results that is consonant with how his work was interpreted by, especially, his sixteenth- and early seventeenth-century successors. For an alternate interpretation that may be closer to the context of the third century CE, see Thomaidis, 2005.

[16] See Heath, 1964, p. 161 for this and the quotations in the next sentence.

[17] For this and the following quote, see Heath, 1964, p. 146.

the problem. Note that the first option yields $x + 2 = \frac{225}{64}$, a square; the second gives $x + 3 = \frac{289}{64}$, a square; and both result in the same "required" number $x = \frac{97}{64}$.

It is important to recognize, however, that Diophantus glossed over one subtle point here. Since his solutions must always be positive and rational, he had to assign the numerical values to the product carefully. By choosing 3 and 2 as his initial numbers, the "difference between the two expressions" $x + 3$ and $x + 2$ was 1, which he then chose to write as the product of 4 and $\frac{1}{4}$. These numerical choices resulted in the positive rational solution $x = \frac{97}{64}$. Will any pair of reciprocals do the trick? Notice that if we replace 4 and $\frac{1}{4}$ by 2 and $\frac{1}{2}$ in the calculation above, the result is $x = -\frac{23}{16}$, a negative—hence impossible—solution for Diophantus. In order to set this problem up so as to yield a legitimate result, Diophantus had to select the reciprocals r and $\frac{1}{r}$ such that, in particular,

$$x = \left(\frac{2\mathbb{Y}}{2} \right)^2 - 2 = \left(\frac{\left(r - \frac{1}{r} \right)}{2} \right)^2 - 2 > 0,$$

that is,

$$\left(\frac{\left(r - \frac{1}{r} \right)}{2} \right)^2 > 2.$$

A SAMPLING FROM THE *ARITHMETICA*: THE ARABIC BOOKS

By the end of Book III, Diophantus had exposed his readers to many scenarios involving equations of the first and especially the second degree. A good teacher, he acknowledged in opening Book IV that he had "done that according to categories which beginners can memorize and grasp the nature of."[18] Book IV and subsequent books, however, were intended not only to continue to hone the students' skills in those types of problems but also to take them to the next level, that is, to problems involving cubes in addition to squares. In presenting the remaining material, Diophantus assured his readers that he would "follow the same

---

[18] See Sesiano, 1982, p. 87 for this and the following quotation.

path and advance you along it from one step to another and from one kind to another for the sake of experience and skill. Then," he continued, "when you are acquainted with what I have presented, you will be able to find the answer to many problems which I have not presented, since I shall have shown to you the procedure for solving a great many problems and shall have explained to you an example of each of their types."

Before entering this next phase of their mathematical training, however, Diophantus's students needed to have at least two additional tools in their algebraic toolbox: they would have to understand how to manipulate higher powers of x, and they would need always to reduce higher-order equations to lowest terms. To this end, Diophantus laid out two more rules before launching into the next series of problems.

The first—revisiting the powers of x from the first through the sixth degree that Diophantus had defined in Book I—detailed how to divide these powers by lower ones. Diophantus worked through each power separately, beginning with x^2 and noting that "every square multiplied by its side gives an x^3. When I then divide x^3 by x^2, the result is the side of x^3; if x^3 is divided by x, namely the root of the said x^2, the result is x^2."[19] He concluded with the analogous scale for x^5, explaining that "when x^5 is then multiplied by x, the result is the same as when x^3 is multiplied by itself and when x^2 is multiplied by x^4, and it is called x^6. If x^6 is divided by x, namely the root of x^2, the result is x^5; if it is divided by x^4, the result is x^2; if it is divided by x^5, the result is x, namely the root of x^2."[20]

The second new rule treated the issue of how properly to reduce higher-order expressions. As in the earlier books, problems would still need to be rendered ultimately in the form $ax^m = b$ in order to be solved, so one further step might be required. Routine completion and balancing could reduce a given problem to an expression of the form $ax^m = x^n$, for $1 \leq n < m \leq 6$, so, Diophantus explained, "it will then be necessary to divide the whole by a unit of the side having the lesser degree in order to obtain one species equal to a number." With these preliminaries out of the way, Diophantus led his students into, what were for them, previously uncharted mathematical waters.

---

[19] Sesiano, 1982, p. 87. For the next quotation and those in the next two paragraphs, see pp. 88–89.

[20] As noted, later in Book IV, Diophantus also presented x^8 and x^9 analogously, although, at least in the surviving books, he never considered x^7.

As in Book II where he had first introduced problems involving quadratic expressions, Diophantus approached the matter of cubics gently, gradually building up to more and more complicated manipulations in terms of combinations of squares and cubes. In IV.1, for example, he asked his readers simply "to find two cubic numbers the sum of which is a square number." The approach was straightforward, especially in light of the techniques he had presented earlier in the text. He took one of the cubes to be x^3 and the other to be the cube of "an arbitrary number of x's, say $2x$." Given this setup, $x^3 + 8x^3 = 9x^3$ needed to be a square. Again, "mak[ing] the side of that square any number of x's we please, say $6x$," the equation to be satisfied was $9x^3 = 36x^2$. Performing the enhanced sort of reduction he had detailed in the preface to Book IV yielded $x = 4$, so the two desired cubes were 64 and 512, and their sum, 576, "is a square with 24 as its side."

In IV.3, he turned the tables and asked for "two square numbers the sum of which is a cube," but he demonstrated that the problem-solving technique was analogous to that of IV.1 mutatis mutandis. Taking x^2 and $4x^2$ as the two squares and x^3 as the cube, he needed to solve $5x^2 = x^3$, which he did by dividing both sides of the expression by x^2 to find that $x = 5$. The two squares were thus 25 and 100, and their sum, 125, is a cube.

Both IV.1 and IV.3 translate easily into modern notation. In the first, it is a matter of solving $(1 + a^3)x^3 = b^2x^2$ so that $x = \frac{b^2}{1+a^3}$, while in the second, $(1 + a^2)x^2 = b^3x^3$ reduces to $x = \frac{1+a^2}{b^3}$. Problems IV.2 and IV.4 asked analogous questions for differences.

Interestingly, here Diophantus did not, as one might have expected, pose problems representative of the other possible combinations of squares, cubes, sums, and differences, namely, to find two squares the sum of which is a square, or to find two squares the difference of which is a square, or to find two cubes the sum of which is a cube, or to find two cubes the difference of which is a cube. Recall, though, that he had already done the first in II.8, so there would have been no real reason to repeat it here. As for the other three, the second had, for all intents and purposes, already been solved in II.10. There, Diophantus showed the reader how "to find two square numbers having a given difference," and nothing precludes that "given difference" itself from being a square.[21]

---

[21] See Heath, 1964, p. 146.

The third and fourth cases, however, are truly unsolvable, with the third being a special case of what would come to be known as Fermat's last theorem after the seventeenth-century French mathematician, Pierre de Fermat (see chapter 13); namely, no three positive integers a, b, c can satisfy the equation $a^n + b^n = c^n$, for n an integer greater than 2. While it is unlikely that Diophantus would have realized the theorem in its full generality, it may well have been the case that since he could not find an integer solution to what we would write as the equation $a^3 + b^3 = c^3$, he wisely chose not to broach the example in his text.

From sums and differences of squares and cubes, Diophantus moved on to products of them. After seeking in IV.5 "two square numbers which comprise [that is, whose product is] a cubic number," he proceeded in IV.6 and IV.7 to look for "two numbers, one square and the other cubic, which comprise a square number" and "which comprise a cubic number," respectively.[22] In IV.6, in particular, he chose x^2 and $(2x)^3$ as the square and cube and required their product $8x^5$ to be a square. Taking the square simply to be x^2 and performing the enhanced reduction, the problem became x^3 "equal to units," but what is needed is a final expression in x. To achieve that, instead of taking the latter square to be x^2, he exploited the properties of powers as laid out in the prefatory definitions, and took it instead to be $(4x^2)^2$. Then $8x^5 = 16x^4$, $x = 2$, and the desired square and cube were 4 and 64, respectively, with product 256, a square.

While the solutions to these examples played out with few surprises, such was not the case for IV.8, where Diophantus's students learned a valuable lesson. There, he challenged them "to find two cubic numbers which comprise a square number," and he led them down the calculational path to which they had become accustomed, namely, he posited x^3 for the first cube and $(ax)^3 = (2x)^3$ for the second. The product was then $8x^6$, and so $8x^6$ had to be "a square."[23] But, as Diophantus warned, this "is not correct." In other words, an equation of the form $8x^6 = (bx)^2$ cannot be satisfied in rational numbers, since it would require $x^4 = \frac{b^2}{8}$, and 8 is not a square. The choice of $a = 2$ had clearly been misguided. Here, then, is an example of the sort of conundrum with which Diophantus would have been faced had he played out all of the possible permutations

---

[22] Sesiano, 1982, pp. 90–91. For problem IV.6 and the quotations that follow in this paragraph, see p. 90.

[23] Sesiano, 1982, p. 91. For the quotations that follow in this and the next paragraph, see pp. 91–92.

arising in the context of problems IV.1 through IV.4, except in this case, he could achieve a rational solution by making different initial assumptions, that is, he could solve this problem within his mathematical universe.

As in IV.6, instead of taking the square to be $(bx)^2$, take it of the form $(bx^2)^2$. The equation to be solved then became $a^3x^6 = b^2x^4$, and performing the enhanced reduction yielded $a^3x^2 = b^2$, namely, the problem of finding a cube and a square, the product of which is a square, or the *very problem* Diophantus had solved in IV.6. Therefore, as in IV.6, take the square to be 4 and the cube to be 64, that is, take $x = 2$ and $a = 4$ (*not* $a = 2$). Their product, 256, was thus a square, and $b = 16$.

This solved the auxiliary problem to which Diophantus had reduced IV.8, but it still remained explicitly to produce the two cubes with product a square, and Diophantus did that in IV.9. Since $x = 2$, the first cube was $2^3 = 8$, the second $(ax)^3 = (4 \cdot 2)^3 = 64 \cdot 8 = 512$, and the product $8 \cdot 512 = 2096 = 64^2$, a square.

Myriad problems involving more and more complicated relations between squares and cubes fill the rest of Book IV as well as Books V, VI, and VII. In all, Diophantus challenged his readers with 100 problems in these four books. To get just a sense of the degree of complexity (and convolution) that these problems could entail, consider VII.15 where Diophantus asked his readers "to divide a given square number into four parts such that two of the four parts each leave, when subtracted from the given square number, a square, and the other two also each give, when added to the given square number, a square number."[24] In modern notation, the problem was to find a, b, c, d such that, given a square A^2, it is the case that $A^2 = a + b + c + d$ at the same time that $A^2 - a$, $A^2 - b$, $A^2 + c$, and $A^2 + d$ are all squares.

Diophantus opened by taking 25 as the "given square number," but noted more abstractly that for arbitrary x, $x^2 + (2x + 1)$ and $x^2 + (4x + 4)$ are both squares as are $x^2 - (2x - 1)$ and $x^2 - (4x - 4)$. Thus, taking $c = 2x + 1$, $d = 4x + 4$, $a = 2x - 1$, and $b = 4x - 4$, the last four conditions were satisfied, but the first condition became $x^2 = 12x$ or $x = 12$. In other words, had the initially given square been 144, we would be done at this point, and the four values sought would be $c = 25$, $d = 52$, $a = 23$, and $b = 44$. As Diophantus put it, "we have completed the requisite search for the said square number; but we have not reached the desired end of

---

[24] See Sesiano, 1982, p. 167 for this and the quotations in the following paragraph.

the problem." To get there, "we have to multiply each of the parts of 144 by 25 and divide the results by 144." Relative to the square 25, then, the four desired numbers were $c = \frac{625}{144}$, $d = \frac{1300}{144}$, $a = \frac{575}{144}$, and $b = \frac{1100}{144}$. In modern notation, this is clear. Since $144 = 25 + 52 + 23 + 44$, we have $25 = \frac{144 \times 25}{144} = \frac{(25+52+23+44) \times 25}{144} = \frac{25 \times 25}{144} + \frac{52 \times 25}{144} + \frac{23 \times 25}{144} + \frac{44 \times 25}{144}$. Given what he knew about expressing quadratics as squares, all Diophantus had to do to accommodate any "given square" in this problem was to rationalize appropriately in the end.

A SAMPLING FROM THE *ARITHMETICA*: THE REMAINING GREEK BOOKS

Although it is not clear whether the remaining Greek books—G-IV, G-V, and G-VI—are Diophantus's original Books VIII–X or whether they are otherwise dispersed through what were Diophantus's Books VIII–XIII, it does seem clear on the basis of their content and their relative difficulty that they followed the seven books we have examined thus far.[25] In particular, like Books IV–VII, Books G-IV and G-V also primarily treat squares and cubes, but instead of finding two or three numbers that satisfy given conditions as in the earlier books, Diophantus now began to challenge his readers to find three or four numbers in various relations to one another. In Book G-VI, however, he diverged from the earlier books in an interesting way. It treated relations among squares, but in the geometric context of finding right triangles with sides satisfying certain algebraic conditions in addition, of course, to the Pythagorean theorem. Unlike problems in any of the other books, those in Book G-VI explicitly linked Diophantus's *algebraic* techniques with *geometrical* problems of triangle construction. To get a sense of the novel twists and turns that the problems in these later books could take—and of the insights that their solution techniques could provide—consider three problems, one each from Books G-IV, G-V, and G-VI.

In G-IV.31, Diophantus charged his students "to divide unity into two parts such that, if given numbers are added to them respectively, the product of the two sums gives a square."[26] He offered two different

[25] Compare the argument and evidence for this in Sesiano, 1982, pp. 4–8.
[26] Heath, 1964, p. 189.

solutions, the second involving his so-called "method of limits."[27] Take the two "given numbers" to be 3 and 5 and let the two parts into which 1 has been divided be $x - 3$ and $4 - x$. Consider the product of $3 + (x - 3)$ and $5 + (4 - x)$, that is, $x(9 - x)$; it must be a square, say, $x(9 - x) = 4x^2$. Under these assumptions, $x = \frac{9}{5}$. "But," Diophantus explained,"I cannot take 3 from $\frac{9}{5}$, and x must be > 3 and < 4."[28] So, how had he arrived at the problematic value for x? It had come "from 9/(square+1), and, since $x > 3$, this square+1 should be < 3, so that the square must be less than 2; but, since $x < 4$, the square $+1$ must be $> \frac{9}{4}$, so that the square must be $> \frac{5}{4}$." In other words, Diophantus had initially taken the square too big. By analyzing the problem more closely, he was able to reduce it to one of finding a square between two limits, namely, $\frac{5}{4}$ and 2. Consider, then, $\frac{5}{4} < x^2 < 2$. To generate some numbers that are easier to think about and to introduce a numerical square, multiply through by 1 in the form of $\frac{64}{64}$ to get $\frac{80}{64} < x^2 < \frac{128}{64}$. It is now easy to see that $x^2 = \frac{100}{64} = \frac{25}{16}$ fills the bill. To finish the problem, Diophantus used this new square instead of 4 to get the equation $x(9 - x) = \frac{25}{16}x^2$. This then yielded $x = \frac{144}{41}$ and the associated division of unity $\left(\frac{21}{41}, \frac{20}{41}\right)$. A simple check shows that $\left(\frac{21}{41} + 3\right)\left(\frac{20}{41} + 5\right) = \frac{32400}{41^2} = \frac{180^2}{41^2}$, that is, a square, as desired.

This problem illustrated at least one technique for introducing what we might call a "mid-course correction" into a Diophantine analysis. The specific choice of the "given numbers" as well as that of the division of unity, introduced numerical constraints on subsequent choices of which Diophantus had to be aware if he was going to avoid, as he had to, negative numbers. It was sometimes the case that he did not have the freedom to choose any value whatsoever at a given stage of his analysis. The "method of limits" gave him at least one means to extricate himself from such situations by generating appropriate bounds within which his choice should be made.

Degrees of freedom also presented problems in Book G-V. Consider, for example, G-V.29, where Diophantus sought "to find three squares such that the sum of their squares is a square."[29] Taking x^2, 4, and 9 as the three squares, the problem thus required that $x^4 + 97$ be a square, and Diophantus set it equal to $\left(x^2 - 10\right)^2$ to get $x^2 = \frac{3}{20}$. "If the ratio

[27] Heath, 1964, p. 94.
[28] Heath, 1964, p. 189. For the quotations that follow in this paragraph, see p. 190.
[29] Heath, 1964, p. 224. The next quotation is also from this page.

of 3 to 20 were the ratio of a square to a square," he continued, "the problem would be solved; but it is not." What to do? In a classic example of "back-reckoning,"[30] Diophantus realized that he needed to take three different squares as his starting point, but how was he to find them? As he explained, "I have to find two squares (p^2, q^2, say) and a number (m, say) such that $m^2 - p^4 - q^4$ has to $2m$ the ratio of a square to a square."[31]

To see why this is the case, consider what went wrong with his original choice. If we set $p^2 = 4$, $q^2 = 9$, and $m = 10$, then the equation Diophantus had to solve was $(x^2)^2 + (p^2)^2 + (q^2)^2 = (x^2 - m)^2$. This reduced to $x^2 = \frac{m^2 - p^4 - q^4}{2m}$, the right-hand side of which needed to be a square but was not. He just needed better choices of p, q, and m. To get them, he "let $p^2 = z^2$, $q^2 = 4$, and $m = z^2 + 4$." Then $m^2 - p^4 - q^4 = 8z^2$, and he needed $\frac{8z^2}{2z^2 + 8}$ or $\frac{4z^2}{z^2 + 4}$ to be the ratio of two squares. Since the numerator is a square, he had to ensure that the denominator was, too, so he set $z^2 + 4 = (z + 1)^2$. Solving for z yielded $z = \frac{3}{2}$, and so $p^2 = \frac{9}{4}$, $q^2 = 4$, and $m = \frac{25}{4}$. Rationalizing to eliminate the fractions, he took $p^2 = 9$, $q^2 = 16$, and $m = 25$. In his words, "starting again, we put for the squares x^2, 9, and 16; then the sum of the squares $= x^4 + 337 = (x^2 - 25)^2$, and $x = \frac{12}{5}$. The required squares are $\frac{144}{25}$, 9, 16."

There are at least two things to notice here. First, Diophantus was actually dealing with a *fourth-degree*—as opposed to a quadratic or cubic— equation, although it immediately reduced to a quadratic that he should, in principal, have been able to solve. Second, because he could only work with rational numbers, he had, by hook or by crook, to set up his problems so that rational solutions would result. In this case, when his arbitrary initial choices did not produce a rational answer, he showed how to employ the technique of back-reckoning to relaunch the problem successfully, another valuable lesson for his students.

He also put back-reckoning to good use in Book G-VI in the context of his problems on right triangles. In G-VI.6, for example, he asked his readers "to find a right-angled triangle such that the area added to one of the perpendiculars makes a given number."[32] Taking the "given

---

[30] The term is due to the nineteenth-century historian of Greek mathematics Georg H. F. Nesselmann in his classic work, *Die Algebra der Griechen*. See the section "Methode der Zurückrechnung und Nebenaufgabe" in Nesselmann, 1842, pp. 369–381.

[31] Heath, 1964, p. 224. The quotations in the next paragraph are also from this page.

[32] Heath, 1964, p. 228. The quotations that follow in this and the next paragraph are also from this page.

number" to be 7, he considered the right triangle of sides $(3x, 4x, 5x)$. The condition to be satisfied was then $6x^2 + 3x = 7$, but as Diophantus noted, "in order that this might be solved, it would be necessary that (half coefficient of $x)^2$ + product of coefficient of x^2 and absolute term should be a square." Clearly, then, Diophantus was aware of the same sort of solution algorithm for quadratics of the form $ax^2 + bx = c$ that the ancient Mesopotamians employed in tablets such as MS 5112 (recall the discussion in chapter 2) and to which Euclid's construction of II.6 reduced in the *Elements* (recall chapter 3). (In both of those earlier cases, $a = 1$.) However, the problem Diophantus encountered in G-VI.6, at least as he had set it up, was that the numerical choices in place yielded $\left(\frac{3}{2}\right)^2 + 6 \cdot 7$ which is not a square. "Hence," he continued, "we must find, to replace $(3, 4, 5)$, a right-angled triangle such that ($\frac{1}{2}$ one perpendicular $)^2 + 7$ times area = a square." In other words, we need to apply the technique of back-reckoning in order to generate starting values that will yield a legitimate numerical solution.

This new problem was not hard. Diophantus chose as the triangle's perpendicular sides m and 1. Thus, the condition to be satisfied became $\left(\frac{1}{2} \cdot 1\right)^2 + 7 \cdot \frac{1}{2}m$ = a square or, rationalizing, $1 + 14m$ = a square. Since the triangle is right, however, the Pythagorean theorem also needed to be satisfied, so $1 + m^2$ also had to be square. Thus, Diophantus had to solve another instance of a "double equation," although whereas the first such example he had treated (recall the discussion of II.11 above) was linear, this one was of the second degree. To solve it, though, he followed the same algorithm that he had established in II.11. He took the difference $1 + m^2 - (1 + 14m) = m^2 - 14m$ of the two squares and realized it as the product $m(m - 14)$ of the two factors m and $m - 14$. Taking the difference $m - (m - 14)$ of those two factors, he then set the second of the squares $1 + 14m$ equal to the square of half the difference of the two factors, that is, $\left(\frac{1}{2}(m - (m - 14))\right)^2 = 7^2$ and solved for m to get $m = \frac{24}{7}$. Thus, the Pythagorean triple generated was $(\frac{24}{7}, 1, \frac{25}{7})$ or, rationalizing, $(24, 7, 25)$. Using this as the initial assumption on the right triangle, that is, considering a Pythagorean triple of the form $(24x, 7x, 25x)$, the condition to be satisfied was $84x^2 + 7x = 7$, "and $x = \frac{1}{4}$." The desired right triangle thus had sides 6, $\frac{7}{4}$, and $\frac{25}{4}$.

Notice here that although Diophantus had insisted in the preliminary definitions that each problem should ultimately be reduced to one of the form $ax^m = b$, G-VI.6 actually reduced to an expression "where two terms are left equal to one term,"[33] precisely the type of case he had promised in Book I to deal with systematically at some point in the *Arithmetica*. That promise went unfulfilled here; Diophantus seemed merely to pull the answer out of the hat. Without too much trial and error, he could have hit upon the solution $x = \frac{1}{4}$, or he could well have used the Mesopotamian algorithm to which he had alluded earlier in the problem. Neither here nor elsewhere in the extant books of the *Arithmetica*, however, did he provide us with enough evidence to say for sure.

THE RECEPTION AND TRANSMISSION OF THE *ARITHMETICA*

Much as Euclid had systematized and codified the geometrical results known in his day (in addition perhaps to adding his own), Diophantus may be seen as having systematized and codified a certain set of problem-solving techniques (in addition perhaps to adding his own) that had been evolving since ancient Egyptian times and that were already in evidence in texts such as the *Rhind Papyrus*.[34] His was primarily a didactic mission, as he stated more than once in the extant books of his text; he wanted to teach others how to employ the techniques that had evolved for determining relations between rational numbers by providing them with a series of problems solved and explained. By mastering his examples, his readers would be able to go on to solve problems of their own.

The mathematical style reflected in Diophantus's work thus contrasted markedly with that of Euclid's *Elements*, despite their seemingly common aim of codification. Writing a generation after Diophantus, and perhaps in reaction to the kind and style of mathematics that his work represented, Pappus of Alexandria (ca. 290–350 CE) sought to revive classical Greek geometry in his eight-book overview of the subject entitled the *Mathematical Collection*.[35] A work in the didactic, handbook tradition characteristic of Roman Alexandria, the *Mathematical Collection* surveyed

---

[33] Heath, 1964, p. 131

[34] Since the nineteenth century, there has been much disagreement among historians of mathematics as to the originality and purpose of Diophantus's *Arithmetica*. See, for example, the literature review reflected in the opening six chapters of Heath, 1964. The interpretation presented here follows Klein, 1968; Sesiano, 1999; and Cuomo, 2001.

[35] For a readable and comprehensive survey of Pappus's mathematical contributions, see Heath, 1921, 2:355–439.

the various texts written and topics pursued by ancient Greek geometers, topics like arithmetic (Book I), Apollonius's method for handling large numbers (Book II), the five regular polyhedra (Book III), properties of special curves like the Archimedean spiral and the quadratrix of Hippias (Book IV), the thirteen semiregular solids treated by Archimedes (Book V), astronomy (Book VI), analysis and synthesis (Book VII), and mechanics (Book VIII). In Book VII, in particular, Pappus compared and contrasted the two types of analysis and synthesis that had evolved, one characteristic of geometry and especially of Euclid's *Elements*, the other characteristic of Diophantus's *Arithmetica*. As used in geometry, analysis and synthesis demonstrated a proposition by first, in the analysis phase, supposing true what was to be proved and reducing that to an identity or other known proposition and, then, in the synthesis phase, reversing the process. As used in a context like that of the *Arithmetica*, however, analysis and synthesis treated a given problem by, first, in the analysis phase, supposing the problem to be solved, establishing some relations between known and unknown quantities, and reducing to some relation in terms of the smallest number of unknowns possible and, then, in the synthesis phase, checking that the solution found is correct.[36] Although Pappus's intent in the *Mathematical Collection* was to highlight and reemphasize the former sort of analysis and synthesis and thereby once again to stimulate geometrical research, his juxtaposition of the two types suggests that the latter—as exemplified by Diophantus's work—may have gained a certain primacy in later Roman Alexandria.

In the next and final generation of scholars at the Museum, Theon of Alexandria (ca. 335–405 CE) continued in the scholarly tradition that so highly valued new editions of and commentaries on the great books to be found in the Library. Like Pappus, Theon focused on mathematical texts—particularly those of Euclid and the second-century-CE astronomer Claudius Ptolemy—and he imparted his mathematical erudition to Alexandria's intellectual elite. Among his students was his own daughter, Hypatia (ca. 355–415 CE), who not only aided him in the production of his scholarly editions but also wrote her own commentaries on Apollonius's *Conics* and on Diophantus's *Arithmetica*.[37] Although no

[36] Compare Heath, 1921, vol. 2, pp. 400–401, and Sesiano, 1982, p. 48.
[37] For a modern biography of Hypatia that analyzes and attempts to unravel the various myths and legends that have been perpetuated about her life, see Dzielska, 1995.

copies of her Diophantine commentary are extant, it is not impossible that at least the four Arabic books—Books IV–VII discussed above—that have come down to us may have stemmed from her work. Those books— as opposed to the six extant Greek books, that is, Books I–III and G-IV through G-VI—possess "all the characteristics of a commentary made at about the time of the decline of Greek mathematics," that is, the fifth century CE.[38]

The decline of Greek mathematics, indeed the decline of Alexandria as an intellectual center, roughly coincided with Hypatia's brutal death in 415 CE. An influential teacher in her own right, Hypatia founded a school in which she taught ethics, ontology, astronomy, and mathematics—in addition to philosophy "including the ancient pagan ideas of Pythagoras, Plato, and Aristotle as well as the Neoplatonism of Ammonius Saccas and Plotinus"—to men who would ultimately number among Alexandria's leaders.[39] Because of her influence and what were viewed as her pagan leanings, Hypatia ultimately found herself caught up in a power play orchestrated by the Alexandrian patriarch, (later Saint) Cyril (ca. 376–444 CE), to expel non-Christians from the city and thereby to challenge the secular civil authority with which Hypatia had prominently aligned herself. Her role in this ended with her murder at the hands of a mob, and with her death, "her city also began to die."[40] Just over two centuries later in 646 CE, Alexandria was destroyed at the hands of the conquering Muslims.

The Muslim conquest of Alexandria marked a cultural and intellectual shift in the medieval world order away from the West and toward the East. As Islam grew and its kingdoms developed, there was an increasing demand in the Middle East for the "lost" knowledge of the conquered. In a complex process of transmission (which we shall discuss more fully in chapter 8), manuscripts such as Euclid's *Elements* and Diophantus's *Arithmetica* ultimately made their way into newly created libraries in Baghdad and elsewhere where they were translated into Arabic for further study. We owe the existence of Books IV–VII of the *Arithmetica*—and hence our understanding of the problem-solving techniques so different in flavor and form, if not in actual fact, from the mathematics to be

---

[38] Sesiano, 1982, p. 71.
[39] Pollard and Reid, 2006, p. 268.
[40] Pollard and Reid, 2006, p. 280.

found in Book II of Euclid's *Elements*—to this process and to the Arabic translation that Qusṭā ibn Lūqā made of them around the middle of the ninth century. Before turning to an examination of medieval Islamic contributions to the history of algebra, however, we consider developments that took place first in China and then in India.

5

Algebraic Thought in Ancient and Medieval China

Although archaeological evidence supports the fact that the Chinese engaged in numerical calculation as early as the middle of the second millennium BCE, the earliest detailed written evidence of the solution of mathematical problems in China is the *Suan shu shu* (or *Book of Numbers and Computation*), a book written on 200 bamboo strips discovered in 1984 in a tomb dated to approximately 200 BCE.[1] Like Mesopotamian texts much earlier, the *Suan shu shu* was a book of problems with their solutions.

China was already a highly centralized bureaucratic state by the time the *Suan shu shu* was composed, having been unified under Qin Emperor Shi Huangdi by 221 BCE. On his death, his heirs were overthrown, and the Han dynasty, which lasted some 400 years, came to power. That dynasty instituted a system of examinations for entrance into the civil service, and it appears that the *Suan shu shu*—as well as the somewhat later and more famous *Jiuzhang suanshu* (or *Nine Chapters on the Mathematical Art*)—were required reading for civil service candidates. The *Suan shu shu*, for example, treated such highly practical issues as area and volume determination, the pricing of goods, and the distribution of taxes, while the *Nine Chapters*[2] contained problems not only of this kind but also of a less purely practical nature that were, nevertheless, cast as real-life situations.

During the thousand years after the fall of the Han dynasty, China witnessed alternating periods characterized either by unity under a native

---

[1] New archaeological evidence of two slightly earlier Chinese books on arithmetic has also recently come to light. See Volkov, 2012.

[2] In what follows, we have generally used the English translation of the *Jiuzhang suanshu* by Shen et al., 1999. However, the slightly newer French translation, Chemla and Guo, 2004, sometimes contains better readings.

or foreign dynasty or by warring small kingdoms. Regardless of these broader political contingencies, however, the civil service examinations—and so the need for mathematical texts to use in preparation for them—remained a social constant. Chinese mathematicians—who seemingly worked in isolation and in widely disparate parts of the country—gradually developed new methods for treating various problems that their works needed to contain.[3] In particular, they derived a numerical procedure for solving equations of arbitrary degree and a method for solving simultaneous congruences, now called the Chinese remainder problem. Interestingly, the mathematical texts that survive were largely written without reference to earlier texts, although frequently, later texts contained similar problems side by side with problems illustrating new methods. Texts continued to have a decidedly practical bent, but many of the newer methods and problems—like the Chinese remainder problem that probably arose in the solution of astronomical and calendrical problems—clearly went well beyond any practical necessity. It is evident, then, that in China, as in the other civilizations with which we have dealt, there were creative mathematicians interested in following ideas wherever they led, and, in some cases, those ideas led to techniques that we consider algebraic.

PROPORTIONS AND LINEAR EQUATIONS

As a case in point, although the *Suan shu shu* was largely devoted to calculational techniques and geometry, a few of its problems—involving proportionality—have an algebraic cast. Consider, for example, problem 37: "In collecting hemp, 10 *bu* (of volume) give 1 bundle [of circumference] 30 *cun*. Now one dries it [and it becomes] 28 *cun*. Question: how many *bu* yield one [standard] bundle?"[4] Since drying was assumed to shrink the circumference of the bundle in the ratio 28:30, and since the cross-sectional area of a bundle was proportional to the square of the circumference, the question, in modern terms, was to determine the

---

[3] For a good summary of the idea of mathematics in early imperial China, see Cullen, 2009.

[4] Cullen, 2004, p. 60. The next quotation is also on this page. Cullen has posted his translation of the entire *Suan shu shu* at http://www.nri.org.uk/suanshushu.html. For another translation, see Dauben, 2008.

new volume V such that $10 : V = 28^2 : 30^2$. The author simply gave the algorithm: "Let the dried multiply itself to make the divisor. Let the fresh multiply itself, and further multiply by the number of *bu* [through] the dividend accommodating the divisor." In other words, letting the divisor be 28^2 (the dried) and the dividend be 30^2 (the fresh), the answer is 10×30^2 divided by 28^2 or $11\frac{47}{98}$.

The *Suan shu shu* also contained a work problem similar to the one about the cobbler found in the Egyptian *Moscow Papyrus* (recall chapter 2): "1 man in 1 day makes 30 arrows; he feathers 20 arrows. Now it is desired to instruct the same man to make arrows and feather them. In one day, how many does he make?"[5] As in the previous problem, the author presented a solution procedure without further explanation: "Combine the arrows and feathering to make the divisor [30 + 20 = 50]; take the arrows and the feathering multiplied together to make the dividend [30 × 20 = 600]." Therefore, the solution is 600 ÷ 50 = 12.

How did the author arrive at this? He may have noted that if one can make 30 arrows in one day and feather 20, then it would take $\frac{1}{30}$ of a day to make one arrow and $\frac{1}{30} + \frac{1}{20} = \frac{20+30}{30 \cdot 20} = \frac{50}{600} = \frac{1}{12}$ of a day to make one feathered arrow. Thus, the number of arrows that can be made in one day is the inverse of this value, namely, $\frac{600}{50} = 12$, just as required by the solution.

A deeper grasp of proportionality was reflected in several other problems, in which the author used what the Chinese termed the "method of excess and deficit" to solve what we can interpret as a system of two linear equations in two unknowns. This method is a version of what later became known as the "rule of double false position," namely, "guess" two possible solutions and then adjust the guess to get the correct solution. One problem of this type reads, "*Li* hulled grain is 3 coins for 2 *dou*; *bai* is 2 coins for 3 *dou*; Now we have 10 *dou* of *li* and *bai*. Selling it, we get 13 coins. Question: how much each of *li* and *bai*?"[6] In modern terms, if x is the number of *dou* of *li* and y that of *bai*, then the problem reduces to these two equations in two unknowns:

$$x + y = 10,$$

$$\frac{3}{2}x + \frac{2}{3}y = 13.$$

[5] For this and the next quotation, see Cullen, 2004, p. 79.
[6] Cullen, 2004, pp. 85–86. The next quotation may also be found here.

To solve them, the author of the *Suan shu shu* provided his reader with this algorithm: "Making it all *li*, then the coins are 2 in excess. Making it all *bai*, then the coins are $6\frac{1}{3}$ in deficit. Put the excess and deficit together to make the divisor; Multiply the 10 *dou* by the deficit for the *bai*; multiply the 10 *dou* by the excess for the *li*. In each case, accommodate the divisor to obtain 1 *dou*." In other words, the author noted that, by first guessing that all of the grain was *li*, the total price would be $10 \times \frac{3}{2} = 15$, which is 2 in excess of the required 13 coins. On the other hand, if one guessed that all of the grain was *bai*, then the total price would be $10 \times \frac{2}{3} = 6\frac{2}{3}$, which is $6\frac{1}{3}$ in deficit. He then explained that to determine the amount of *li*, multiply 10 by the deficit $6\frac{1}{3}$ and divide by the sum of 2 and $6\frac{1}{3}$ to get $7\frac{3}{5}$ *dou*. Similarly, to find the amount of *bai*, multiply 10 by the excess 2 and divide by the same divisor to get $2\frac{2}{5}$.

How might this algorithm have arisen? Note that a change from the correct, but unknown, value x of *li* to the first guessed value of 10 involved a change in the total price of 2, while a change from the second guessed value of 0 to the correct value x involved a change in the total price of $6\frac{1}{3}$. Since the price "function," expressed in modern terms as $\frac{3}{2}x + \frac{2}{3}(10 - x)$ or as $\frac{5}{6}x + \frac{20}{3}$, is what we now call a linear function, the ratio of the two pairs of changes must be equal:

$$\frac{10 - x}{2} = \frac{x}{6\frac{1}{3}} \qquad \text{or} \qquad \left(2 + 6\frac{1}{3}\right)x = 6\frac{1}{3} \times 10.$$

The given algorithm for the *li* then follows, and the argument for the algorithm for the *bai* is analogous.

Problems similar to these, as well as many more, also appear in the *Nine Chapters*, written perhaps two hundred years after the *Suan shu shu*. For example, the second chapter of the *Nine Chapters* was entirely devoted to proportionality problems. It opened with a table of exchange rates involving 20 different types of grain and then used that table in numerous problems, among which problem 6 was typical: "Now 9 *dou* 8 *sheng* of millet is required as coarse crushed wheat [where 1 *dou* = 10 *sheng*]. Tell how much is obtained."[7] Since the table gave the rate of exchange between millet and coarse crushed wheat as 50 to 54, the method for solving the problem was stated as "multiply by 27, divide by 25," that is,

---

[7] Shen et al., 1999, p. 145.

multiply the 9 *dou* 8 *sheng* by $\frac{27}{25}$ to get 10 *dou* $5\frac{21}{25}$ *sheng* of coarse crushed wheat. This represented a special case of the general "rule of three," namely, "take the given number to multiply the sought rate. The product is the dividend. The given rate is the divisor. Divide."[8] The author of the *Nine Chapters* had frequent recourse to the rule of three in this form.

The seventh chapter of the *Nine Chapters* treated a different kind of algebraic problem, those solved by the rule of double false position, called, as noted earlier, the rule of excess and deficit.[9] It opened with several examples directly couched in the latter language and gave explicit solution rules. For example, problem 1 stated, "Now an item is purchased jointly; if everyone contributes 8, the excess is 3; if everyone contributes 7, the deficit is 4. Tell the number of people, the item price, what is each?"[10] The author used the following rule to arrive at the answer: "Display the contribution rates; lay down the corresponding excess and deficit below. Cross-multiply by the contribution rates; combine them as dividend; combine the excess and deficit as divisor ....To relate the excess and the deficit for the articles jointly purchased, lay down the contribution rates. Subtract the smaller from the greater; take the remainder to reduce the divisor and the dividend. The reduced dividend is the price of an item. The reduced divisor is the number of people."[11] In other words, he formed the following table:

$$8 \quad 7$$
$$3 \quad 4.$$

Cross-multiplication and addition yielded $8 \times 4 + 7 \times 3 = 53$ for the dividend, and the divisor was $3 + 4 = 7$. Finally, subtracting the smaller from the larger rate—that is, $8 - 7 = 1$—and dividing the divisor and dividend each by 1 gave the number of people as 7 and the price as 53. Here, as elsewhere in Chinese mathematics, directions such as "display" and "lay down" refer to the use of counting rods on a counting board. The rods, generally not more than 10 cm long, were made of bamboo or

---

[8] Shen et al., 1999, p. 141.
[9] Hart, 2011 contains a detailed treatment of problems of this type as well as problems in several linear equations in several unknowns from chapter eight of the *Nine Chapters*.
[10] Shen et al., 1999, p. 358.
[11] Shen et al., 1999, p. 359.

wood or ivory and were laid out in columns and rows on some convenient surface. There is no evidence of any standard board with a preexisting grid; mathematicians probably just used whatever surface was available.[12]

Although the algorithm in this problem was explicit, again, its origin was obscure. In this case, however, as in other parts of the *Nine Chapters*, light is shed by the commentator, Liu Hui. Liu lived in the third century CE and produced the edition of the *Nine Chapters* that has come down to us. In relation to this problem, he noted that the cross-multiplication step made the excess and the deficit equal. That is, taking four contributions of 8 by each person, the excess was 12, and taking three contributions of 7 by each person, the deficit was also 12. The total contribution in 7 steps was then 53, and this gave neither excess nor deficit; equivalently, a contribution of $\frac{53}{7}$ gave neither excess nor deficit. To determine the correct number of people and the actual price, Liu next noted that each change in payment of $8 - 7 = 1$ coin per person caused the total contributed to swing by $4 + 3 = 7$ coins. Thus, the number of people was actually the 7 in the denominator of $\frac{53}{7}$ divided by $1 = 8 - 7$ (the difference in the rates of contribution). It followed that also dividing the numerator by 1 gave 53 as the total price. This fully explained the rule as presented by the author of the *Nine Chapters*.

Much of the remainder of the seventh chapter involved problems that can be thought of as systems of two linear equations in two unknowns, but they can be recast in terms of excess and deficit and solved by the algorithm. For example, problem 17 specified that "the price of 1 acre of good land is 300 pieces of gold; the price of 7 acres of bad land is 500. One has purchased altogether 100 acres; the price was 10,000. How much good land was bought and how much bad?"[13]

The solution procedure was similar to that of the *Suan shu shu* problem above; the author simply took two guesses for the amount of good land and applied the excess–deficit procedure to the linear expression for the price, which, in modern terminology, can be expressed as $300x + \frac{500}{7}y$, where x and y are the amount of good and bad land, respectively. Explicitly, he said, "suppose there are 20 acres of good land and 80 of bad. Then the surplus is $1714\frac{2}{7}$. If there are 10 acres of good land and 90 of bad, the deficiency is $571\frac{3}{7}$." To complete the problem,

---

[12] For more on counting boards, see Lam and Ang, 1992, particularly pp. 21–27.

[13] For this and the quotation that follows in the next paragraph, see Shen et al., 1999, p. 376.

multiply 20 by $571\frac{3}{7}$ and 10 by $1714\frac{2}{7}$, add the products, and finally divide this sum by the sum of $1714\frac{2}{7}$ and $571\frac{3}{7}$. The result, $12\frac{1}{2}$ acres, was the amount of good land. The amount of bad land, $87\frac{1}{2}$ acres, followed easily.

We can express the algorithm used here symbolically as

$$x = \frac{b_1 x_2 + b_2 x_1}{b_1 + b_2},$$

where b_1 is the surplus determined by guess x_1 and b_2 is the deficiency determined by guess x_2. This slightly generalizes the algorithm discussed above in connection with the similar problem from the *Suan shu shu*, where x_2 was taken to be 0. As before, the basic principle of linearity was clearly involved in deriving the algorithm, but the author's exact mode of thought remains unknown.

The theme of two linear equations in two unknowns continued in the eighth chapter of the *Nine Chapters*. There, the author laid out an alternative method for solving such systems that can also be used to solve larger systems and that is virtually identical, even down to the matrix form, to the modern method of Gaussian elimination. Given that adding and subtracting matrix columns, as in Gaussian elimination, may well produce negative quantities, even if the original problem and its solution only deal with positive quantities, the Chinese also developed, in this context, the basic methods for adding and subtracting such quantities.

By way of example, consider problem 8 of the eighth chapter: "Now sell 2 cattle and 5 sheep to buy 13 pigs. Surplus 1000 coins. Sell 3 cattle and 3 pigs to buy 9 sheep. There is exactly enough money. Sell 6 sheep and 8 pigs. Then buy 5 cattle. There is a 600 coin deficit. Tell what is the price of a cow, a sheep, and a pig, respectively."[14] In modern symbolism, if x represents the price of a cow, y the price of a sheep, and z the price of a pig, then we have the following system of equations:

$$2x + 5y - 13z = 1000,$$

$$3x - 9y + 3z = 0,$$

$$-5x + 6y + 8z = -600.$$

[14] Shen et al., 1999, p. 409.

The Chinese author displayed the coefficients in a rectangular array:

$$
\begin{array}{rrr}
-5 & 3 & 2 \\
6 & -9 & 5 \\
8 & 3 & -13 \\
-600 & 0 & 1000
\end{array}
$$

As in Gaussian elimination, the idea was to multiply the columns by appropriate numbers and then add or subtract such multiplied columns in order to arrive at a "triangular" system that was then easy to solve.

   In this problem, begin by multiplying the middle column by 2 and then subtracting 3 times the right column to get

$$
\begin{array}{rrr}
-5 & 0 & 2 \\
6 & -33 & 5 \\
8 & 45 & -13 \\
-600 & -3000 & 1000
\end{array}
$$

Then multiplying the left column by 2 and adding 5 times the right column yields

$$
\begin{array}{rrr}
0 & 0 & 2 \\
37 & -33 & 5 \\
-49 & 45 & -13 \\
3800 & -3000 & 1000
\end{array}
$$

Finally, multiplying the left column by 33 and adding it to 37 times the middle column gives

$$
\begin{array}{rrr}
0 & 0 & 2 \\
0 & -33 & 5 \\
48 & 45 & -13 \\
14,400 & -3000 & 1000
\end{array}
$$

The new left-hand column translates into the equation $48z = 14,400$, thus the price z of a pig is 300. Next, "back substituting" using the equation from the new middle column gives $-33y + 45 \cdot 300 = -3000$ so that $y = 500$. Finally, substituting the values for y and z into the new right-hand column results in $x = 1200$.[15]

Although, as usual, the original author did not explain the derivation or the reasoning behind his algorithm, Liu Hui justified it in his commentary: "If the rates in one column are subtracted from those in another, this does not affect the proportions of the remainders."[16] In other words, Liu essentially invoked the "axiom" to the effect that when equals are subtracted from equals, equals remain. The same is obviously true for addition.

A final example—problem 13—from the eighth chapter reflects the level of complexity of the algebraic scenarios treated in the *Nine Chapters*. This problem translated into the following system of 5 equations in 6 unknowns:[17]

$$
\begin{aligned}
2x + y \qquad\qquad\qquad\qquad &= s, \\
3y + z \qquad\qquad\qquad &= s, \\
4z + u \qquad\qquad &= s, \\
5u + v &= s, \\
x \qquad\qquad\qquad + 6v &= s.
\end{aligned}
$$

The standard solution method led ultimately to the equation $v = \frac{76s}{721}$. So if $s = 721$, then $v = 76$. This was the single answer given. Unfortunately, we do not know whether the Chinese of the Han period considered other possibilities for s or whether they appreciated the implications of an infinite number of solutions. In general, they only dealt with problems with an equal number of equations and unknowns, and no records remain of any discussion as to why that situation produced a unique solution or what happened in other situations.

---

[15] The actual back-substitution instructions in the original are not clear, and Liu Hui's commentary gives two versions, including the continuation of "Gaussian elimination" to eliminate the two entries below the "2" in the right column and the entry below the "−33" in the middle column. See Hart, 2011, pp. 78–81, 93–109 for more details.

[16] Shen et al., 1999, p. 400.

[17] Shen et al., 1999, p. 413.

POLYNOMIAL EQUATIONS

Over the centuries, the Chinese devised methods for solving quadratic and higher-degree polynomial equations. Unlike Mesopotamian—or as we shall see in the next chapter, Indian—mathematicians, the Chinese did not develop a procedure for solving quadratic equations that can be translated into our quadratic formula. Instead, theirs was a numerical procedure that approximated solutions to quadratic and higher-degree equations to any desired degree of accuracy. For example, the *Nine Chapters* provided techniques for finding square and cube roots, that is, for solving equations of the form $x^2 = a$ and $x^3 = b$ numerically. The technique for finding square roots was essentially the same as the one taught in modern-day schools until the advent of calculators, based on the binomial formula $(x + y)^2 = x^2 + 2xy + y^2$. That is, to extract the square root of, for example, $106,929$, apply the algorithm expressed in the following diagram:

$$
\begin{array}{r}
\begin{array}{ccc} 3 & 2 & 7 \end{array} \\
\hline
|1\ 0\ 6\ 9\ 2\ 9 \\
9 \\
\hline
6\ 2|\quad 1\ 6\ 9 \\
1\ 2\ 4 \\
\hline
6\ 4\ 7|\quad 4\ 5\ 2\ 9 \\
4\ 5\ 2\ 9\ . \\
\hline
\end{array}
$$

The idea is that we first note that the root n is a three-digit number beginning with 3, that is $n = 300 + y$, where y is a two-digit number. We enter the 3 on the top line and then square it and subtract 9. But because $n^2 = 90,000 + 600y + y^2$, this subtraction of 9 is really the subtraction of $90,000$ from $106,929$ leaving $16,929$. We then need to find y such that $600y + y^2 = 16,929$. To accomplish this, we put 6 ($= 2 \times 3$) in the fourth line and make a trial division of 6 into 16, the first two digits of $16,929$. Since the trial gives us "2," this means that we are assuming that y begins with 2, or $y = 20 + z$. To check that "2" is correct, we then multiply 2 by the 62, which is equivalent to substituting $y = 20 + z$ into the equation for y. The substitution gives us that $12,400 + 640z + z^2 = 16,929$, or that $640z + z^2 = 4529$, thus confirming that "2" is the correct value for the second digit. We then repeat the steps by doubling the 32 already found to get 64 and make a trial division of 4529 by 64 to find the last digit z of $64z$. In this case, we determine that $z = 7$ works as a solution to the equation for z. Thus, the square root of 106,929 is equal to 327. The cube root technique is a generalization of the square root technique, using instead the binomial expansion $(x + y)^3 = x^3 + 3x^2y + 3xy^2 + y^3$.

Interestingly, although there are several problems in the *Nine Chapters* that require finding a square or cube root, there is only one mixed quadratic equation—problem 20 of the ninth chapter, the chapter on right triangles based on what we call the Pythagorean theorem: "Given a square city of unknown side, with gates opening in the middle, 20 *bu* from the north gate there is a tree, which is visible when one goes 14 *bu* from the south gate and then 1775 *bu* westward. What is the length of each side?"[18] To answer this question, the author instructed, "Take the distance from the north gate to multiply the westward distance $[20 \times 1775 = 35, 500]$. Double the product as the *shi* [71,000]. Take the sum of the distance from the north and south gates as the linear coefficient $[20 + 14 = 34]$." In other words, the problem resulted in the mixed quadratic equation $x^2 + 34x = 71, 000$, where x was the unknown side of the square city, 71,000 was the constant term [*shi*], and 34 the linear coefficient. (In figure 5.1, it is clear that the upper triangle and the

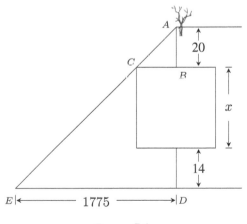

Figure 5.1.

[18] For this and the next quotation, see Shen et al., 1999, p. 507.

entire large triangle are similar. Thus, the proportion $20 : \frac{x}{2} = (34 + x) :$ 1775 is equivalent to the given quadratic equation.) However, the rule at this point just concluded, "Extract the root to obtain the side of the city," without explaining how to do it. The answer was simply given as 250 *bu*, and although Liu in his commentary showed how to derive the equation, he, too, failed to indicate how to solve it. Evidently, then, the Chinese had a method for solving a mixed quadratic equation, but one that was not explicitly written down at this time.

Quadratic and cubic equations continued to crop up occasionally in Chinese works over the course of the next thousand years but, again, with solutions and no indication of methodology. In the *Zhang Qiujian suan jing* (or *The Mathematical Classic of Zhang Quijian*) of the late fifth century CE, for example, the reader was given a segment of a circle with chord $68\frac{3}{5}$ and area $514\frac{31}{45}$ and asked to find the height.[19] The solution was given as $12\frac{2}{3}$, but the description of the method is missing from the manuscript. Presumably, the author used the ancient segment area approximation formula $A = \frac{1}{2}h(h + c)$, where c is the chord length and h is the height, that is, the perpendicular distance from the chord to the circumference. The equation for h is therefore $\frac{1}{2}h^2 + 34\frac{3}{10}h = 514\frac{31}{45}$, which was somehow solved. A cubic equation, derived from a problem involving earthworks for irrigation, occurred in a work by Wang Xiaotong (early seventh century CE), but again no method of solution was given other than a cryptic reference to the rule of cube root extraction.[20] Evidently, then, a method also existed for solving such cubic equations during the first millennium of the common era.

We know more specifically that in the mid-eleventh century, Jia Xian realized that the numbers in what would later be termed the Pascal triangle in the West could be used to generalize the square and cube root procedures of the *Nine Chapters* to higher roots and also that this idea could be extended to solve polynomial equations of any degree. Jia's work is lost, but an improved version of his method for solving equations appeared in the *Shushu jiuzhang* (or *Mathematical Treatise in Nine Sections*) (1247) of Qin Jiushao (1202–1261). Given a particular polynomial equation $p(x) = 0$, the first step, just as in the problem of finding a

---

[19] Shen et al., 1999, pp. 128 and 509.
[20] Li and Du, 1987, pp. 100–104.

square root, was to determine the number $n+1$ of decimal digits in the integer part of the answer and then to find the first digit. For example, Qin considered the equation $-x^4 + 763{,}200x^2 - 40{,}642{,}560{,}000 = 0$ and determined that the answer was a three-digit number beginning with 8. This determination was presumably made through a combination of experience and trial. Once the first digit a was determined, Qin's approach was, in effect, to set $x = 10^n a + y$, substitute this value into the equation, and then derive a new equation in y, the solution of which had one fewer digits than the original. In this case, setting $x = 800 + y$ and substituting produces the equation $-y^4 - 3200y^3 - 3{,}076{,}800y^2 - 826{,}880{,}000y + 38{,}205{,}440{,}000 = 0$. The next step, then, was to guess the first digit of y and repeat the process. Since the Chinese used a decimal system, they could apply this algorithm as often as necessary to achieve an answer to a particular level of accuracy. Frequently, however, once they found the integer value of the answer, they just approximated the rest of the answer by a fraction.[21]

The Chinese mathematicians did not, of course, use modern algebra techniques to "substitute" $x = 10^n a + y$ into the original equation as did William Horner in the similar nineteenth-century method that is named after him.[22] Instead, they set up the problem on a counting board with each row standing for a particular power of the unknown. The results of the "substitution" then came from what we now call the repeated synthetic division of the original polynomial by $x - 10^n a \ (= y)$. Once the coefficients of the equation for y had been determined, the process started anew with a guess for the first digit b of y. Although the numbers from the Pascal triangle are apparent in the process because of the repeated taking of powers of binomials, Qin himself did not mention the triangle. He did, however, solve twenty-six different equations in the *Nine Sections* by the method, and since several of his contemporaries solved similar equations by the same method, it is evident that he and the broader Chinese mathematical community had a correct algorithm for treating these problems.

Several points about these solutions are noteworthy. First, for Qin, equations generally stemmed from geometric or other problems that had

---

[21] For more details and a full description of the solution of this particular equation, see Libbrecht, 1973, pp. 180–189; another example is found in Dauben, 2007, pp. 319–322.

[22] See Clapham and Nicholson, 2005, p. 212 for a description of Horner's method.

only one solution; that solution was thus the one given in the text, even when other positive solutions were possible. For example, Qin gave the solution 840 for the equation discussed above, although 240 is also a solution. As for negative numbers, the Chinese were perfectly fluent in their use, but while they often showed up in the course of a calculation, they never appeared as a final solution. Still, although numerical methods for finding solutions to equations of degree higher than 2 existed in both India and Islam, it is interesting that the Chinese never found the quadratic formula, through geometry or any other method. Instead, they generalized their procedure for finding square roots to an intricate numerical procedure that enabled them to approximate roots of any polynomial equation.

Qin Jiushao had three contemporaries who also contributed significantly to the mathematics involved in solving equations: Li Ye (1192–1279), Yang Hui (second half of the thirteenth century), and Zhu Shijie (late thirteenth century). The first of these was most concerned with teaching how to set up algebraic equations from geometric problems. He assumed that his readers would know the actual solution procedure, presumably the method explained by Qin. By way of example, consider problem 17 of the eleventh chapter of Li Ye's *Ceyuan haijing* (or *Sea Mirror of Circle Measurements*) (1248), a problem reminiscent of problem 20 of the ninth chapter of the *Nine Chapters*: "There is a tree 135 *bu* from the southern gate of a circular walled city. The tree can be seen if one walks 15 *bu* [north] from the northern gate and then 208 *bu* in the eastward direction. Find the diameter of the circular walled city."[23] Li set up his equation by exploiting the similarity of triangles ADO and ACB (figure 5.2). Since $AB : AO = BC : DO$, we have, in modern notation,

$$AB = \frac{208(x+135)}{x},$$

where x is the radius of the circle. But $AB^2 = AC^2 + BC^2$, so $AB^2 - (AC - BC)^2 = 2(AC \cdot BC) =$ twice the area of the rectangle determined by AC and BC. Symbolically,

$$AB^2 - (AC - BC)^2 = -4x^2 + 232x + 39,900 + \frac{11,681,280}{x} + \frac{788,486,400}{x^2}.$$

[23] Dauben, 2007, p. 325.

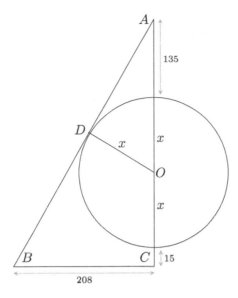

Figure 5.2.

Since twice the area of the rectangle is also equal to $416(2x + 150)$, setting these two expressions equal and simplifying gives the quartic equation

$$-4x^4 - 600x^3 - 22{,}500x^2 + 11{,}681{,}280x + 788{,}486{,}400 = 0.$$

Li then just instructed his readers to "extract the root" and gave the answer $x = 240$.

Of Li's younger contemporaries, Yang Hui primarily publicized Jia Xian's work on using Pascal's triangle to solve polynomial equations. This method was later employed both in Islam and in early modern Europe, but whether there was any transmission from China is unknown. Zhu Shijie, however, developed an important new technique for systematically solving systems of polynomial equations in several unknowns. He successfully worked with up to four unknowns, by associating to each combination of powers of one or two of the unknowns, a region of the counting board in which to place its coefficient. He then manipulated the coefficients of his equations to reduce the system to a single equation in one unknown, an equation that could be solved by Qin's method.

Figure 5.3. Problem 3 from Zhu Shijie's *Precious Mirror* (1303), reproduced in Guo Shuchun, *Shuxue juan* (or *Mathematical Volumes*) (1993), 1:1209b–1210a, in the series *Zhongguo kexue jishu dianji tonghui*. Ed. Ren Jiyu. Zhengzhou: Henan jiaoyu chubanshe. The statement of the problem is in columns a–b, the answer is in column c, and columns d–l are the "working" of the problem.

To see Zhu's method in action, consider problem 3 from his *Siyuan yujian* (or *Precious Mirror of the Four Elements*) (1303): "The sum of the base, altitude and hypotenuse [of a right triangle] divided by the difference of the hypotenuse and altitude is equal to the area of the rectangle. It is also given that the sum of the hypotenuse and the difference of the altitude and base divided by the difference of the hypotenuse and base is equal to the base. Find the hypotenuse. Answer: 5 *bu*."[24] In modern notation, if the base is x, the altitude y, and the hypotenuse z, then the given data yield these equations:

$$\frac{x+y+z}{z-y} = xy \qquad \text{and} \qquad \frac{z+y-x}{z-x} = x.$$

---

[24] Dauben, 2007, p. 351. For more details on this and similar problems, see Dauben, 2007, pp. 343–365.

Clearing these equations of fractions and adding in the equation derived from the Pythagorean theorem produces the following system:

$$xyz - xy^2 - x - y - z = 0, \tag{5.1}$$

$$xz - x^2 + x - y - z = 0, \tag{5.2}$$

$$x^2 + y^2 - z^2 = 0. \tag{5.3}$$

Begin by eliminating y^2 from (5.1) and (5.3) by multiplying the latter equation by x and adding it to the former. The resulting equation is

$$xyz - y + x^3 - xz^2 - x - z = 0. \tag{5.4}$$

Next, eliminate y from (5.2) and (5.4) by first rewriting them in the form $A_1 y + A_2 = 0$ and $B_1 y + B_2 = 0$, where A_i and B_i do not contain y. Then multiply the first equation by B_1, the second equation by A_1, and subtract. What remains is a polynomial without y: $A_2 B_1 - A_1 B_2 = 0$. Specifically, (5.2) becomes $(-1)y + (-x^2 + xz + x - z) = 0$, while (5.4) becomes $(xz - 1)y + (x^3 - xz^2 - x - z) = 0$. Equation $A_2 B_1 - A_1 B_2 = 0$ is then $(xz - 1)(-x^2 + xz + x - z) - (-1)(x^3 - xz^2 - x - z) = 0$, which simplifies to $x^2 - x^2 z + xz^2 + xz - 2z^2 + x - z - 2 = 0$.

For reasons connected with the appearance of the counting board, Zhu replaced z by x and x by y in the latter equation to get

$$y^2 - y^2 x + yx^2 + yx - 2x^2 + y - x - 2 = 0, \tag{5.5}$$

an equation in the two unknowns x and y. To get another equation in these two unknowns, he repeated the basic elimination technique on (5.3) and (5.2) and eliminated y^2 by writing the two equations in the forms $A_1 y + A_2 = 0$ and $B_1 y + B_2 = 0$, where A_i and B_i do not contain y^2. Thus, $A_1 = y$, $A_2 = x^2 - z^2$, $B_1 = -1$, and $B_2 = xz - x^2 + x - z$. By computing $A_1 B_2 - A_2 B_1 = 0$ as before, the y^2 terms are eliminated. The resulting equation is $-x^2 y + xyz + xy - yz + x^2 - z^2 = 0$. Again, writing this new equation in the form $A_1 y + A_2 = 0$ with $A_1 = -x^2 + xz + x - z$ and $A_2 = x^2 - z^2$ and writing (5.2) in the same form $B_1 y + B_2 = 0$ as before, he calculated $A_1B_2 - A_2B_1 = 0$ to eliminate the y. This equation, after some simplification, became $x^3 - 2x^2z - 2x^2 + 4xz + xz^2 - 2z^2 + 2x - 2z = 0$.

As before, Zhu replaced z by x and x by y. Together with equation (5.5) this yielded a system of two equations in two unknowns:

$$y^2 - y^2x + yx^2 + yx - 2x^2 + y - x - 2 = 0,$$

$$y^3 - 2y^2x - 2y^2 + 4yx + yx^2 - 2x^2 + 2y - 2x = 0.$$

Zhu finally applied the same basic methodology to eliminate first y^3, then y^2, and finally y from both equations. The end result was a fourth-degree equation in x: $x^4 - 6x^3 + 4x^2 + 6x - 5 = 0$. Having reduced the original system to a single equation in one unknown, he assumed that the solution process for this equation was familiar and asserted that the solution, that is, the hypotenuse of the triangle, was equal to 5.

We should note that Zhu's description of his method was very cryptic. He only wrote down a few of the many equations he needed to complete the solution; the description of the solution just given is the result of extensive modern scholarship. It appears that no Chinese mathematician in the years immediately following Zhu could extend his method, or even, perhaps, understand it at all. Furthermore, although Zhu and Qin exploited the counting board to the fullest, the physical board had limitations. Equations remained numerical, thereby precluding the development of a theory of equations like that which emerged several centuries later in the West. In addition, the political shifts in China associated with the change in dynasties around the fourteenth century resulted in a decline in mathematical activity, so that, soon, not even the thirteenth-century works we have discussed here were studied.

INDETERMINATE ANALYSIS

In addition to developing techniques for solving determinate equations and systems of equations, the Chinese also worked out methods for solving indeterminate systems of equations. Zhang Quijian's *Mathematical Classic* contained, for instance, the initial appearance of the problem of the "hundred fowls," famous because it also occurred in various guises in mathematics texts in India, the Islamic world, and Europe. Zhang posed the problem this way: "Now one rooster is worth 5 *qian*, one hen 3 *qian*, and 3 chicks 1 *qian*. It is required to buy 100 fowls with 100 *qian*. In each case, find the number of roosters, hens and chicks bought."[25] In modern notation, this translates—with x the number of roosters, y the number of hens, and z the number of chicks—into a system of two equations in three unknowns:

$$x + y + z = 100,$$

$$5x + 3y + \frac{1}{3}z = 100.$$

Zhang gave three answers—4 roosters, 18 hens, 78 chicks; 8 roosters, 11 hens, 81 chicks; and 12 roosters, 4 hens, 84 chicks—but he only hinted at a method: "Increase the roosters every time by 4, decrease the hens every time by 7, and increase the chicks every time by 3." In other words, he noted that changing the values in this way preserved both the cost and the number of birds. It is also possible to solve this problem by a modification of the "Gaussian elimination" method found in the *Nine Chapters* and to get as a general solution $x = -100 + 4t, y = 200 - 7t, z = 3t$ from which Zhang's description follows, but Zhang gave no clue as to his actual method. Nevertheless, his answers are the only ones in which all three values are positive.

In the thirteenth century, Yang Hui repeated this problem and then gave a similar one: "100 coins buy Wenzhou oranges, green oranges, and golden oranges, 100 in total. If a Wenzhou orange costs 7 coins, a green orange [costs] 3 coins, and 3 golden oranges cost 1 coin, how many oranges of the three kinds will be bought."[26] This problem can be translated into the system

$$x + y + z = 100,$$

$$7x + 3y + \frac{1}{3}z = 100,$$

[25] For this and next quotation, see Dauben, 2007, p. 307.
[26] For this and the following quotation, see Dauben, 2007, p. 333.

where x is the number of Wenzhou oranges, y the number of green oranges, and z the number of golden oranges. Yang, however, presented a solution method more detailed than Zhang's: "From 3 times 100 coins subtract 100 coins; from 3 times the cost of a Wenzhou orange, i.e., 21, subtract 1; the remainder is 20. From 3 times the cost of a green orange, i.e., 9, subtract 1, the remainder is 8. The sum of the remainders is 28. Divide 200 by 28, we have the integer 6. These are the numbers to be found: 6 Wenzhou oranges and 6 green oranges, respectively. And then $\frac{200-6\times28}{8} = 4$; this is the difference of the number of Wenzhou oranges and green oranges. hence the sum of them is 16, whereas the number of golden oranges to be found is 84."

Yang's method, while better articulated than Zhang's, was still not entirely clear. In modern terms, however, the beginning of the solution is straightforward. We multiply the second equation by 3 and then subtract the first equation, giving $20x + 8y = 200$. Note that this equation implies that x must be even. And since Yang writes of the "difference," we can take t as the difference $y - x$ and substitute. We get $28x + 8t = 200$ or $t = \frac{200-28x}{8}$. Certainly, the largest even number making t positive is $x = 6$. In that case, $t = 4$, $y = 10$, and $z = 84$, the solution given by Yang. However, this is only one possible solution, and Yang did not give any of the other three solutions. No better explanation of a method for solving these problems appeared elsewhere in medieval Chinese mathematics.

THE CHINESE REMAINDER PROBLEM

In contrast to the problem of solving indeterminate systems of equations, that of solving simultaneous congruences—the Chinese remainder problem—had been handled thoroughly by the thirteenth century. This type of problem had arisen most likely in the context of calendrical calculations like the following.[27] The Chinese assumed that at a certain point in time, the winter solstice, the new moon, and the beginning of the sixty-day cycle used in Chinese dating all occurred simultaneously. If, in a later year, the winter solstice occurred r days into a sixty-day cycle and s days after the new moon, then that year was N years after the initial date,

---

[27] Needham, 1959, p. 119 offers another possible explanation involving prognostication for the original interest in the Chinese remainder problem.

where N satisfied the simultaneous congruences

$$aN \equiv r \ (\mathrm{mod}\, 60), \qquad aN \equiv s \ (\mathrm{mod}\, b),$$

for a the number of days in the year and b the number of days from new moon to new moon. Unfortunately, the extant records of ancient calendars give no indication as to how Chinese astronomers actually determined N.

Simpler versions of congruence problems occur, however, in various mathematical works, with the earliest example appearing in the *Sunzi suanjing* (or *Mathematical Classic of Master Sun*), a work probably written late in the third century CE. There, Master Sun explained that "We have things of which we do not know the number; if we count them by threes, the remainder is 2; if we count them by fives, the remainder is 3; if we count them by sevens, the remainder is 2. How many things are there?"[28] In modern notation, the problem sought an integer N simultaneously satisfying

$$N = 3x + 2, \qquad N = 5y + 3, \qquad N = 7z + 2,$$

for integral values x, y, z, or (what amounts to the same thing) satisfying the congruences

$$N \equiv 2 \ (\mathrm{mod}\, 3), \qquad N \equiv 3 \ (\mathrm{mod}\, 5), \qquad N \equiv 2 \ (\mathrm{mod}\, 7).$$

Master Sun gave not only the answer but also his method of solution: "If you count by threes and have the remainder 2, put 140. If you count by fives and have the remainder 3, put 63. If you count by sevens and have the remainder 2, put 30. Add these numbers and you get 233. From this subtract 210 and you get 23." He further explained that "for each unity as remainder when counting by threes, put 70. For each unity as remainder when counting by fives, put 21. For each unity as remainder when counting by sevens, put 15. If the sum is 106 or more, subtract 105 from this and you get the result."

---

[28] Dauben, 2007, p. 299. The two quotations that follow in this paragraph may also be found here.

In modern notation, Master Sun apparently noted that

$$70 \equiv 1 \ (\mathrm{mod}\, 3) \equiv 0 \ (\mathrm{mod}\, 5) \equiv 0 \ (\mathrm{mod}\, 7),$$

$$21 \equiv 1 \ (\mathrm{mod}\, 5) \equiv 0 \ (\mathrm{mod}\, 3) \equiv 0 \ (\mathrm{mod}\, 7),$$

and

$$15 \equiv 1 \ (\mathrm{mod}\, 7) \equiv 0 \ (\mathrm{mod}\, 3) \equiv 0 \ (\mathrm{mod}\, 5).$$

Hence $2 \times 70 + 3 \times 21 + 2 \times 15 = 233$ satisfied the desired congruences. Because any multiple of 105 is divisible by 3, 5, and 7, subtracting 105 twice yielded the smallest positive value. This problem was, however, the only one of its type that Master Sun presented, so we do not know whether he had developed a general method for finding integers congruent to 1 modulo m_i but congruent to 0 modulo m_j, $j \neq i$, for given integers $m_1, m_2, m_3, \ldots, m_k$, the most difficult part of the complete solution. The numbers in this particular problem were easy enough to find by inspection.

In the thirteenth century, Yang Hui repeated this problem and gave similar ones, but Qin Jiushao first published a general solution method in his *Mathematical Treatise in Nine Sections*. There, he described what he called the *ta-yen* rule for solving simultaneous linear congruences, congruences that, in modern notation, may be written $N \equiv r_i \ (\mathrm{mod}\, m_i)$, for $i = 1, 2, \ldots, n$. To get a feel for his method, consider problem 5 of the first chapter of his treatise: There are three farmers, each of whom grew the same amount of rice (measured in *sheng*). Farmer A sold his rice on the official market of his own prefecture, in which rice was sold in lots of 83 *sheng*. He was left with 32 *sheng* after selling an integral number of lots. Farmer B was left with 70 *sheng* after selling his rice to the villagers of Anji, where rice was sold in lots of 110 *sheng*. Farmer C was left with 30 *sheng* after selling his rice to a middleman from Pingjiang, where rice was sold in lots of 135 *sheng*. How much rice had each farmer produced and how many lots did each sell?[29]

---

[29] Adapted from Dauben, 2007, p. 318.

In modern notation, this Chinese remainder problem requires that we find N so that

$$N \equiv 32 \ (\mathrm{mod}\,83), \qquad N \equiv 70 \ (\mathrm{mod}\,110), \qquad N \equiv 30 \ (\mathrm{mod}\,135).$$

To do this, first reduce the moduli to relatively prime ones. The second congruence shows that N is divisible by 5. Therefore, if N satisfies $N \equiv 3 \ (\mathrm{mod}\,27)$, then $N = 3 + 27t$ or $N = 30 + 27u$. Since this implies that u is a multiple of 5, we can rewrite this last equation as $N = 30 + 135v$ and conclude that the third congruence in the original set is satisfied. We must thus solve the following set of congruences with relatively prime moduli:

$$N \equiv 32 \ (\mathrm{mod}\,83), \qquad N \equiv 70 \ (\mathrm{mod}\,110), \qquad N \equiv 3 \ (\mathrm{mod}\,27).$$

The next step is to determine M, the product of the moduli $m_1 = 83$, $m_2 = 110$, and $m_3 = 27$. Thus, $M = 83 \times 110 \times 27 = 246{,}510$. Since any two solutions of the system will be congruent modulo M, once Qin found one solution, he generally determined the smallest positive solution by subtracting this value sufficiently many times.

Qin continued by dividing M by each of the moduli m_i in turn to get values we shall designate by M_i. Here $M_1 = M \div m_1 = 246{,}510 \div 83 = 2970$; $M_2 = 246{,}510 \div 110 = 2241$; and $M_3 = 246{,}510 \div 27 = 9130$. Each M_i satisfies $M_i \equiv 0 \ (\mathrm{mod}\,m_j)$ for $j \neq i$. Subtracting from each of the M_i as many copies of the corresponding m_i as possible yields the remainders of M_i modulo m_i. These remainders, labeled P_i, are $P_1 = 2970 - 35 \times 83 = 65$; $P_2 = 2241 - 20 \times 110 = 41$; and $P_3 = 9130 - 338 \times 27 = 4$. Of course, $P_i \equiv M_i \ (\mathrm{mod}\,m_i)$, for each i, so P_i and m_i are relatively prime.

To complete the problem, it is necessary actually to solve congruences, in particular, the congruences $P_i x_i \equiv 1 \ (\mathrm{mod}\,m_i)$. It is then easy to see that one answer to the problem is

$$N = \sum_{i=1}^{n} r_i M_i x_i,$$

in analogy to the solution to Master Sun's problem. Because each m_i divides M, any multiple of M can be subtracted from N to get other solutions.

In this example, Qin needed to solve the three congruences $65x_1 \equiv 1 \pmod{83}$, $41x_2 \equiv 1 \pmod{110}$, and $4x_3 \equiv 1 \pmod{27}$. He used what he called the "technique of finding one," essentially the Euclidean algorithm applied to P_i and m_i, to determine the solutions $x_1 = 23$, $x_2 = 51$, and $x_3 = 7$. It followed that

$$N = \sum_{i=1}^{3} r_i M_i x_i = 32 \cdot 2970 \cdot 23 + 70 \cdot 2241 \cdot 51 + 3 \cdot 9130 \cdot 7$$

$$= 2{,}185{,}920 + 8{,}000{,}370 + 191{,}730$$

$$= 10{,}378{,}020.$$

Qin finally determined the smallest positive solution to the congruence by subtracting $42M = 42 \cdot 246{,}510 = 10{,}353{,}420$ to get the final answer that each farmer produced $N = 24{,}600$ *sheng*. The rest of the problem's solution then followed easily, namely, A sold 296 lots, B sold 223 lots, and C sold 182 lots.

****

Chinese mathematicians developed much interesting mathematics prior to the beginning of the fourteenth century, especially in topics that we consider algebraic. Most of this work stemmed from efforts to solve certain practical problems. Some techniques—such as the method for solving equations numerically and the use of the Euclidean algorithm in solving congruences—are similar to ideas developed later elsewhere. There is little evidence, however, to suggest that these ideas were transmitted from China to India or Islam or the West.

There is also little evidence of the further indigenous development of mathematics in China after the thirteenth century. Classical texts were still read and used as a basis for civil service examinations, but the methods they expounded were not extended to new problems, and new ideas did not emerge. Following the introduction of western mathematics into China by the Jesuits in the late sixteenth century, the Chinese attempted to reconcile new western ideas and their indigenous thought. Gradually, however, western mathematics took hold, and indigenous mathematics—including indigenous algebraic techniques—disappeared.

6

Algebraic Thought in Medieval India

Unlike China, India was never unified completely during the ancient or medieval period.[1] Frequently, rulers in one or another of the small kingdoms on the Indian subcontinent conquered larger areas and established empires, but these rarely endured longer than a century. For example, shortly after the death of Alexander the Great in 323 BCE, Chandragupta Maurya (ca. 340–298 BCE) unified northern India, while his grandson Ashoka (304–232 BCE) spread Mauryan rule to most of the subcontinent, leaving edicts containing the earliest written evidence of Indian numerals engraved on pillars throughout the kingdom. After Ashoka's death, however, his sons fought over the inheritance, and the empire soon disintegrated. Early in the fourth century CE, northern India was again united under the dynasty of the Guptas, who encouraged art and medicine as well as the spread of Hindu culture to neighboring areas in southeast Asia. Just two generations after its height under Chandra Gupta II (reigned ca. 375–415 CE), this empire too collapsed in the wake of assaults from the northwest. Despite changes in dynasties and sizes of kingdoms, many Indian rulers encouraged the study of astronomy in order to ensure accurate answers to calendrical and astrological questions.

The earliest mathematical ideas that we can definitely attribute to Indian sources are found in the *Śulba-sūtras*, works dating from sometime in the first millennium BCE, that describe the intricate sacrificial system of the brahmins. Most of the ideas contained in these texts are geometrical, but many of the problems can be thought of as geometrical algebra in the sense of chapter 3 (see below for an example). Indian mathematics in the first millennium CE, however, did not refer to the geometrical algebra of

---

[1] See Plofker, 2009, pp. 4–10 for a summary of the history of the Indian subcontinent.

the *Śulba-sūtras*. In fact, what we think of as algebra in the Indian context is purely algorithmic.

The first glimpse of such ideas in the work of an identifiable Indian mathematician is in the *Āryabhaṭīya* (ca. 500 CE), which, like most works containing mathematics (*gaṇita*) over the next thousand years, was fundamentally astronomical and so belonged to the genre of *siddhānta* (or astronomical treatises). Its author, Āryabhaṭa (b. 476 CE), lived near the Gupta capital of Pāṭaliputra (modern Patna) on the Ganges River in Bihar in northern India, and while he did not claim originality in his work, he seems to have participated in a well-established mathematical tradition. Although the Indian origins of the algebraic ideas Āryabhaṭa presented are unknown (Āryabhaṭa himself claimed they were divinely revealed), we do know that he and his successors were creative mathematicians, who went beyond the practical problem solving of *siddhānta* to develop new areas of mathematics—and particularly algebra—that piqued their interest. Another key feature of this mathematical work is its form. Like most medieval Indian texts, it was composed in Sanskrit verse. Originally a spoken language, Sanskrit had become, by the middle of the first millennium, a shared language of Indian scholarship, even though most Indians spoke one of numerous other languages. In this regard, Sanskrit and medieval Latin served similar purposes. Since Sanskrit was venerated as sacred speech, moreover, the texts were designed to be recited and memorized. The requirements of poetic form thus often superseded those of technical clarity and resulted in the production, ultimately, of supplementary prose commentaries by the author himself or by subsequent scholars.[2]

The central mathematical chapter of the *Āryabhaṭīya*, the text's second chapter, consists of thirty-three verses, each containing a brief rule for a particular kind of calculation. The requirements of poetry make some of the rules cryptic, but with the help of a commentary by Bhāskara I (early seventh century CE) and the two mathematical chapters of the *Brāhma-sphuṭa-siddhānta* of Brahmagupta (early seventh century CE), both written around 628 CE, it is generally possible to understand Āryabhaṭa's meaning.

This chapter will explore medieval Indian algebraic thought in the works of Āryabhaṭa and Brahmagupta before moving on to consider the

---

[2] For more discussion of this idea, see Plofker, 2009, pp. 10–12. Indeed, this book is the best one available on the history of Indian mathematics.

contributions of later Indian mathematicians to such topics as solving determinate and indeterminate equations, including the so-called Pell equation $Dx^2 + 1 = y^2$, where D, x, and y are positive integers. It will close with a look at some of the more advanced algebraic topics in the work of the Kerala school, which began in the fourteenth century in the south of India and lasted for over two hundred years.

PROPORTIONS AND LINEAR EQUATIONS

As we have seen, one important method for determining unknown quantities was the rule of three, a rule also central to Indian mathematics as exemplified in the *Āryabhaṭīya*.[3] For example, verse twenty-six of this work's second chapter (denoted 2.26) reads, "Now when one has multiplied that fruit quantity of the rule of three by the desire quantity, what has been obtained from that divided by the measure should be this fruit of the desire."[4] As Bhāskara I explained, we are given both a "measure quantity" m that produces a "fruit quantity" p and a "desire quantity" q for which we want to know the "fruit of the desire" x. In modern terminology, this boils down to the proportion $m : p = q : x$, and the solution is just what Āryabhaṭa claimed, namely, $x = \frac{pq}{m}$. It is interesting to note that Āryabhaṭa stated the "rule of three" as if it were a common procedure. It is thus reasonable to assume that solving proportions in this way was well known to his readers.

To clarify the rule of three further, Bhāskara I included several examples in his commentary, among them the following: "A *pala* and a half of musk have been obtained with eight and one third *rūpakas*. Let a powerful person compute what I should obtain with a *rūpaka* and one fifth."[5] Here, $8\frac{1}{3}$ is the measure, $1\frac{1}{2}$ the fruit, and $1\frac{1}{5}$ the "desire quantity." Thus, the "fruit of the desire" is $(1\frac{1}{2} \times 1\frac{1}{5}) \div 8\frac{1}{3} = \frac{27}{125}$.

Bhāskara I also generalized the rule of three to a rule of five, seven, and so on. For example, suppose "the interest of one hundred in a month should be five. Say how great is the interest of twenty invested for six months."[6] Here, the original fruit $p_1 = 5$ is produced by two measures

---

[3] For much more on the rule of three in India, see Sarma, 2002.
[4] Plofker, 2007, p. 414.
[5] Keller, 2006, p. 110.
[6] Keller, 2006, p. 112.

$m_1 = 100$ and $m_2 = 1$. We want to know the new "fruit" when the desires are $q_1 = 20$ and $q_2 = 6$. According to Bhāskara I, we can treat this as two linked rule-of-three problems, that is, we calculate p_2 (the fruit of the desire q_1, leaving the time period fixed) by solving the proportion $m_1 : p_1 = q_1 : p_2$, so $p_2 = \frac{p_1 q_1}{m_1}$. We then change the time period and calculate the fruit x by solving $m_2 : p_2 = q_2 : x$, so $x = \frac{p_2 q_2}{m_2} = \frac{p_1 q_1 q_2}{m_1 m_2} = 6$. In any case, we have in the rule of five, just as in the original rule of three, a simple algorithm involving multiplication and division to determine the answer. Neither Bhāskara I nor Brahmagupta, who gave the same rule of three as well as an inverse rule of three, provided any further explanation of the method. We can therefore safely assume that the understanding of proportion was so basic that no real explanation was necessary, even in a textbook.

Verse 2.30 of the *Āryabhaṭīya* gave a rule for solving more general linear equations, although it seems to be very specific to a particular type of problem: "One should divide the difference of coins belonging to two men by the difference of beads. The result is the price of a bead, if what is made into money for each man is equal."[7] The idea here, again explained by Bhāskara I, was that "bead" stood for an object of unknown price, and the problem aimed to find that price. For example, "the first tradesman has seven horses with perpetual strength and auspicious marks and a hundred *dravyas* are seen by me in his hand. Nine horses and the amount of eighty *dravyas* belonging to the second tradesman are seen. The price of one horse, and the equal wealth of both tradesmen should be told by assuming the same price for all the horses."[8] Here, since the wealth of the two men was the same, the equation to solve was $7x + 100 = 9x + 80$, where x is the price of one horse. To do this, divide the "difference of coins" $(100 - 80)$ by the "difference of beads" $(9 - 7)$ to get $20 \div 2 = 10$ *dravyas* as the value of each horse. In general, Āryabhaṭa's method found the solution of the linear equation $ax + c = bx + d$ to be $x = \frac{d-c}{a-b}$.

Brahamagupta had a rule very similar to that of Āryabhaṭa, although he specifically stated it as a rule for solving equations in one unknown: "The difference between *rūpas*, when inverted and divided by the difference of the unknowns, is the unknown in the equation."[9] Brahmagupta

[7] Plofker, 2007, p. 415.
[8] Plofker, 2007, p. 415.
[9] Plofker, 2007, p. 430, verse 18.43.

did not do much with this rule, but Mahāvīra (ca. 850) considered many linear equations of various types in his *Gaṇita-sāra-saṅgraha*, the first independent mathematical text in Sanskrit to have survived. Mahāvīra's problems were generally very intricate and written in a flowery style that became common in Indian mathematics during the next several centuries. For example, "Of a collection of mango fruits, the king took $\frac{1}{6}$, the queen took $\frac{1}{5}$ of the remainder, and three chief princes took $\frac{1}{4}, \frac{1}{3}, \frac{1}{2}$ [of what remained at each stage]; and the youngest child took the remaining three mangos. O you, who are clever in working miscellaneous problems on fractions, give out the measure of that collection of mangos."[10] In modern notation, this problem may be translated into the equation

$$x = \frac{x}{6} + \frac{1}{5} \cdot \frac{5}{6}x + \frac{1}{4} \cdot \frac{4}{6}x + \frac{1}{3} \cdot \frac{3}{6}x + \frac{1}{2} \cdot \frac{2}{6}x + 3,$$

where it is evident that each person ended up with exactly the same number of mangos. Mahāvīra gave rules for solving this and several other types of linear equations, but, unlike Brahmagupta, he did not give a general rule for solving all linear equations in one unknown. General rules for solving quadratic equations were, however, given by several Indian mathematicians.

QUADRATIC EQUATIONS

The *Baudhāyana-śulba-sūtra* (ca. 600 BCE) uses geometrical algebra in treating what we recognize as quadratic equations. In particular, we find there the problem of transforming a rectangle into a square, that is, in modern notation, the problem of solving $x^2 = cd$, where c and d are the length and width of the given rectangle. The same problem may be found in Euclid's *Elements* II.14, although Euclid and the author of the *Śulba-sūtras* effected the solution differently. The problem and solution are given as follows in verse 2.5: "If it is desired to transform a rectangle into a square, its breadth is taken as the side of a square [and this square on the breadth is cut off from the rectangle]. The remainder [of the rectangle] is divided into two equal parts and placed on two sides

[10] Mahāvīra, 1912, verses 4.29–30 or Plofker, 2007, p. 445.

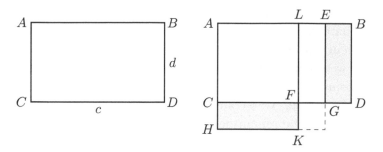

Figure 6.1.

[one part on each]. The empty space [in the corner] is filled up with a [square] piece. The removal of it [of the square piece from the square thus formed to get the required square] has been stated [in 2.2]."[11] (See figure 6.1.) In other words, we cut off a square of side $CF = BD = d$ from the original rectangle of length $c = CD$, divide the remaining rectangle in half, and place one piece on each side of the square. We therefore have transformed the rectangle into a gnomon of the same area. We then complete the square with a new square of side $GF = \frac{c-d}{2}$. Thus, we have demonstrated essentially the same result as *Elements* II.6, namely, that

$$cd + \left(\frac{c-d}{2}\right)^2 = \left(\frac{c+d}{2}\right)^2,$$

or that the rectangle has been expressed as the difference of two squares.

The *Śulba-sūtras*, however, had already stated "the removal of it," that is, how to express the difference of two squares as a single square in verse 2.2: "If it is desired to remove a square from another, a [rectangular] part is cut off from the larger [square] with the side of the smaller one to be removed; the [longer] side of the cut-off [rectangular] part is placed across so as to touch the opposite side; by this contact [the side] is cut off. With the cut-off [part], the difference [of the two squares] is obtained [as a square]."[12] (See figure 6.2.) Thus, if $BC = a$ and $PS = b$, we set $DL = AK = b$, draw LK, and find M so that $KM = LK = a$. By the Pythagorean theorem, well known in India by this time, we have $AM^2 = MK^2 - KA^2 = a^2 - b^2$. Thus, AM is the side of the square that

[11] Plofker, 2007, p. 391.
[12] Plofker, 2007, p. 390.

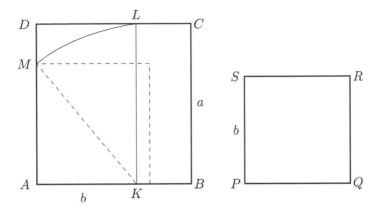

Figure 6.2.

is the difference of the two original squares, and we have found a square equal in area to the original rectangle.

This method of completing the square geometrically seems to have had no effect on the solution of quadratic equations a millennium later. For example, consider verses 2.19 and 2.20 from the *Āryabhaṭīya*, both of which ostensibly deal with the sum of an arithmetic progression. In verse 2.19, Āryabhaṭa stated that "the desired number of terms decreased by one, halved, . . . , having the common difference for multiplier, [and] increased by the first term, is the mean [value]. [The result], multiplied by the desired, is the value of the desired [number of terms]. Or else, the first and last added together multiplied by half the number of terms is the value."[13] Thus, he presented two algorithms for the sum S_n of an arithmetic progression with initial term a and common difference d, namely,

$$S_n = n\left[\left(\frac{n-1}{2}\right)d + a\right] = \frac{n}{2}[a + (a + (n-1)d)].\tag{6.1}$$

In the next verse, he specified that "the square root of the value of the terms multiplied by eight and the common difference, increased by the square of the difference of twice the first term and the common difference, decreased by twice the first term, divided by its common difference, increased by one and halved [is the number of terms]." In

---

[13] Keller, 2006, p. 93. The next quote appears on p. 97.

other words, knowing the sum S_n, the number n of terms was given by the algorithm

$$n = \frac{1}{2} \left[\frac{\sqrt{8S_n d + (2a - d)^2} - 2a}{d} + 1 \right]. \tag{6.2}$$

Rewriting (6.1) as a quadratic equation in n, namely, $dn^2 + (2a - d)n - 2S_n = 0$, the value for n in (6.2) follows from the quadratic formula. Thus, it is reasonable to assume that Āryabhaṭa knew how to solve quadratic equations by using that formula.

We also see this in verse 2.24: "The square root of the product of two quantities with the square of two for multiplier, increased by the square of the difference of the two, is increased or decreased by the difference, and halved, this will produce the two multipliers of that product."[14] In modern notation, Āryabhaṭa claimed that if we know the product $xy = c$ of two quantities and their difference $x - y = b$, then the two quantities are

$$x = \frac{\sqrt{4c + b^2} + b}{2} \quad \text{and} \quad y = \frac{\sqrt{4c + b^2} - b}{2}.$$

In other words, he provided an algorithm for solving the quadratic system $xy = c, x - y = b$, again equivalent to a version of the quadratic formula.

Although Bhāskara I gave examples to illustrate these verses from Āryabhaṭa's text, he neither discussed how his predecessor might have arrived at these solutions nor generalized the procedure to that of finding the solution of an arbitrary quadratic. Furthermore, in his commentary on verse 2.24, he made mention neither of the similarities between the procedure there and in verse 2.20 nor of the Śulba-sūtras.

In the Brāhma-sphuṭa-siddhānta, however, Brahmagupta did present a general algorithm for solving quadratic equations, in particular, equations we would write as $ax^2 + bx = c$. As he put it, "diminish by the middle [number] the square root of the rūpas multiplied by four times the square and increased by the square of the middle [number]; divide [the remainder] by twice the square. [The result is] the middle [number]."[15]

---

[14] Keller, 2006, p. 104.
[15] Plofker, 2007, p. 431, verse 18.44.

Here, the "middle number" is the coefficient b (and also the unknown itself); the *rūpas* is the constant term c and the "square" is the coefficient a. Brahmagupta's words thus easily translate into the formula

$$x = \frac{\sqrt{4ac + b^2} - b}{2a}.$$

While he also gave the same rule as Āryabhaṭa for determining two numbers if their product and difference are known, he gave only rules, not numerical examples.

A numerical example was given, however, by the ninth-century commentator, Pṛthūdakasvāmin, who followed the quadratic formula rule in solving the equation $x^2 - 10x = -9$. Still, neither Brahmagupta nor his commentator noted that there could well be two positive solutions to a given quadratic equation. They also gave no clues as to how the formula was discovered. For example, there is no indication, as in ancient Mesopotamia, of a geometric basis for the solution, nor is there any purely algebraic derivation. We are simply presented with the algorithm.[16]

Mahāvīra also considered a quadratic equation in his *Gaṇita-sāra-saṅgraha*: "One-third of a herd of elephants and three times the square root of the remaining part of the herd were seen on a mountain slope; and in a lake was seen a male elephant along with three female elephants [constituting the ultimate remainder]. How many were the elephants here?"[17] In his answer, Mahāvīra gave no general rule for determining the solution of problems of this type. Rather, he laid out a very specific rule—albeit one equivalent to the quadratic formula—for solving a problem dealing with square roots of an unknown collective quantity and remainders after certain elements are removed. Like his predecessors, he provided no hints as to how he determined the method of solution, perhaps because the poetic format left no space for any derivation.

Several hundred years later, Bhāskara II (1114–1185) also dealt with quadratic equations in his *Bīja-gaṇita*, written around 1160. He, however, gave an explicit solution method involving completing the square and

---

[16] It is, of course, possible that Āryabhaṭa or Brahmagupta derived the result by adapting the geometrical algebra of the *Śulba-sūtras*, but there is no evidence that they even knew of these works.

[17] Mahāvīra, 1912, verse 4.41 or Plofker, 2007, p. 445.

not just an algorithm similar to the quadratic formula. Moreover, he dealt with multiple roots, at least when both are positive. Thus, to solve a quadratic equation of the form $ax^2 \pm bx = c$, he suggested multiplying each side by an appropriate number and also adding the same number to or subtracting the same number from both sides so that the left-hand side became a perfect square and the equation became $(rx - s)^2 = d$. He then solved the equation $rx - s = \sqrt{d}$ for x, but he noted that "if the square-root of the side having the manifest number is less than the negative *rūpa* in the root of the side having the unknown, making that negative and positive, the measure of the unknown in two values is sometimes to be obtained."[18] In other words, if $\sqrt{d} < s$, then there are two values for x, namely, $\frac{s+\sqrt{d}}{r}$ and $\frac{s-\sqrt{d}}{r}$. To get a better sense of his use of "sometimes," consider two of his examples.

In one, he set the following scene: "The eighth part of a troop of monkeys, squared, is dancing joyfully in a clearing in the woods; twelve are seen on a hill enjoying making noises and counter-noises. How many are there?"[19] In effecting a solution here and elsewhere in his work, Bhāskara II used not symbols but rather a standard term *yāvattāvat* (meaning "as much as so much") for the unknown quantity, which he often abbreviated *yā*. Similarly, for our x^2, he used *yā va*, short for *yāvattāvat varga* or the "square of the unknown." Rather than using Bhāskara II's shorthand, however, we shall cast this problem in modern symbols as $(\frac{1}{8}x)^2 + 12 = x$. Bhāskara II multiplied every term by 64 and subtracted to get $x^2 - 64x = -768$. To complete the square, he next added 32^2 to each side, yielding $x^2 - 64x + 1024 = 256$. Taking square roots gave $x - 32 = 16$. He then noted that "the number of the root on the absolute side is here less than the known number, with the negative sign, in the root on the side of the unknown."[20] Therefore, 16 can be made positive or negative. So, he concluded, "a two-fold value of the unknown is thence obtained, 48 and 16."

The next problem he considered only had one meaningful solution, though: "The shadow of a twelve-digit gnomon diminished by a third of its hypotenuse became fourteen digits. Tell me that shadow quickly,

---

[18] Bhāskara II, 2000, verse 121.

[19] Bhāskara II, 2000, verse 130.

[20] Colebrooke, 1817, p. 216 for this and the next quotation. This statement is part of Bhāskara II's own commentary on verse 130.

oh calculator."[21] To solve the problem—one basically involving a right triangle formed by a pole (the gnomon), its shadow, and the line connecting the tip of the shadow with the tip of the pole—we see that if the shadow is equal to x and the hypotenuse equal to h, then $x - \frac{1}{3}h = 14$, so $h = 3x - 42$. Therefore, the problem boiled down to the equation $(3x - 42)^2 = x^2 + 144$. Expanding and simplifying gave $8x^2 - 252x = -1620$, and Bhāskara II noted that if he multiplied by 2 and added 63^2, the left-hand side would be a perfect square, that is, he would have the equation $16x^2 - 504x + 63^2 = 729$, or $(4x - 63)^2 = 27^2$. Therefore, $4x - 63 = \pm 27$, and there were two solutions, $\frac{45}{2}$ and 9. But, he noted, "the second value of the shadow is less than fourteen. Therefore, by reason of its incongruity, it should not be taken."[22] In other words, the second (positive) solution to the equation imposed a negative in the calculation of the problem and therefore made no sense in his practical context.

Bhāskara II thus at least considered two solutions of a quadratic. In the case of a quadratic with a positive and a negative root, however, he only found the positive one. His examples of quadratic equations also never had two negative roots or two nonreal roots or irrational roots. All of his problems, in fact, had rational solutions. Given that he knew how to solve quadratics algorithmically—and that he clearly understood what a perfect square binomial looked like—it is rather surprising that he never gave examples with irrational answers. He surely would have run into such problems in his scratch work, but perhaps he wanted to publish only problems with rational answers. After all, there could not be an irrational number of monkeys in a troop!

In addition to solving quadratic equations, Bhāskara II gave a number of cubic and quartic equations that could be treated by completing the third or fourth power. For example, "what quantity, the square of the square of which, diminished by the quantity multiplied by two hundred, increased by the square of the quantity, and multiplied by two, is ten thousand diminished by one? Tell this quantity if you know the operations of algebra."[23] In modern notation, this equation can be written as $x^4 - 2(x^2 + 200x) = 9999$, or $x^4 - 2x^2 - 400x = 9999$. If, Bhāskara II noted,

[21] Bhāskara II, 2000, verse 131.
[22] Colebrooke, 1817, p. 218.
[23] Bhāskara II, 2000, verses 128–129.

we add $400x + 1$ to each side, the left-hand—but not the right-hand—side becomes a square. "Hence, ingenuity is in this case called for."[24] The ingenuity consisted of adding $4x^2 + 400x + 1$ to each side, making the left-hand side $(x^2 + 1)^2$ and the right-hand side $[2(x + 50)]^2$. Taking square roots yields $x^2 + 1 = 2x + 100$, or $x^2 - 2x + 1 = 100$, or $(x - 1)^2 = 100$, or, finally, $x = 11$. (Bhāskara II did not comment on why he only took positive square roots, but it is clear that taking the negative ones would not have resulted in a positive answer.)

Although Bhāskara II certainly understood the rationale he described for solving quadratic equations, proof did not become a central part of Indian mathematics until the Kerala school began its work in the fourteenth century.[25] For example, Citrabhānu (early sixteenth century), a member of that school, wrote a brief treatise in which he gave proofs, both algebraic and geometric, of twenty-one rules generalizing Āryabhaṭa's rule for finding two quantities if their product and difference are known. That is, given the seven equations $x + y = a$, $x - y = b$, $xy = c$, $x^2 + y^2 = d$, $x^2 - y^2 = e$, $x^3 + y^3 = f$, and $x^3 - y^3 = g$, Citrabhānu presented rules and justifications for solving each of the twenty-one combinations of these equations taken two at a time, given known values of the constants. For instance, although Āryabhaṭa's verse 2.24 gave the rule for solving the second and third equations simultaneously, Citrabhānu proved it by showing that $(x + y)^2 = 4xy + (x - y)^2$. Therefore, $x + y = \sqrt{4c + b^2}$, and, by Citrabhānu's rule for solving the first and second equations, since the sum and difference of the two quantities are known,

$$x = \frac{\sqrt{4c + b^2} + b}{2} \quad \text{and} \quad y = \frac{\sqrt{4c + b^2} - b}{2},$$

exactly as claimed by Āryabhaṭa.

Interestingly, Citrabhānu justified all of his quadratic problems algebraically by essentially deriving the formulas, but in many of his cubic problems, he gave a geometrical rather than an algebraic derivation. Consider the case in which b and g are known, that is, where we want to solve the system $x - y = b$, $x^3 - y^3 = g$. Citrabhānu's rule specified,

---

[24] Colebrooke, 1817, p. 215.
[25] The work of the Kerala school is discussed in Plofker, 2007, chapter 7. In particular, see pp. 247–248 for a discussion of "proof."

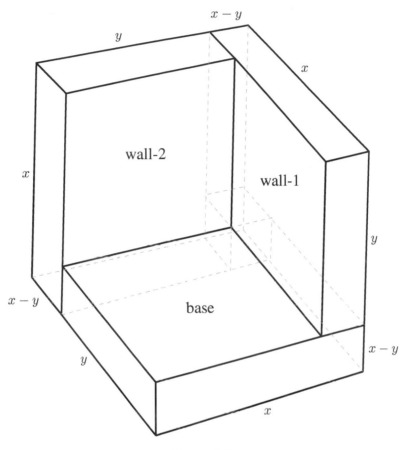

Figure 6.3.

"When the difference between the cubes of the two quantities is decreased by the cube of the difference and divided by three times the difference, it [the quotient] will be the product of the two [quantities]."[26] In modern symbols, this translates as $xy = c = \frac{g-b^3}{3b}$. The previous rule then allows for the determination of x and y. But how did Citrabhānu come up with this?

He described a figure formed by two walls, a base, and a small cube, the outer three dimensions of which were all equal to x and the inner three dimensions of which all equaled y. Thus, the region between the outer surface and the inner empty space is exactly $x^3 - y^3$. (See figure 6.3.)

---

[26] Hayashi and Kusuba, 1998, p. 11. This article discusses all of Citrabhānu's rules and derivations.

This region is formed, however, by three rectangular parallelepipeds, each of height x, width y, and depth $x - y$, as well as a small cube of side $x - y$. Thus, $x^3 - y^3 = 3xy(x - y) + (x - y)^3$, or $g = 3cb + b^3$, from which Citrabhānu's rule followed.

INDETERMINATE EQUATIONS

Besides treating linear and quadratic equations in one unknown, Indian mathematicians also handled indeterminate equations in several unknowns. For example, Mahāvīra presented a version of the hundred fowls problem in his treatise: "Doves are sold at the rate of 5 for 3 coins, cranes at the rate of 7 for 5, swans at the rate of 9 for 7, and peacocks at the rate of 3 for 9. A certain man was told to bring at these rates 100 birds for 100 coins for the amusement of the king's son and was sent to do so. What amount does he give for each?"[27]

While Mahāvīra gave a rather complicated rule for the solution, Bhāskara II presented the same problem with a procedure showing explicitly why there are multiple solutions. He set his unknowns—which we label d, c, s, and p, but which he distinguished by different color names—equal to the number of "sets" of doves, cranes, swans, and peacocks, respectively. From the prices and the numbers of birds, he derived the two equations

$$3d + 5c + 7s + 9p = 100,$$
$$5d + 7c + 9s + 3p = 100,$$

and proceeded to solve them. He first solved each equation for d, equated the two expressions, and found the equation $c = 50 - 2s - 9p$. Taking an arbitrary value 4 for p, he reduced the equation to the form $c + 2s = 14$, for which the solution is $s = t$, $c = 14 - 2t$, for t arbitrary. It followed that $d = \frac{1}{3}(100 - 5c - 7s - 9p) = \frac{1}{3}(100 - 5(14 - 2t) - 7t - 36) = \frac{1}{3}(3t - 6) = t - 2$. Then setting $t = 3$, he calculated that $d = 1$, $c = 8$, $s = 3$, and $p = 4$ and hence that the number of doves is 5, of cranes 56, of swans 27, and of peacocks 12, their prices being,

[27] Mahāvīra, 1912, p. 134.

respectively, 3, 40, 21, and 36. He noted further that other choices of t gave different values for the solution. Thus, "by means of suppositions, a multitude of answers may be obtained."[28]

LINEAR CONGRUENCES AND THE PULVERIZER

Like the equations just discussed, pairs of what we call linear congruences also have multiple solutions. Unlike the situation relative to indeterminate equations, however, Indian mathematicians originated a unique method, called the "pulverizer," for solving these systems, a method that does not appear in any other tradition. In modern notation, the problem can be posed in various equivalent ways:[29]

(1) Find N satisfying $N \equiv a \pmod{r}$ and $N \equiv b \pmod{s}$;

(2) Find x and y such that $N = a + rx = b + sy$;

(3) Find x and y such that $a + rx = b + sy$; or

(4) Find x and y such that $rx + c = sy$, where $c = a - b$.

Āryabhaṭa stated the pulverizer method for solving the first version of this problem; Bhāskara I expanded somewhat on Āryabhaṭa's brief verses and gave examples; Brahmagupta essentially repeated the rules with further examples. The central idea of the procedure was to apply what is now called the Euclidean algorithm to the two numbers r and s. Each step produced a new indeterminate equation with smaller coefficients than the previous one until the final equation was solvable by inspection. Going back through the steps allowed for the determination of each unknown in turn until the original equation was reached. Evidently, this method was called the "pulverizer" because, by repeating the division process, the original numbers were "pulverized" until they were small enough to provide an easy solution.

By way of example, consider one problem given by the commentator Pṛthūdakasvāmin on the work of Brahmagupta:[30] $N \equiv 10 \pmod{137}$ and $N \equiv 0 \pmod{60}$. Since this problem can be rewritten (see (4) above) as the

[28] Colebrooke, 1817, pp. 234–235.
[29] See Plofker, 2007, pp. 134–136 and p. 150 for more details.
[30] Colebrooke, 1817, p. 336.

single equation $60y = 137x + 10$, start by dividing 137 by 60 and making the step-by-step substitutions in accordance with the successive quotients appearing in the Euclidean algorithm:

$$y = \frac{137x + 10}{60} = 2x + \frac{17x + 10}{60} = 2x + z,$$

$$x = \frac{60z - 10}{17} = 3z + \frac{9z - 10}{17} = 3z + u,$$

$$z = \frac{17u + 10}{9} = 1u + \frac{8u + 10}{9} = 1u + v,$$

$$u = \frac{9v - 10}{8} = 1v + \frac{1v - 10}{8} = 1v + w,$$

$$v = 8w + 10.$$

At this point, the last equation is solvable by inspection as $w = 1, v = 18$. To find the remaining unknowns, use substitution to work back up the list from bottom to top:

$$u = 1v + w = 1 \cdot 18 + 1 = 19, \quad z = 1u + v = 1 \cdot 19 + 18 = 37,$$

$$x = 3z + u = 3 \cdot 37 + 19 = 130, \quad y = 2x + z = 2 \cdot 130 + 37 = 297.$$

Thus, $x = 130$, $y = 297$ is a solution of the original equation in two unknowns.

The problem, however, actually called for determining N and, in fact, the *smallest* N that solved the original problem. To find that, note that $N = 10 + 137x = 17{,}820$, but since any two solutions for N are equivalent modulo the product $137 \cdot 60 = 8220$, we can subtract 8220 twice from 17,820 to get 1380 as the smallest solution for N. This also gives smaller solutions for x and y, for, since $1380 = 10 + 137x = 60y$, we have $x = 10$, $y = 23$ as the smallest solution of the linear equation in two unknowns.

Sometimes, however, the mathematicians were not interested in N but rather in solving a single linear equation in two unknowns. Bhāskara II, for example, incorporated several such problems in his *Bīja-gaṇita*, including the following: "Oh calculator! Tell me quickly that multiplier which, multiplied by a pair of hundreds combined with twenty-one, joined with sixty-five, and divided by a pair of hundreds diminished by five, arrives at the state of being reduced without a remainder."[31] In modern terms, the problem was to solve the equation $221x + 65 = 195y$.

We have already seen that, beginning in the third century, various Chinese authors were interested in solving systems of linear congruences. A close comparison of their method with that of Indian mathematicians, however, shows that the two were quite different. Indian authors usually treated a system of two congruences, while the Chinese dealt with a larger system. When an Indian author did treat a problem similar to a "Chinese remainder problem," he solved the congruences two at a time, rather than all at once as the Chinese did. The only similarity between the Indian and Chinese methods would seem to be that both hinged on the Euclidean algorithm. The more interesting questions— unanswerable given current evidence—then become, did either culture learn the algorithm from the Greeks?, did all three learn it from an earlier culture?, or, did the two Asian cultures simply discover the algorithm independently?

There is good evidence that Brahmagupta and Āryabhaṭa were orig- inally interested in congruence problems for the same basic reason as the Chinese, namely, for use in astronomy, because many of the problems in their texts have astronomical settings.[32] The Indian astro- nomical system of the fifth and sixth century, like the ancient Chinese system, hinged on the notion of a large astronomical period at the beginning and end of which all the planets (including the sun and moon) had longitude zero. The idea was that all worldly events would recur with this same period. For Āryabhaṭa, the fundamental period was the *Mahāyuga* of 4,320,000 years, while for Brahmagupta, it was the *Kalpa* of 1,000 *Mahāyugas*. In any case, calculations with heavenly bodies required knowing their average motions. Since it was impossible to determine those motions empirically with any accuracy, it became necessary to calculate them on the basis of current observations and on the fact that all the planets were at approximately the same place at the beginning of the period. These calculations involved solving linear congruences.

---

[31] Bhāskara II, 2000, verses 56–57.
[32] See Plofker, 2007, pp. 72–79 for a discussion of Indian astronomy.

THE PELL EQUATION

The ability to solve single linear equations in two unknowns turned out to be important in the solution of another type of indeterminate equation: the quadratic equation of the form $Dx^2 + 1 = y^2$. This is usually called the Pell equation today after the seventeenth-century Englishman John Pell (1611–1685), even though he actually had very little to do with his eponymous equation. The detailed study of these equations, in fact, first took place in India and ultimately marked the high point of medieval Indian algebra.

To get a sense of the Indian approach to equations of this type, consider the following example from the work of Brahmagupta: "He who computes within a year the square of [a number]. . . multiplied by ninety-two. . . and increased by one [that is] a square, [he] is a calculator."[33] This problem was thus to solve $92x^2 + 1 = y^2$, and Brahmagupta began this way: "Put down twice the square root of a given square multiplied by a multiplier and increased or diminished by an arbitrary [number]."[34] In other words, start by putting down any number, say 1, and determining by inspection a value b_0 such that $92 \cdot 1^2 + b_0$ is a square. Choosing $b_0 = 8$ gives $92 \cdot 1^2 + 8 = 100 = 10^2$ or three numbers x_0, b_0, y_0 satisfying the equation $Dx_0^2 + b_0 = y_0^2$. For convenience, we represent this by writing that (x_0, y_0) is a solution for additive b_0. Thus, $(1, 10)$ is a solution for additive 8.

Brahmagupta next wrote this solution in two rows as

$$x_0 \quad y_0 \quad b_0$$
$$x_0 \quad y_0 \quad b_0$$

or

$$1 \quad 10 \quad 8$$
$$1 \quad 10 \quad 8.$$

As he explained, "the product of the first [pair], multiplied by the multiplier, with the product of the last [pair], is the last computed.

---

[33] Plofker, 2007, p. 433, verse 18.75.

[34] Plofker, 2007, p. 432, verses 18.64 and 18.65. The quotation in the next paragraph may also be found here.

The sum of the thunderbolt products [cross multiplication] is the first. The additive is equal to the product of the additives. The two square roots, divided by the [original] additive or subtractive, are the additive *rūpas*." In other words, he found a new value for the "last root" y by setting $y_1 = Dx_0^2 + y_0^2$, specifically, $y_1 = 92(1)^2 + 10^2 = 192$; a new value for the "first root" x, namely, $x_1 = x_0y_0 + x_0y_0$ or $x_1 = 2x_0y_0$, so $x_1 = 20$; and a new additive $b_1 = b_0^2 = 64$. Therefore, $(x_1, y_1) = (20, 192)$ is a solution for additive $b_1 = 64$, or $92(20)^2 + 64 = 192^2$. Finally, to get the results for additive 1, Brahmagupta divided x_1 and y_1 by 8 to get $(\frac{5}{2}, 24)$. Here, however, since x_1 is not an integer, the answer was unsatisfactory, so he used the same array technique as before to get the integral solution $(120, 1151)$ for additive 1, that is, $92 \cdot 120^2 + 1 = 1151^2$.

Although Brahmagupta provided neither a proof for his methods nor even a hint as to its derivation, his result is easy enough to verify.[35] In fact, he even generalized the array method to assert that if (u_0, v_0) is a solution for additive c_0 and (u_1, v_1) is a solution for additive c_1, then $(u_0v_1 + u_1v_0, Du_0u_1 + v_0v_1)$ is a solution for additive c_0c_1. Calling the new solution the *composition* of the two original solutions, it is straightforward to show that the composition is a solution in light of the identity

$$D(u_0v_1 + u_1v_0)^2 + c_0c_1 = (Du_0u_1 + v_0v_1)^2,$$

and given that $Du_0^2 + c_0 = v_0^2$ and $Du_1^2 + c_1 = v_1^2$.

This example not only illustrates Brahmagupta's method, but also highlights its limitations. The solution for additive 1 in the general case is the pair $(\frac{x_1}{b_0}, \frac{y_1}{b_0})$, but there is no guarantee either that these will be integers or that integers would necessarily result by combining this solution with itself. Brahmagupta merely gave rules that helped find solutions. For example, he noted that composition allowed him to get other solutions for any additive, provided he knew one solution for the given additive as well as a solution for additive 1. Second, if he had a solution (u, v) for additives ± 4, he showed how to find a solution for additive 1. Thus, if

[35] For an extensive discussion of Brahmagupta's rules, including the terminology used, see Datta and Singh, 1935–1938, 2:141–161.

(u, v) is a solution of $Dx^2 - 4 = y^2$, then

$$(u_1, v_1) = \left(\frac{1}{2}uv(v^2 + 1)(v^2 + 3), (v^2 + 2)\left[\frac{1}{2}(v^2 + 1)(v^2 + 3) - 1 \right] \right)$$

is a solution of $Dx^2 + 1 = y^2$, with a similar rule for additive 4. And, although Brahmagupta did not mention this explicitly, Indian mathematicians also knew how to reach additive 1, if they had a solution for additives ± 2 or -1.[36]

Why this problem interested Indian mathematicians remains a mystery. Some of Brahmagupta's examples used astronomical variables for x and y, but there is no indication that the problems actually came from real-life situations. In any case, the Pell equation became a tradition in Indian mathematics. Studied through the next several centuries, it was solved completely by the otherwise unknown Acarya Jayadeva (ca. 1000).[37] Here, however, we shall consider the solution Bhāskara II presented in his twelfth-century text, *Līlāvatī*.

Bhāskara II began by recapitulating Brahmagupta's procedure. In particular, he emphasized that once one solution pair had been found, indefinitely many others could be found by composition. More importantly, however, he discussed the so-called cyclic method (*cakravāla*), which, once a solution was found for some additive, enabled him eventually to reach a solution for additive 1. For the general case, Bhāskara II explained that "making the smaller and greatest roots and the additive into the dividend, the additive, and the divisor, the multiplier is to be imagined. When the square of the multiplier is subtracted from the 'nature' or is diminished by the 'nature' so that the remainder is small, that divided by the additive is the new additive. It is reversed if the square of the multiplier is subtracted from the 'nature.' The quotient of the multiplier is the smaller square root; from that is found the greatest root."[38]

To see how this works, consider one of Bhāskara II's specific examples: $67x^2 + 1 = y^2$. Begin, as before, by choosing a solution pair (u, v) for any additive b. For example, take $(1, 8)$ as a solution for additive -3. Next, solve the indeterminate equation $um + v = bn$ for m by the pulverizer

[36] Such problems are solved by commentators on Brahmagupta's work.
[37] Shukla, 1954.
[38] Plofker, 2007, p. 473, verse 75.

method—the equation here is $1m + 8 = -3n$—with the square of m as close to D (the "nature") as possible. Then take $b_1 = \pm\frac{D-m^2}{b}$ (which may be negative) for the new additive. The new first root is $u_1 = \frac{um+v}{b}$, while the new last root is $v_1 = \sqrt{Du_1^2 + b_1}$. In this case, to have m^2 as close as possible to 67, we choose $m = 7$ so that

$$\frac{D - m^2}{b} = \frac{67 - 49}{-3} = -6.$$

Because the subtraction is of the square from the coefficient, however, the new additive is 6. Thus, the new first root is $u_1 = \frac{1\cdot7+8}{-3} = -5$, but given that these roots are always squared, take $u_1 = 5$ so that $v_1 = \sqrt{67 \cdot 25 + 6} = \sqrt{1681} = 41$, and $(5, 41)$ is a solution for additive 6. "Then," according to Bhāskara II, "[it is done] repeatedly, leaving aside the previous square roots and additives. They call this the *cakravāla* (cycle). Thus there are two integer square roots increased by four, two or one. The supposition for the sake of an additive one is from the roots with four and two as additives."[39] In other words, he noted, if the above operation is repeated, eventually a solution for additive or subtractive four, two, or one will be reached. As we have seen, in those cases, we can find a solution for additive 1. But why does this method always give integral values at each stage? And why does repeating the method eventually give a solution pair for additives ± 4, ± 2, or ± 1?

To answer the first question, we note that Bhāskara II's method can be derived by composing the first solution (u, v) for additive b with the obvious solution $(1, m)$ for additive $m^2 - D$. It follows that $(u', v') = (mu + v, Du + mv)$ is a solution for additive $b(m^2 - D)$. Dividing the resulting equation by b^2 gives the solution $(u_1, v_1) = (\frac{mu+v}{b}, \frac{Du+mv}{b})$ for additive $\frac{m^2-D}{b}$. It is then clear why m must be found so that $mu + v$ is a multiple of b. And, although Bhāskara did not present a proof, one can show that if $\frac{mu+v}{b}$ is integral, so are $\frac{m^2-D}{b}$ and $\frac{Du+mv}{b} = \pm\sqrt{Du_1^2 + b_1}$. The reason that $m^2 - D$ is chosen "small" is so that the second question can be answered, but neither Bhāskara nor, probably, Jayadeva, presented a proof that the cycle method eventually leads to a solution to the original problem. It may well have been that Bhāskara had simply done enough examples

---

[39] Plofker, 2007, p. 474, verse 75.

to convince himself of its truth. An actual proof was not given until the nineteenth century.[40]

To complete our numerical example, then, recall that we had derived the result $67 \cdot 5^2 + 6 = 41^2$. The next step requires that we solve $5m + 41 = 6n$, with $|m^2 - 67|$ small. The appropriate choice is $m = 5$, so $(u_2, v_2) = (11, 90)$ is a solution for additive -7, or $67 \cdot 11^2 - 7 = 90^2$. Next, solve $11m + 90 = -7n$. The value $m = 9$ works, and $(u_3, v_3) = (27, 221)$ is a solution for additive -2, or $67 \cdot 27^2 - 2 = 221^2$. At this point, since additive -2 has been reached, compose $(27, 221)$ with itself. This gives $(u_4, v_4) = (11{,}934,\ 97{,}684)$ as a solution for additive 4. Dividing by 2 yields the solution $x = 5967$, $y = 48{,}842$ for the original equation $67x^2 + 1 = y^2$.

SUMS OF SERIES

Indian mathematicians also considered algebraic problems of kinds other than the solution of equations. One particular algebraic result that was discussed beginning in the time of Āryabhaṭa was a formula for determining the sum of integral powers. Recall that verse 2.19 of the *Āryabhaṭīya* gave, as a special case, a formula for the sum of the integers from 1 to n. Verse 2.22 then presented the formulas for the sum of the squares and the cubes: "One sixth of the product of three quantities which are, in due order, the number of terms, that increased by one, and that increased by the number of terms; that will be the solid made of a pile of squares, and the square of a pile is the solid made of a pile of cubes."[41]

These two statements easily translate into modern notation as

$$\sum_{i=1}^{n} i^2 = \frac{1}{6}n(n+1)(2n+1) \quad \text{and} \quad \sum_{i=1}^{n} i^3 = (1+2+\cdots+n)^2.$$

The first result was known, in essence, to Archimedes, while the second is almost obvious, at least as a hypothesis, after working a few numerical examples. As usual, Āryabhaṭa gave no indication as to how he discovered

[40] See Srinavasiengar, 1967, pp. 110–117 for more details on the *cakravāla* method, including the proof that all solutions produced by the method are integral.
[41] Keller, 2006, p. 100.

or proved these results, and Bhāskara I provided little insight in his commentary. The problem, however, became more important nine centuries later in the work of the Kerala school in the south of India.

This school, which began with the fourteenth-century mathematician Mādhava, is most famous for determining the power series for the sine, cosine, and arctangent functions. As part of the derivation of these series, it was necessary to know formulas not only for the sums of first powers, squares, and cubes, but also for the sums of higher integral powers. Since the members of this school, as noted above, were more concerned with proof than most earlier Indian mathematicians, we find Nīlakaṇṭha (ca. 1445–1545), a member of Mādhava's school, providing geometric proofs of Āryabhaṭa's square and cube formulas in his own commentary on the *Āryabhaṭīya*.

Consider, for example, Nīlakaṇṭha's proof of the formula for the sum of the squares, a proof reminiscent of Citrabhānu's argument discussed earlier: "Being that this [result on the sum of the squares] is demonstrated if there is equality of the total of the series of squares multiplied by six and the product of the three quantities, their equality is to be shown. A figure with height equal to the term-count, width equal to the term-count plus one, [and] length equal to the term-count plus one plus the term-count is [equal to] the product of the three quantities. But that figure can be made to construct the total of the series of squares multiplied by six."[42] Nīlakaṇṭha constructed the figure in this way. At each stage k, he used three "dominoes" of thickness 1, width k, and length $2k$. Thus, the total volume of these dominoes was $6k^2$. From the largest set ($k = n$), he constructed four walls of the figure (figure 6.4). One of the dominoes formed one wall, while a second formed the floor. The third domino was broken into two pieces, one of length $n + 1$ and one of length $n - 1$. These formed the two ends. Thus, the total length of the box was $2n + 1$, its width $n + 1$, and its height n. The inner space of the box has length $2n - 1 = 2(n - 1) + 1$, width $n = (n - 1) + 1$, and height $n - 1$. Thus, the walls of this new space can be built with three dominoes of thickness 1, width $n - 1$, and length $2(n - 1)$. Repeating this process until the entire box is filled, the result follows.

To get formulas for higher powers, another technique was needed, and this was provided by Jyeṣṭhadeva, another member of Mādhava's school,

---

[42] Plofker, 2007, p. 494.

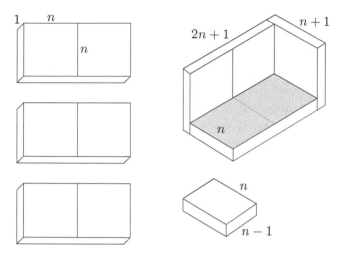

Figure 6.4.

in his *Gaṇita-yukti-bhāṣā* (ca. 1530). As it turned out, the development of sine and cosine series did not actually require an exact formula for higher powers. All that was required was the basic approximation for large n, namely,

$$\sum_{i=1}^{n-1} i^k \approx \frac{n^{k+1}}{k+1}.$$

To find it, Jyeṣṭhadeva proceeded inductively.[43] He knew from the exact formula for $k = 1$ that, for large n, $\sum i \approx \frac{n^2}{2}$. Now note that

$$n\sum_{i=1}^{n} i - \sum_{i=1}^{n} i^2 = n[n + (n-1) + \cdots + 1] - [n^2 + (n-1)^2 + \cdots + 1^2]$$

$$= 1(n-1) + 2(n-2) + 3(n-3) + \cdots + (n-1)1.$$

This can be rewritten in the form

$$n\sum_{i=1}^{n} i - \sum_{i=1}^{n} i^2 = \sum_{i=1}^{n-1} i + \sum_{i=1}^{n-2} i + \cdots + \sum_{i=1}^{1} i = \sum_{p=1}^{n-1} \left(\sum_{i=1}^{p} i \right).$$

---

[43] For more details relative to the technical discussion that follows, see Sarma, 2008, pp. 192–197.

For large n, however, each of the summands in the double summation can be approximated as half the square of the largest term. That is

$$n \sum_{i=1}^{n} i - \sum_{i=1}^{n} i^2 \approx \frac{(n-1)^2}{2} + \frac{(n-2)^2}{2} + \cdots + \frac{1^2}{2} = \frac{1}{2} \sum_{i=1}^{n-1} i^2.$$

Therefore, for n large, we have

$$n \sum_{i-1}^{n} i \approx \frac{3}{2} \sum_{i=1}^{n} i^2 - \frac{1}{2} n^2 \qquad \text{or} \qquad \frac{3}{2} \sum_{i-1}^{n} i^2 \approx n \sum_{i-1}^{n} i + \frac{1}{2} n^2$$

or, finally,

$$\sum_{i=1}^{n} i^2 \approx \frac{2}{3} n \cdot \frac{n^2}{2} + \frac{1}{3} n^2 \approx \frac{n^3}{3}.$$

Jyeṣṭhadeva continued in the same manner for $k = 3$, showing that

$$n \sum_{i=1}^{n} i^2 - \sum_{i=1}^{n} i^3 = \sum_{p=1}^{n-1} \left(\sum_{i=1}^{p} i^2 \right).$$

Then, assuming the approximation for sums of squares and because, for large n, it does not matter whether one sums to n or $n - 1$, he concluded that

$$n \sum_{i=1}^{n} i^2 - \sum_{i=1}^{n} i^3 \approx \frac{1}{3} \sum_{i=1}^{n} i^3$$

and therefore that

$$\sum_{i=1}^{n} i^3 \approx \frac{n^4}{4}.$$

Similarly, in general,

$$n \sum_{i=1}^{n} i^{k-1} - \sum_{i=1}^{n} i^k = \sum_{p=1}^{n-1} \left(\sum_{i=1}^{p} i^{k-1} \right),$$

and this result, given the approximation for the sum of the $(k-1)$st powers, implies that

$$\sum_{i=1}^{n} i^k \approx \frac{n^{k+1}}{k+1}.$$

As we shall see in the next chapter, these formulas were also demonstrated by Islamic mathematicians in order to calculate volumes. Later on, of course, they proved essential in the development of the integral calculus in Europe.

****

Many questions surround the development of Indian algebraic ideas. For example, we have seen that as early as the fifth century CE, Indian mathematicians knew and used the algorithm that we call the quadratic formula for solving quadratic equations, but whether they developed it independently or learned of it from other cultures is not at all clear. Regardless, in light of the development of this formula in both ancient Mesopotamia and medieval Islam, it is particularly surprising that there is no evidence of a geometric basis for it—or, indeed, for any procedure for solving problems of quadratic type—in the Indian context. Indian mathematicians certainly considered geometric models when they dealt with cubic problems and even with summation problems, but they seemingly determined their solutions of quadratic problems completely algorithmically. Thus, although it is conceivable that Islam could have learned Indian algebra during the ninth and tenth centuries, the available evidence suggests that Indian algebra did not provide a model for the more sophisticated, geometrical algebra that emerged during that time period in the Islamic world.

A deeper mystery surrounds the development of the procedure for solving the Pell equation. As far as we can tell, this was a totally indigenous development; no trace of the Pell equation has been found to date in medieval Islamic mathematics. Moreover, since the equation's eventual solution in Europe is quite independent of the Indian context, it would appear that knowledge of the Indian solution never left the Indian subcontinent.

Like China, India was eventually opened to western mathematical ideas, but, unlike their Chinese counterparts, Indian mathematicians continued to develop their own ideas, particularly in Kerala. It was only in the nineteenth century, after the British conquest of the country, that the influx of western mathematics managed to supplant the strong indigenous Indian mathematical tradition.[44] From that point on, Indian mathematicians, in general, and Indian algebraists, in particular, came to share a mathematical vernacular with the West.

---

[44] Even though there were Portuguese Jesuit missionaries in India by the seventeenth century, there is no evidence from that time of the transmission and actual transplantation of western mathematical ideas there.

7

Algebraic Thought in Medieval Islam

I n the first half of the seventh century a new civilization arose on the Arabian Peninsula. Under the inspiration of the prophet Muḥammad, the new monotheistic religion of Islam quickly drew adherents and pushed outward. In the century following Muḥammad's capture of Mecca in 630 CE, Islamic armies conquered an immense territory while propagating the new religion both among the previously poly-theistic tribes of the Middle East and among the followers of other faiths.[1] Syria and then Egypt were taken from the Byzantine Empire, while Persia was conquered by 642 CE. Soon, the victorious armies reached India and central Asia in the east and then North Africa and Spain in the west. Their western march was eventually halted at Tours in France by the army of Charles Martel in 732, while in the Caucasus, the Khazars successfully repelled the Islamic invaders several times between the mid-seventh century and 737, defeats that probably prevented the destruction of the Byzantine Empire in the eighth century.

By the mid-700s, issues of governing an immense empire gradually replaced problems of conquest. Muḥammad's successors, the Umayyad caliphs, originally set up their capital in Damascus, but after almost a century of wars, the caliphate split into several parts. In the east, the 'Abbāsids overthrew the Umayyads in 750. A key source of support for their new dynasty was Persia, whose rulers had been overthrown by the Umayyads, but from whom the 'Abbāsids learned how to rule an empire. Favorable conditions for the development of a new culture soon obtained, particularly in this eastern Islamic domain.

---

[1] Lindberg, 1992, pp. 166–170 has a brief treatment of the rise and expansion of Islam, while Saliba, 2007 has a very detailed treatment and argument about the growth and eventual decline of Islamic science.

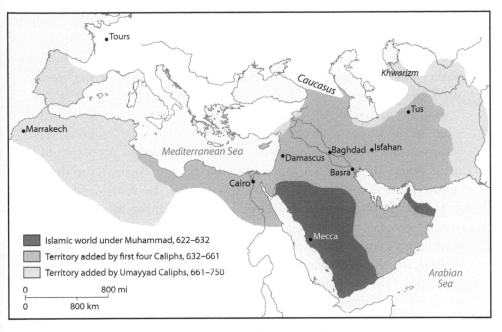

Figure 7.1. Map of medieval Islam.

In 762 the caliph al-Manṣūr founded a new capital at Baghdad, a city that quickly became a flourishing commercial and intellectual center. As the initial impulses of Islamic orthodoxy were replaced by a more tolerant atmosphere, the intellectual accomplishments of all residents of the caliphate were welcomed and encouraged. The caliph Hārūn al-Rashīd, who ruled from 786 to 809, began a program of translation of manuscripts from various languages into Arabic. Texts—among them many of the classic works in the original Greek, including, in particular, many mathematical and scientific texts—were collected from academies in the Near East that had been established by scholars fleeing from persecution at the ancient academies in Athens and Alexandria. Sources in Persia and India were also secured and included treatises in Persian, Syriac versions of Greek works, and texts in Sanskrit. Hārūn's successor, the caliph al-Ma'mūn (reigned 813–833), established a library, the *Bayt al-Ḥikma* (or House of Wisdom), where the manuscripts were stored and where the long procedure of copying and translating them took place. By the end of the ninth century, translators had produced Arabic versions of many of the principal works of Euclid, Archimedes, Apollonius,

Diophantus, Ptolemy, and other Greek mathematicians. Islamic scholars had also somehow absorbed the ancient mathematical traditions of the Mesopotamian scribes and, in addition, had learned some of the mathematics of the Indians, especially trigonometry and the decimal place-value system.

In addition to receiving these "scientific" texts from two major cultures, Islam learned from what are often called "subscientific" sources. Thus, examples of recreational problems—such as the famous problem of how many grains of wheat are on a chessboard if there is 1 on the first square, 2 on the second, 4 on the third, 8 on the fourth, and so on, up to 64 doublings—are to be found in Islam. Similar problems exist in virtually every culture, including ancient Mesopotamia and China at the beginning of our era. While it is probably fruitless to look for the origins of these problems, transmission along the Silk Routes remains a likely possibility. There were also sources for Islamic mathematics in mercantile arithmetic and in the geometry of surveyors and architects.

Islamic scholars during the first few hundred years of Islamic rule did more than just bring these sources together, however. They amalgamated them into a new whole and infused their mathematics with what they felt was divine inspiration. As we have noted frequently, creative mathematicians had always carried investigations well beyond the dictates of immediate necessity, but many in Islam felt that this was actually a requirement of God.[2] At least early on, Islamic culture regarded "secular knowledge" as a way to "holy knowledge" and not as in conflict with it. Learning was therefore encouraged, and those who had demonstrated sparks of creativity were often supported by rulers—both secular and religious authorities—so that they could pursue their ideas as far as possible. Furthermore, since those in power were naturally interested in the needs of daily life, Islamic mathematicians, unlike most of their Greek predecessors, nearly all contributed to both theory and practical applications. They also sought to assure that "practitioners"—timekeepers, lawyers, architects, artisans, and others—understood the mathematics and that mathematicians understood the perspective of the practitioners. Since Islamic law regulates every aspect of Muslim life—religious and ritual, political and legal, social and behavioral—mathematicians needed to concern themselves with much else besides

---

[2] See Høyrup, 1987, pp. 300–308 for more detail on the ideas here.

"pure" mathematics. Abū Rayḥān al-Bīrūnī (973–1055) conveyed the basic Islamic attitude toward mathematics when he wrote,

> You know well...for which reason I began searching for a number of demonstrations proving a statement due to the ancient Greeks concerning the division of the broken line in an arbitrary circular arch by means of the perpendicular from its center...and which passion I felt for the subject...so that you reproached me my preoccupation with these chapters of geometry, not knowing the true essence of these subjects, which consists precisely in going in each matter beyond what is necessary....Whatever way he [the geometer] may go, through exercise will he be lifted from the physical to the divine teachings, which are little accessible because of the difficulty [in understanding] their meaning...and because of the circumstance that not everybody is able to have a conception of them, especially not the one who turns away from the art of demonstration. You would be right, God give you strength, to reproach me, had I neglected to search for these ways, and used my time where an easier approach would suffice; or if the work had not arrived at the point which constituted the fundamentals of astronomy, that is to the calculation of the chords in the circle and the ratio of their magnitude to that supposed for the diameter.[3]

By the time al-Bīrūnī wrote these words, however, the status of mathematical thought in Islam was changing. Areas of mathematics more advanced than basic arithmetic had been classified as "foreign sciences"—in contrast to the "religious sciences" that included religious law and speculative theology—but many of the early Islamic leaders had nevertheless encouraged their study.[4] As more orthodox Islamic leaders came to power in certain parts of the Islamic world, the feeling grew that the foreign sciences were potentially subversive to the faith and certainly superfluous to the needs of life, either here or hereafter. More and more, the *madrasas*, the Islamic institutions of higher learning, tended to concentrate on the teaching of Islamic law. There were therefore

---

[3] From the preface to the *Book on Finding the Chords in the Circle* (ca. 1030), quoted in Høyrup, 1987, pp. 306–307.

[4] For a brief treatment of these issues, see Lindberg, 1992, pp. 170–175. Huff, 1993 has a more extended, if controversial, discussion.

fewer students studying advanced topics in mathematics. Thus, although there were significant mathematical achievements in Islam through the fifteenth century, the creation of new mathematics gradually waned in importance. In fact, the same al-Bīrūnī who articulated the sense of the mission of the mathematician recognized the impediments to the further development of science in Islam. As he put it,

> The number of sciences is great, and it may be still greater if the public mind is directed towards them at such times as they are in the ascendancy and in general favor with all, when people not only honor science itself, but also its representatives. To do this is, in the first instance, the duty of those who rule over them, of kings and princes. For they alone could free the minds of scholars from the daily anxieties for the necessities of life, and stimulate their energies to earn more fame and favor, the yearning for which is the pith and marrow of human nature. The present times, however, are not of this kind. They are the very opposite, and therefore it is quite impossible that a new science or any new kind of research should arise in our days. What we have of sciences is nothing but the scanty remains of bygone better times.[5]

Still, it is useful to remember, when dealing with the subject of Islamic mathematics, that the mathematical enterprise in Islam lasted longer than both the era of classical Greek mathematics and the age of "modern mathematics." With varying degrees of intensity, Islamic mathematics developed for over 600 years. In that period, mathematics arose in various centers linked by networks of communication, primarily using the Arabic language, that persisted despite the general absence of political unity. Baghdad, the first major center, supported many of the earliest Islamic mathematicians. Cairo and Persia emerged as foci of mathematical activity after the ninth century, while Spain supported a good deal of activity between the eighth and twelfth centuries. Finally, Marrakech in North Africa had a number of mathematicians in the twelfth and thirteenth centuries. While it is generally not easy to trace the mutual influences between these centers, it is possible to gain some insight into the paths that the mathematical ideas took in medieval Islam by tracking the use of particular theorems or examples.

[5] Al-Bīrūnī, 1910, 1:152.

QUADRATIC EQUATIONS

Islamic mathematicians made their most important contributions in the area of algebra. In fact, it is fair to say that they turned the earlier study of topics that we recognize as algebraic—from both ancient Mesopotamia and classical Greece—into a new science called *al-jabr*. Interestingly, the quadratic-equation-solving procedures of the Mesopotamian scribes seem only to have reached Islamic scholars in the same geographic area some two and a half millennia after their creation. How did this delayed transmission take place?

The question still remains impossible to answer, but we do have a few hints. In particular, a work, probably of the ninth century, written by the otherwise unknown Islamic scholar, Abu Bekr, and translated into Latin in the twelfth century as *Liber mensurationum*, contains quadratic problems treated in language reminiscent of old Mesopotamian tablets. That is, the words are all geometric—square, side, and so on—and not, as we shall see shortly in an "algebraic" language, where the unknown quantity is called a "thing" (*shay*) and the square of the quantity is called a "wealth" (*māl*). In fact, some of the problems are presented with two solutions, the first using only the geometric words and following the Mesopotamian procedure, the second reflecting the newer "algebra" procedures as detailed below in the work of al-Khwārizmī. There are also several instances in this text of problems involving the sum or difference of the area and the four sides of a square, a problem that occurred earlier in problem 23 of the Mesopotamian text BM 13901 discussed in chapter 2, but only later in texts that still used the earlier geometric language, such as the tenth-century work of the Spanish Islamic writer, ibn Abdun.[6]

Around the same time that the Arabic original of the *Liber mensurationum* was being written, the chief Greek mathematical classics were being translated into Arabic, and Islamic scholars were studying and writing commentaries on them. The most important idea that these scholars learned from their study of Greek works was that of proof based on explicit axioms. They absorbed the ethos that a mathematical problem could not be considered solved until its solution was proved valid. What

---

[6] See Høyrup, 1996 for more details. This type of problem and language also occurs in the twelfth-century Hebrew work of Abraham bar Hiyya, a work evidently based on some earlier Islamic sources.

does such a proof look like for an algebraic problem? The answer seemed clear. Geometry, not algebra, was found in Greek texts, so the only real proofs were geometric. Islamic scholars thus set themselves the task of justifying algebraic rules—whether ancient Mesopotamian or newly discovered—through geometry. Islamic algebra was thereby rooted both in the ancient material of the Mesopotamian scribes and in the classical Greek heritage of geometry.

One of the earliest Islamic algebra texts, entitled *Al-kitāb al-mukhtaṣar fī ḥisāb al-jabr wa'l-muqābala* (or *The Compendious Book on the Calculation of al-Jabr and al-Muqābala*), was written around 825 by Muḥammad ibn Mūsā al-Khwārizmī (ca. 780–850) and ultimately had immense influence not only in the Islamic world but also in Europe. The term *al-jabr* can be translated as "restoring" and refers to the operation of "transposing" a subtracted quantity on one side of an equation to the other side where it becomes an added quantity. The word *al-muqābala* can be translated as "balancing" and refers to the reduction of a positive term by subtracting equal amounts from both sides of the equation. Thus, the conversion of $3x + 2 = 4 - 2x$ to $5x + 2 = 4$ is an example of *al-jabr*, while the conversion of the latter to $5x = 2$ is an example of *al-muqābala*. (Of course, these formulations are modern; al-Khwārizmī had no symbols whatsoever.) Our modern word "algebra," a corrupted form of the Arabic *al-jabr*, came to connote the entire science represented by al-Khwārizmī's work.

Why did al-Khwārizmī write the text that helped develop that science? As he explained in its introduction,

> That fondness for science, by which God has distinguished the Imam al-Ma'mūn, the Commander of the Faithful, . . . that affability and condescension which he shows to the learned, that promptitude with which he protects and supports them in the elucidation of obscurities and in the removal of difficulties, has encouraged me to compose a short work on calculating by *al-jabr* and *al-muqābala*, confining it to what is easiest and most useful in arithmetic, such as men constantly require in cases of inheritance, legacies, partition, law-suits, and trade, and in all their dealings with one another, or where the measuring of lands, the digging of canals, geometrical computation, and other objects of various sorts and kinds are concerned.[7]

---

[7] Rosen, 1831, p. 7 and Berggren, 2007, p. 542.

As this makes clear, al-Khwārizmī was primarily interested in writing a practical, not a theoretical, manual. Still, he had been sufficiently influenced by the introduction of Greek mathematics into Baghdad that he felt constrained to give geometric proofs of his algebraic procedures. Those proofs, however, are not Greek. Rather, they strongly resemble the Mesopotamian geometric arguments out of which the algebraic algorithms grew. Like his Mesopotamian predecessors, al-Khwārizmī also gave numerous examples and problems, but Greek influences showed through in the systematic classification of problems as well as in the very detailed explanations of his methods.

Al-Khwārizmī opened his text with the avowal that "what people generally want in calculating... is a number,"[8] that is, the solution of an equation. His text would thus be a manual for equation solving. The quantities he dealt with were generally of three kinds, and he gave them technical names: *māl* (wealth) for the square (of the unknown), *jidhr* (root) for the unknown itself, thought of as the root of the *māl*, and *'adad mufrad* (simple number) to represent the constant in an equation. Later writers often used *shay* (thing) in place of *jidhr* to emphasize that this value was unknown, and they sometimes substituted a monetary term, such as *dirhem*, for the constant, but the main point is that these terms referred to numerical and not geometrical quantities.[9] It is therefore at this time period that algebra evolved from its geometric stage in Mesopotamia and Greece to an algorithmic, equation-solving stage in Islam.

With his terms set, al-Khwārizmī noted that six cases of equations can be written using these three kinds of quantities:

(1) Squares are equal to roots (or $ax^2 = bx$);

(2) Squares are equal to numbers (or $ax^2 = c$);

(3) Roots are equal to numbers (or $bx = c$);

(4) Squares and roots are equal to numbers (or $ax^2 + bx = c$);

(5) Squares and numbers are equal to roots (or $ax^2 + c = bx$); and

(6) Roots and numbers are equal to squares (or $bx + c = ax^2$).

---

[8] Rosen, 1831, p. 5 and Berggren, 2007, p. 543.
[9] See Oaks and Alkhateeb, 2005 for more discussion of the use of these technical words in Islamic algebra.

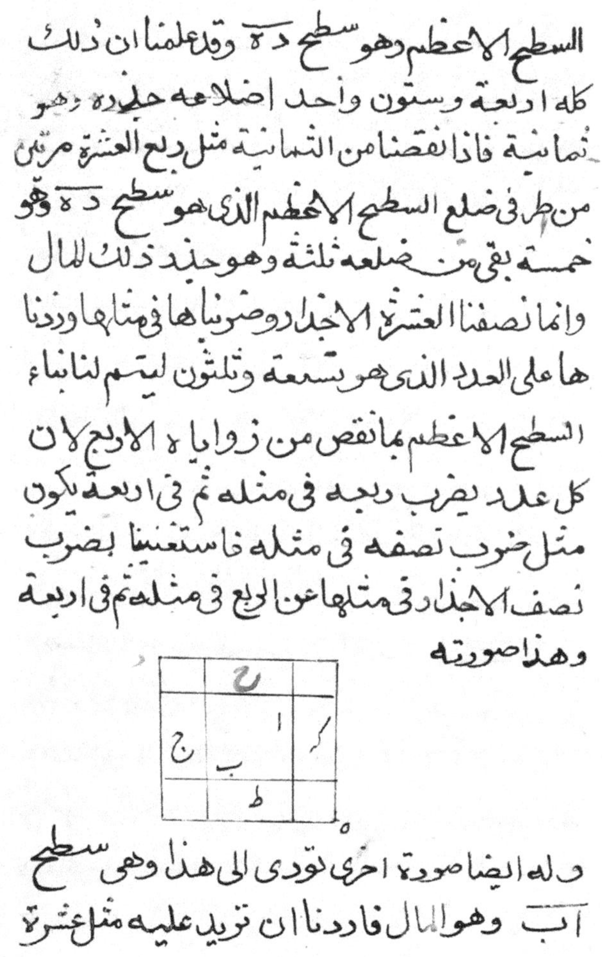

السطح الاعظم وهو سطح د ه وقد علمنا ان ذلك

كله اربعة وستون واحد اضلاعه حذره وهو

ثمانية فاذا نقصنا من الثمانية مثل ربع العشرة مرتين

من طرفى ضلع السطح الاعظم الذى هو سطح د ه وهو

خمسة بقى من ... ضلعه ثلثة وهو جذر ذلك المال

وانما نصفنا العشرة الاجذار وضربناها فى مثلها واردنا

ها على العدد الذى هو تسعة وثلثون ليتم لنا بناء

السطح الاعظم بما نقص من زواياه الاربع لان

كل عدد يضرب ربعه فى مثله ثم فى اربعة يكون

مثل ضرب نصفه فى مثله فاستغنينا بضرب

نصف الاجذار فى مثلها عن الربع فى مثله ثم فى اربعة

وهذا صورته

وله ايضا صورة اخرى تؤدى الى هذا وهى سطح

آب وهو المال فاردنا ان نزيد عليه مثل عشرة

Figure 7.2. Medieval Arabic manuscript of al-Khwārizmī's *Algebra* illustrating a derivation of the algorithm for an equation of case (4), from the Smith Collection of the Rare Book and Manuscript Library, Columbia University.

The six-fold classification—which became and remained standard in Western algebra texts into the seventeenth century—was necessary because Islamic mathematicians, unlike their Indian counterparts, did not deal with negative numbers. (This reflects both the Greek influence and the limitations of geometry.) Coefficients, as well as the roots of equations, must be positive. These types listed are thus the only ones with positive solutions. (Our standard form $ax^2 + bx + c = 0$ would have made no sense to al-Khwārizmī, because if the coefficients are all positive, the roots cannot be.)

Although al-Khwārizmī's solutions of the first three cases were straightforward—we need only note that 0 is not considered a solution of the first type—his rules for the last three were more interesting. Consider his solution of case (5), where squares and numbers are equal to roots:

> A square and twenty-one in numbers are equal to ten roots of the same square. That is to say, what must be the amount of a square, which, when twenty-one *dirhems* are added to it, becomes equal to the equivalent of ten roots of that square? Solution: Halve the number of the roots; the half is five. Multiply this by itself; the product is twenty-five. Subtract from this the twenty-one which are connected with the square; the remainder is four. Extract its root; it is two. Subtract this from the half of the roots, which is five; the remainder is three. This is the root of the square which you required, and the square is nine. Or you may add the root to the half of the roots; the sum is seven; this is the root of the square which you sought for, and the square itself is forty-nine.[10]

Al-Khwārizmī added a further gloss:

> When you meet with an instance which refers you to this case, try its solution by addition, and if that does not serve, then subtraction certainly will. For in this case both addition and subtraction may be employed, which will not answer in any other of the three cases in which the number of the roots must be halved. And know, that, when in a question belonging to this case you have halved the number of the roots and multiplied the half by itself, if the product be less than the number of *dirhems* connected with the square, then the instance is impossible; but if the product be equal to the *dirhems* by themselves, then the root of the square is equal to the half of the roots alone, without either addition or subtraction.

This algorithm, translated into modern symbolism, corresponds to a version of our quadratic formula for solving the equation

---

[10] Rosen, 1831, p. 11 and Berggren, 2007, p. 544. The next quotation is also found in both sources.

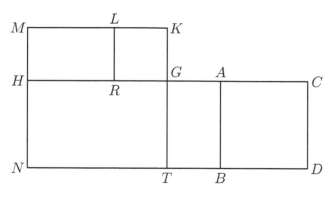

Figure 7.3.

$x^2 + c = bx$:

$$x = \frac{b}{2} \pm \sqrt{\left(\frac{b}{2}\right)^2 - c}.$$

As noted above, even though the equations were between terms representing numbers, the author felt compelled to justify his procedure by giving a geometric argument, where the *māl* became an actual geometric square the side of which was the *jidhr* (figure 7.3):

> When a square plus twenty-one *dirhems* are equal to ten roots, we depict the square as a square surface AD of unknown sides. Then we join it to a parallelogram, HB, whose width, HN, is equal to one of the sides of AD. The length of the two surfaces together is equal to the side HC. We know its length to be ten numbers since every square has equal sides and angles, and if one of its sides is multiplied by one, this gives the root of the surface, and if by two, two of its roots. When it is declared that the square plus twenty-one equals ten of its roots, we know that the length of the side HC equals ten numbers because the side CD is a root of the square figure. We divide the line CH into two halves by the point G. Then you know that line HG equals line GC, and that line GT equals line CD. Then we extend line GT a distance equal to the difference between line CG and line GT and make a quadrilateral. The line TK equals line KM making a quadrilateral MT of equal sides and angles. We know that the line TK and the other sides equal five. Its surface is

twenty-five obtained by the multiplication of one-half the roots by itself, or five by five equals twenty-five. We know that the surface HB is the twenty-one that is added to the square. From the surface HB, we cut off a piece by line TK, one of the sides of the surface MT, leaving the surface TA. We take from the line KM line KL which is equal to line GK. We know that line TG equals line ML and that line LK cut from line MK equals line KG. Then the surface MR equals surface TA. We know that surface HT plus surface MR equals surface HB, or twenty-one. But surface MT is twenty-five. And so, we subtract from surface MT, surface HT and surface MR, both [together] equal to twenty-one. We have remaining a small surface RK, of twenty-five less twenty-one or 4. Its root, line RG, is equal to line GA, or two. If we subtract it from line CG, which is one-half the roots, there remains line AC or three. This is the root of the first square. If it is added to line GC which is one-half the roots, it comes to seven, or line RC, the root of a larger square. If twenty-one is added to it, the result is ten of its roots.[11]

It is clear here that rather than a Greek proof, al-Khwārizmī essentially gave a "cut-and-paste" argument designed to derive his procedure by manipulating appropriate squares and rectangles, a method similar to that of the Mesopotamian scribes in their descriptions of methods of solving similar problems. However, even though al-Khwārizmī gave these geometric justifications of his method, he succeeded in changing the focus of solving quadratic equations away from finding actual sides of squares to finding numbers satisfying certain conditions. That is, once he presented the geometry, he did not refer to it again. He simply applied his algorithm as he calculated solutions for numerous examples. In fact, he made clear that he was not necessarily finding the side of a square when he solved a quadratic equation. A root, he wrote, is "anything composed of units which can be multiplied by itself, or any number greater than unity multiplied by itself, or that which is found to be diminished below unity when multiplied by itself."[12] Furthermore, his procedure for solving quadratic equations like those in cases (4)–(6), when the coefficient of the square term is different from one, was the

[11] Rosen, 1831, pp. 16–18 and Berggren, 2007, p. 545.
[12] Karpinski, 1930, p. 69.

arithmetical method of first multiplying or dividing appropriately to make the initial coefficient one, and then proceeding as before. Al-Khwārizmī even admitted somewhat later in his text, when discussing the addition of the "polynomials" $100 + x^2 - 20x$ and $50 + 10x - 2x^2$, that "this does not admit of any figure, because there are three different species, i.e., squares and roots and numbers, and nothing corresponding to them by which they might be represented. . . [Nevertheless], the eluci-dation by words is easy."[13]

While influential, al-Khwārizmī's was evidently not the only algebra text of his time. A fragment of a contemporaneous work remains, namely, the section *Logical Necessities in Mixed Equations* from a longer work *Kitāb al-jabr wa'l muqābala* (or *Book of al-Jabr and al-Muqābala*) by 'Abd al-Ḥamīd ibn Wāsi ibn Turk al-Jīlī. Very little is known about ibn Turk; sources even differ as to whether he was from Iran, Afghanistan, or Syria.[14] Regardless, the extant chapter of his book deals with quadratic equations of al-Khwārizmī's cases (1), (4), (5), and (6) and includes geometric descriptions of the methods of solution that are much more detailed than al-Khwārizmī's. In particular, in case (5), ibn Turk gave geometric versions for all possible situations. His first example is the same as al-Khwārizmī's, namely, $x^2 + 21 = 10x$, but he began the geometrical demonstration by noting that G, the midpoint of CH, may be either on the line segment AH, as in al-Khwārizmī's construction (figure 7.3), or on the line segment CA (see figure 7.4). In this case, squares and rectangles are completed (similar in form to those in figure 7.3), but the solution $x = AC$ is now given as $CG + GA$, thus using the plus sign from the formula. In addition, ibn Turk discussed what he called the "intermediate case," in which the root of the square is exactly equal to half the number of roots. His example for this situation is $x^2 + 25 = 10x$; the geometric diagram then consists simply of a rectangle divided into two equal squares.

Ibn Turk further noted that "there is the logical necessity of impossibil-ity in this type of equation when the numerical quantity. . . is greater than [the square of] half the number of roots,"[15] as, for example, in the case $x^2 + 30 = 10x$. Again, he resorted to a geometric argument. Assuming

[13] Rosen, 1831, pp. 33–34.
[14] Much of what follows in this paragraph draws from Sayili, 1962.
[15] Sayili, 1962, p. 166.

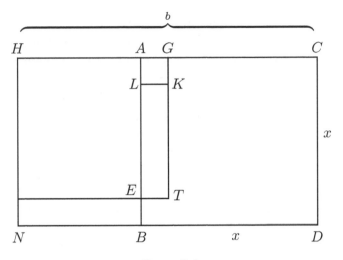

Figure 7.4.

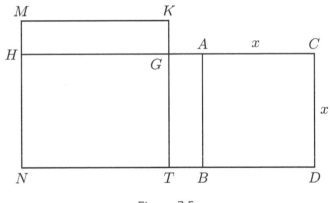

Figure 7.5.

that G is located on the segment AH, the square $KMNT$ is greater than the rectangle $HABN$ (figure 7.5), but the conditions of the problem show that the latter rectangle equals 30, while the former only equals 25. A similar argument works when G is located on CA.

Although the section on quadratic equations of ibn Turk's algebra is the only part still extant, al-Khwārizmī's text contains much else of interest, including an introduction to manipulation with algebraic expressions, explained by reference to similar manipulations with numbers. Al-Khwārizmī thus employed his principles of *al-jabr* and *al-muqābala* to develop laws he could use to reduce complicated equations to one of his standard forms. For example, he explained the distributive law, noting that four multiplications are necessary when $a \pm b$ is multiplied by $c \pm d$. His discussion also made clear that he knew the rules for dealing with multiplication and signs, for, as he stated, "If the units [b and d in our notation]... are positive, then the last multiplication is positive; if they are both negative, then the fourth multiplication is likewise positive. But if one of them is positive and one negative then the fourth multiplication is negative."[16] Beginning in al-Khwārizmī's time, then, these laws of *al-jabr* and *al-muqābala* became standard procedures, even though they were accomplished in words and not symbols. All of his successors used the basic laws of algebraic manipulation without discussing them in any detail.

Having dealt with algebraic manipulation, al-Khwārizmī then presented a large collection of problems that generally involved some manipulation to turn the original "equation" into one of the standard forms so that the appropriate algorithm could be applied. Despite his promise that the text would be useful, however, virtually all of the problems are abstract. This problem is typical, and similar ones continued to appear in algebra texts over the centuries: "I have divided ten into two parts, and having multiplied each part by itself, I have put them together, and have added to them the difference of the two parts previously to their multiplication, and the amount of all this is fifty-four."[17] It is not difficult for us to translate this problem into an equation:

$$(10 - x)^2 + x^2 + (10 - x) - x = 54.$$

Al-Khwārizmī took a bit more space:

> You multiply ten minus thing by itself; it is a hundred and a square minus twenty things. Then multiply also the other thing of the ten by itself; it is one square. Add this together, it will be a hundred plus two squares minus twenty things. It was stated that the difference of the two parts before multiplication should be added to them. You say, therefore, the difference between them is ten minus two things.

[16] Rosen, 1831, p. 33 and Berggren, 2007, p. 546.
[17] Rosen, 1831, p. 43.

The result is a hundred and ten and two squares minus twenty-two things, which are equal to fifty-four *dirhems*.[18]

He then reduced this to the equation "twenty-eight *dirhems* and a square [are] equal to eleven things"—that is, $x^2 + 28 = 11x$—and used his rule for equations as in case (5) to get the solution $x = 4$. He ignored the second root—that is, $x = 7$—since, using it, the sum of the two squares would be 58 and the conditions of the problem would not be met.

There was, however, among al-Khwārizmī's forty problems, one "quasi-real-world" problem: "You divide one *dirhem* among a certain number of men. Now you add one man more to them, and divide again one *dirhem* among them. The quota of each is then one-sixth of a *dirhem* less than at the first time."[19] If x represents the number of men, the equation becomes $\frac{1}{x} - \frac{1}{x+1} = \frac{1}{6}$, which reduces to $x^2 + x = 6$, for which the solution is $x = 2$. And, although later in his algebra text, al-Khwārizmī displayed dozens of involved inheritance problems, the actual mathematics needed in them was never more complicated than the solution of linear equations. One can only conclude that although al-Khwārizmī was interested in teaching his readers how to solve mathematical problems, and especially how to deal with quadratic equations, he could hardly think of any real-life situations that required these equations. Things apparently had not changed in this regard since the time of the Mesopotamians.

Within fifty years of the works by al-Khwārizmī, Islamic mathematicians had generally adopted his algebraic language, but they had also decided that the necessary geometric foundations of the algebraic solution of quadratic equations should be based on the work of Euclid rather than on the ancient traditions. And whether or not Euclid intended this, Islamic writers found the justifications for their quadratic-equation-solving algorithms in propositions II.5 and II.6 of the *Elements*. One of the first of these writers was the Egyptian mathematician Abū Kāmil ibn Aslam (ca. 850–930). In his algebra text, *Kitāb fī al-jabr wa'l-muqābala* (or *Book on al-Jabr and al-Muqābala*), he stated that he would "explain their

[18] Rosen, 1831, pp. 43–44.
[19] Rosen, 1831, p. 63.

rule using geometric figures clarified by wise men of geometry and which is explained in the Book of Euclid."[20]

For example, using al-Khwārizmī's problem $x^2 + 21 = 10x$ and the accompanying figure (figure 7.3), Abū Kāmil noted that since the line HC is divided into equal segments at G and into unequal segments at A, proposition II.5 of Euclid's *Elements* shows that the rectangle determined by CA and HA (the unequal segments of the whole) together with the square on AG (the straight line between the points of section) is equal to the square on CG (the half). Since the rectangle determined by CA and HA is equal to 21 and the square on CG is equal to 25, however, the square on AG must equal 4, making AG itself equal to 2. It then follows, as before, that $x = CA = 3$. Abū Kāmil used a slightly different diagram, but the same proposition from Euclid, to show that x could also equal 7. In his discussion of the equations of cases (4) and (6), moreover, he showed how to justify the algorithms using Euclid's proposition II.6.

Like his predecessor al-Khwārizmī, Abū Kāmil followed his discussion of the different forms of quadratic equations first by a treatment of various algebraic rules, many demonstrated geometrically, and then by a large selection of problems. He did make some advances, though, considering more complicated identities and problems and manipulating quadratic surds. Consider, for example, Abū Kāmil's problem 37, which like so many of al-Khwārizmī's problems, started with dividing 10 into two parts: "If one says that 10 is divided into two parts, and one part is multiplied by itself and the other by the root of 8, and subtracts the quantity of the product of one part times the root of 8 from... the product of the other part multiplied by itself, it gives 40."[21] The equation in this case is, in modern terms, $(10 - x)(10 - x) - x\sqrt{8} = 40$. After rewriting it in words that we represent as $x^2 + 60 = 20x + \sqrt{8x^2}$ $(= (20 + \sqrt{8})x)$, Abū Kāmil carried out the algorithm for the case of "squares and numbers equal roots" to conclude that $x = 10 + \sqrt{2} - \sqrt{42 + \sqrt{800}}$ and that $10 - x$, the "other part," is equal to $\sqrt{42 + \sqrt{800}} - \sqrt{2}$.

Abū Kāmil also showed both how to use substitutions to solve complicated problems and how to deal with "equations of quadratic form."

[20] Levey, 1966, p. 32. For the discussion of the proof of case (5) in the next paragraph, see pp. 40–42.
[21] Levey, 1966, p. 144.

Problem 45 illustrates both ideas: "One says that 10 is divided into two parts, each of which is divided by the other, and when each of the quotients is multiplied by itself and the smaller is subtracted from the larger, then there remains 2."[22] The equation is

$$\left(\frac{x}{10-x}\right)^2 - \left(\frac{10-x}{x}\right)^2 = 2.$$

To simplify this, Abū Kāmil assumed that $10-x$ was smaller than x. He then set a new "thing" y equal to $\frac{10-x}{x}$ and derived the new equation $\frac{1}{y^2} = y^2 + 2$. Multiplying both sides by y^2 gave him a quadratic equation in y^2, namely, $(y^2)^2 + 2y^2 = 1$ for which the solution is $y^2 = \sqrt{2} - 1$. Hence $y = \sqrt{\sqrt{2}-1}$. Then, since $\frac{10-x}{x} = y = \sqrt{\sqrt{2}-1}$, he proceeded to solve for x by first squaring both sides of this equation. Some manipulation produced a new equation of case (5) to which he again applied his algorithm. He wrote out his final result in words that we can translate as $x = 10 + \sqrt{50} - \sqrt{50 + \sqrt{20,000} - \sqrt{5000}}$.

Abū Kāmil even solved systems of equations. Consider his problem 61: "One says that 10 is divided into three parts, and if the smallest is multiplied by itself and added to the middle one multiplied by itself, it equals the largest multiplied by itself, and when the smallest is multiplied by the largest, it equals the middle multiplied by itself."[23] In modern notation, we are asked to find $x < y < z$, where

$$x + y + z = 10, \qquad x^2 + y^2 = z^2, \qquad \text{and} \qquad xz = y^2.$$

Abū Kāmil, evidently realizing that he could use the ancient method of false position, initially ignored the first equation and set $x = 1$ in the second and third equations to get $1 + y^2 = y^4$. Since this is an equation in quadratic form, he could solve it:

$$z = y^2 = \frac{1}{2} + \sqrt{\frac{5}{4}} \qquad \text{and} \qquad y = \sqrt{\frac{1}{2} + \sqrt{\frac{5}{4}}}.$$

[22] Levey, 1966, p. 156.
[23] Levey, 1966, p. 186.

Now, returning to the first equation, he noted that the sum of his three "false" values was

$$1\frac{1}{2} + \sqrt{\frac{5}{4}} + \sqrt{\frac{1}{2} + \sqrt{\frac{5}{4}}},$$

instead of 10. To find the correct values, he needed to divide 10 by this value and multiply the quotient by the "false" values. Since the false value of x was 1, this just meant that the correct value for x was

$$x = \frac{10}{1\frac{1}{2} + \sqrt{\frac{5}{4}} + \sqrt{\frac{1}{2} + \sqrt{\frac{5}{4}}}}.$$

To simplify this was not trivial, but Abū Kāmil began by multiplying the denominator by x and setting the product equal to 10. He ultimately turned this equation into a quadratic equation and succeeded in determining that $x = 5 - \sqrt{\sqrt{3125} - 50}$. To find z by multiplying the false value by this quotient would have been even more difficult, so he chose to find it by beginning the problem anew with the false value $z = 1$. Of course, once he found z, he could determine y by subtraction.

These examples show that Abū Kāmil was willing to use the algebraic algorithms that had been systematized by the time of al-Khwārizmī with more general numbers than rational ones. In particular, he dealt easily with square roots and square roots of expressions with square roots. Thus, he made no distinction between operating with 2 or with $\sqrt{8}$ or even with $\sqrt{\sqrt{2} - 1}$. Moreover, although his solutions would be sides of squares in his geometric derivation, he essentially disregarded the geometry and thought of all of his solutions as "numbers," just as the words for the unknowns would suggest. It did not matter whether a magnitude was technically a square or a fourth power or a root or a root of a root. For Abū Kāmil, the solution of a quadratic equation was not a line segment, as it would be in the interpretation of the appropriate propositions of the *Elements*. It was a "number," even though Abū Kāmil could not perhaps give a proper definition of that term. He therefore had no compunction about combining the various quantities that appeared in the solutions, using general rules.

It is important to remember, however, that Abū Kāmil's algebra, like all Islamic algebra texts of his era, was written without symbols. Thus, the algebraic manipulation—made almost obvious by modern symbolism—is carried out completely verbally. (Of course, in our previous example, the procedure is by no means "obvious" even with symbolism.) It cannot be denied that Abū Kāmil was doing "algebra" using words, but it also seems clear that his readers could easily have gotten lost trying to follow his involved verbal expressions. Thus, in problem 61, the calculation of x as 10 divided by a complicated expression begins: "Multiply the thing by $1\frac{1}{2}$ plus the root of $1\frac{1}{4}$ plus the root of the sum of the root of $1\frac{1}{4}$ and $\frac{1}{2}$; it equals a thing plus $\frac{1}{2}$ a thing plus the root of $1\frac{1}{4}$ squares plus the root of the sum of $\frac{1}{2}$ a square plus the root of $1\frac{1}{4}$ square squares—equal to 10."[24] It is never quite clear, for example, when one reads "root of," to what that phrase applies. Even the scribes who made copies of the manuscripts often left out some numerical phrases because they could not follow the thread of the argument. We return to this question of verbal algebra and its limitations below.

Like al-Khwārizmī, Abū Kāmil also considered "quasi-real-world" problems, in his case, seven of them in all. Although of the same basic type, they were slightly more complicated than the single example al-Khwārizmī had dealt with in his text. Interestingly, in each of his problems, Abū Kāmil showed how to work out the equation by representing the quantities geometrically. For instance, problem 13 reads, "One says that 60 is divided among men equally. Add 3 men and divide 20 among them. Then each portion is less than the first by 26."[25] The equation here is

$$\frac{60}{x} - \frac{20}{x+3} = 26.$$

With $GD = x$, $DH = 3$, $AG = \frac{60}{x}$, rectangle $GB = 60$, and rectangle $GK = 20$ (see figure 7.6), Abū Kāmil demonstrated that $GZ = \frac{20}{x+3}$, and thus $AZ = 26$. Then, rectangle $BZ - 40 = $ rectangle DK. Dividing both sides of this equation by $DH = 3$ and translating into algebra yielded the equation

$$8\frac{2}{3}x - 13\frac{1}{3} = \frac{20}{x+3},$$

[24] Levey, 1966, p. 188.
[25] Levey, 1966, p. 115.

Figure 7.6.

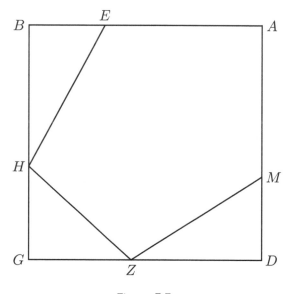

Figure 7.7.

which reduced to the quadratic equation $8\frac{2}{3}x^2 + 12\frac{2}{3}x = 60$ or $x^2 + 1\frac{6}{13}x = 6\frac{12}{13}$. Abū Kāmil then applied the algorithm for case (4) to determine that the original number of men was 2.

In addition to "quasi-real-world" problems, Abū Kāmil also included a number of geometry problems solvable using algebra. For example, he showed how to construct an equilateral (but not equiangular) pentagon inside a square of side 10 (figure 7.7). Beginning by setting one side of the pentagon, AE, equal to a thing [x], he used the Pythagorean theorem to derive a quadratic equation for x and then solved it using the appropriate algorithm to get $x = 20 + \sqrt{200} - \sqrt{200 + \sqrt{320,000}}$.[26]

Abū Kāmil did not state explicitly that this algebraic solution could be constructed geometrically, but since his geometric problems involved

---

[26] See Berggren, 2007, p. 552 for the complete derivation of the solution.

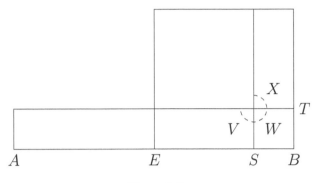

Figure 7.8.

nothing more complicated than taking square roots, a construction was, in fact, always possible. An anonymous Arabic text dating from 1004 actually effected constructions of the solutions for each of the three cases of mixed quadratic equations, using Euclid's propositions VI.28 and VI.29. Once again, this shows that the heritage of these theorems was algebraic and that one could make use of their construction techniques to find a solution to an actual equation stated verbally.[27] For example, case (5) of "squares and numbers equal to roots" ($x^2 + c = bx$) was recast in the form $x(b - x) = c$. In terms of Euclid's application of area techniques, suitably modified for rectangles and squares, we can think of this problem as applying a rectangle with a given area c to a given straight line AB of length b that is deficient by a square. To accomplish this, first find the midpoint E of AB (figure 7.8) and construct the square on EB. Then, assuming in this case that $c < (\frac{b}{2})^2$, construct a square of area $(\frac{b}{2})^2 - c$. Finding a square that is equal to the difference of two rectangles is, of course, a standard Euclidean construction. Finally, choosing the point S on BE so that ES is the side of the constructed square, it is straightforward to show that $x = BS = EB - ES$ yields the solution. The second solution is provided by $x' = AS = AE + ES$.

INDETERMINATE EQUATIONS

Abū Kāmil, both in his algebra text and in a briefer book, *Kitāb al-ṭarā'if fi'l ḥisāb* (or *Book of Rare Things in the Art of Calculation*) considered

[27] Sesiano, 2009, pp. 79–80.

indeterminate equations, some in the style of Diophantus and others reminiscent of problems in Indian texts. Among the "rare things" he solved were six problems involving the buying of one hundred birds for one hundred *dirhems*. As we have seen, similar problems may be found in Indian texts, but, in those, it was only noted that multiple solutions existed without determining their number. Abū Kāmil actually counted the number of solutions. As he wrote about the most complicated problem he tackled, "I found myself before a problem that I solved and for which I discovered a great many solutions; looking deeper for its solutions, I obtained two thousand six hundred and seventy-six correct ones."[28] The problem was this: there were five kinds of birds, with ducks costing two *dirhems* each, chickens costing one *dirhem* each, and with one *dirhem* purchasing two pigeons, three wood pigeons, or four larks. Having no symbolic notation, Abū Kāmil used different words for each of four unknowns ("thing," "seal," and two words representing monetary units); there was no need to name the fifth unknown, since it disappeared quickly from the calculations (see below). In modern notation, the two equations are

$$d + c + p + w + \ell = 100,$$

$$2d + c + \frac{1}{2}p + \frac{1}{3}w + \frac{1}{4}\ell = 100.$$

Abū Kāmil began by solving each equation for c and setting them equal:

$$100 - d - p - w - \ell = 100 - 2d - \frac{1}{2}p - \frac{1}{3}w - \frac{1}{4}\ell \quad \text{or}$$

$$d = \frac{1}{2}p + \frac{2}{3}w + \frac{3}{4}\ell.$$

In other words, choosing values for p, w, and ℓ determined the complete solution. Since the answer must involve positive integers (and never 0), the last equation implies that w must be divisible by 3 and either p is odd and $\ell \equiv 2 \pmod 4$ or p is even and ℓ is divisible by 4. In addition, it must be the case that $d + p + w + \ell = \frac{3}{2}p + \frac{5}{3}w + \frac{7}{4}\ell < 100$. Abū Kāmil then examined each of the two conditions above, in turn, and counted

---

[28] Sesiano, 2009, p. 75. The solution below is adapted from Sesiano's treatment.

possibilities. For example, consider the possible values for p in the first case. Given that the smallest value of w is 3 and the smallest value of ℓ is 2, the inequality gives $\frac{3}{2}p < 92$, which implies that $p < 62$. But if p were equal to 61, with $w = 3$ and $\ell = 2$, then $d = 34$, and the sum of these four quantities would already be 100 before the chickens were considered. It follows that the largest possible value of p is 59. After calculating similarly with w and ℓ, Abū Kāmil concluded that the only possible values for p, w, and ℓ were

$$p = 1, 3, 5, \ldots, 59;$$

$$w = 3, 6, 9, \ldots, 51;$$

$$\ell = 2, 6, 10, \ldots, 50.$$

Since not every triple of p, w, ℓ coming from this list gives a valid solution to the problem, Abū Kāmil had also to provide a systematic method for checking each triple. He began with $w = 3$. When $p = 1$, each of the thirteen values for ℓ gives a valid solution, and the same is true when $p = 3$. When $p = 5$, however, the value $\ell = 50$ does not give a valid solution, so there are only twelve possible values for ℓ. There are also twelve possible values for ℓ when $p = 7$ and $p = 9$, but for $p = 11$, the largest possibility for ℓ is 42, giving only eleven possible values. Continuing in this vein, he determined that there are 212 valid solutions when $w = 3$. He next turned to $w = 6$ and proceeded as before. He reasoned similarly for the other values of w. In the end, for this case, he found that there are 1443 solutions. Unfortunately, he missed two solutions: $w = 54$ is a possibility that leads to two valid solutions: $p = 1$, $\ell = 2$, $d = 38$, $c = 5$ and $p = 3$, $\ell = 2$, $d = 39$, $c = 2$.

In the second case, Abū Kāmil found that the possible values for p, w, and ℓ are

$$p = 2, 4, 6, \ldots, 58;$$

$$w = 3, 6, 9, \ldots, 51;$$

$$\ell = 4, 8, 12, \ldots, 52.$$

Having already given a very detailed analysis in the first case, however, he just presented one solution and mentioned that there are 1233 solutions in all, thus giving his stated total of 2676 (rather than the correct value of 2678) solutions to the problem.

In his algebra text, Abū Kāmil presented thirty-eight indeterminate quadratic problems. Interestingly, although Diophantus's work was first translated into Arabic in the late ninth century by Qusṭā ibn Lūqā, it is virtually certain that Abū Kāmil was unaware of the work, since he neither used Diophantus's system for naming powers nor dealt with negative powers of the unknown. The fact, however, that Abū Kāmil's techniques are similar to those of Diophantus suggests some continuing tradition in Egypt, at least, of solving such problems.

By way of example, consider problem 13, which asks that a given number be divided into two parts such that if one given number is added to one of the parts and the other part is subtracted from another given number, the two results are both squares.[29] As Diophantus often did, Abū Kāmil specified the three given numbers before attempting a solution. The specific system to solve was thus

$$x + y = 10,$$
$$20 + x = z^2,$$
$$50 - y = w^2.$$

Rewriting the second equation as $x = z^2 - 20$, the third equation becomes $w^2 = 50 - y = 50 - 10 + x = 50 - 10 + z^2 - 20 = 20 + z^2$. Because x is positive, we know that $z^2 > 20$, and because $w^2 < 50$, we also have $z^2 < 30$. Making a linear substitution, this time of the form $w = b \pm z$, enables us to find z within these boundaries. With some trials, we find that setting $w = 11 - z$ will work. Thus, $20 + z^2 = (11 - z)^2 = 121 - 22z + z^2$, or $22z = 101$ and $z = \frac{101}{22} = 4\frac{13}{22}$. Then $x = z^2 - 20 = \left(\frac{101}{22}\right)^2 - 20 = \frac{521}{484}$. Therefore, the second part of 10 is $y = 10 - x = \frac{4319}{484}$, and, in fact, $50 - y = \left(\frac{141}{22}\right)^2$, as desired.

After Diophantus's *Arithmetica* was translated into Arabic, other Islamic mathematicians used it in concert with material on indeterminate equations already present in their tradition. In particular, two texts by the Baghdad mathematician, Abū Bakr al-Karajī (953–1029), contain large sections on indeterminate equations. In his major algebra text *al-Fakhrī*[30] (ca. 1012), after discussing the basic ideas of algebra

---

[29] This problem and others are discussed in Sesiano, 1977a, pp. 91–92.
[30] The title of this work is a version of the name of the vizier to whom it was dedicated.

(see below), al-Karajī outlined the method—so frequently used by Diophantus and Abū Kāmil—for making linear substitutions in order to solve certain quadratic indeterminate equations. In a long series of problems, al-Karajī presented most of the problems from Diophantus's Books I and II, sometimes with the same numerical constants and sometimes with different ones but always solved by the Greek author's method. Al-Karajī also incorporated many of the problems found in Abū Kāmil's text, including problem 13 discussed above.

Al-Karajī devoted a second text, al-Badī (or The Marvelous), entirely to the solution of indeterminate equations. There, he gave systematic procedures for solving such equations, again, mostly quadratic ones and, again, using essentially Diophantine methods. Al-Karajī took many of his examples from the Arithmetica, but he drew some from his own al-Fakhrī and some from Abū Kāmil's algebra text. His problem 42, for example, both illustrates a standard Diophantine technique (recall chapter 4) and had implications later on in medieval Europe (see chapter 8). In it, he asked his readers to determine a square such that when 5 is both added to it and subtracted from it, a square is produced.[31] In modern notation, the problem amounts to solving the system

$$x^2 + 5 = u^2,$$
$$x^2 - 5 = v^2.$$

Al-Karajī used Diophantus's method of the double equation, that is, he took the difference of the two equations to get $10 = u^2 - v^2$ and then factored the right-hand side as $(u+v)(u-v)$. Since the sum of the two factors is $2u$, if $d = u + v$, then $u - v = \frac{10}{d}$ and $u = \frac{1}{2}\left(d + \frac{10}{d}\right)$. In this case, d must be chosen so that

$$u^2 - 5 = \left[\frac{1}{2}\left(d + \frac{10}{d}\right)\right]^2 - 5$$

is a square, namely, x^2. Al-Karajī just suggested the use of "trial and error,"[32] but, in order to do this, he probably would, as in earlier

---

[31] Sesiano, 1977b, p. 335, where Sesiano only states the problem symbolically.
[32] Sesiano, 1977b, p. 335.

problems, have first made an appropriate linear substitution. Note that

$$u^2 - 5 = \frac{1}{4}\left[d^2 + 20 + \frac{100}{d^2}\right] - 5 = \frac{25}{d^2} + \frac{d^2}{4} = \frac{100 + d^4}{4d^2}.$$

This will be a square provided $100 + d^4$ is a square. Setting $100 + d^4 = (md^2 - 10)^2$ reduces this equation to $20md^2 = m^2d^4 - d^4$, so $d^2 = \frac{20m}{m^2-1}$. In order for the right-hand side of this expression to be a square, m must be chosen accordingly, by "trial and error." We find that $m = 9$ works, giving $d = \frac{3}{2}$. Thus, the appropriate factorization of 10 is $\frac{3}{2} \cdot \frac{20}{3}$ and $u = \frac{49}{12}$. It follows that $x = \frac{41}{12}$ and $v = \frac{31}{12}$.

THE ALGEBRA OF POLYNOMIALS

The process that began in the writings of al-Khwārizmī of relating arithmetic to algebra, that is, of considering algebra as "generalized arithmetic," continued in the Islamic world with the work of al-Karajī around the turn of the eleventh century and a century later with that of al-Samaw'al ben Yahyā al-Maghribī (1125–1174). These mathematicians were instrumental in showing that the techniques of arithmetic could be fruitfully applied in algebra and, reciprocally, that ideas originally developed in algebra could be important in dealing with numbers.

Al-Karajī set out the aim of algebra in general in *al-Fakhrī*, namely, "the determination of unknowns starting from knowns."[33] In pursuing this goal, he used the basic techniques of arithmetic, converted into techniques for dealing with unknowns. Thus, he began by making a systematic study of the algebra of exponents. Al-Karajī repeated Diophantus's naming scheme, using square-square for x^4, square-cube for x^5, up to cube-cube-cube for x^9, and noted explicitly that one could continue this process indefinitely. He also named the reciprocals of the powers "parts." Since he defined the powers of an unknown x, analogously to the powers of 10, as x times the previous power, it followed

---

[33] Woepcke, 1853, p. 45. This article contains a partial French translation of and much commentary by Woepcke on *al-Fakhrī*. The English translation here is ours.

that there was an infinite sequence of proportions

$$1 : x = x : x^2 = x^2 : x^3 = \cdots$$

and similarly that

$$\frac{1}{x} : \frac{1}{x^2} = \frac{1}{x^2} : \frac{1}{x^3} = \frac{1}{x^3} : \frac{1}{x^4} = \cdots .$$

Once the powers were understood, al-Karajī established general procedures for adding, subtracting, and multiplying monomials and polynomials, noting, for example,[34] that the product of 2 "things" with 3 "parts of
squares" was equal to 6 "parts of a thing," that is, $2x \times \frac{3}{x^2} = \frac{6}{x}$. Since he did
not discuss the zeroth power of an unknown, however, these procedures
were incomplete. He could thus not state a general "rule of exponents,"
partly because he was unable to incorporate rules for negative numbers
into his theory and partly because of his verbal means of expression.
Similarly, although he developed an algorithm for calculating square
roots of polynomials, it was only applicable in limited circumstances. Still,
he did apply arithmetic operations to irrational quantities, both in the
context of solving quadratic equations and in the manipulation of surds.

Consider, for example, what may be the earliest example of a travel
problem in an algebra text: Of two travelers going in the same direction,
the first goes 11 miles per day, while the second, leaving five days later,
goes successively each day 1 mile, 2 miles, 3 miles, and so on. In how
many days will the second traveler overtake the first?[35] Given that the
sum of the integers from 1 to x is $\frac{1}{2}(x^2 + x)$ and given that the first traveler
has a fifty-five-mile head start, the required equation is

$$\frac{1}{2}\left(x^2 + x\right) = 11x + 55 \qquad \text{or} \qquad x^2 = 21x + 110,$$

and its solution is the irrational number $x = \sqrt{220\frac{1}{4}} + 10\frac{1}{2} \approx 25.34$.
Curiously, al-Karajī was apparently bothered neither by the irrational
number of days nor by how it actually fit the conditions of the problem;
it is not at all clear what the solution would mean for the distance

---

[34] Woepcke, 1853, p. 49.
[35] This problem is discussed in Woepcke, 1853, p. 82.

the second traveler would travel on the twenty-sixth day. Al-Karajī also developed various formulas involving surds, such as

$$\sqrt{A - B} = \sqrt{\frac{A + \sqrt{A^2 - B^2}}{2}} - \sqrt{\frac{A - \sqrt{A^2 - B^2}}{2}}$$

and

$$\sqrt[3]{A} + \sqrt[3]{B} = \sqrt[3]{3\sqrt[3]{A^2 B} + 3\sqrt[3]{A B^2} + A + B},$$

relationships that could be used in certain circumstances for simplifying algebraic expressions.

Following al-Karajī by a century, al-Samaw'al was born in Baghdad of Jewish parents from Morocco. By the twelfth century, there were no longer mathematicians in the city who could teach him, so he traveled elsewhere in the Islamic world to study. He apparently learned well enough to write an algebra text, *Al-Bāhir fi'l-ḥisāb* (or *The Shining Book of Calculation*), when he was nineteen. There, he carried the work of algebraic manipulation further than had al-Karajī, especially with his introduction of negative coefficients. Al-Samaw'al expressed his rules for dealing with such coefficients quite clearly: "If we subtract an additive number from an empty power, $[0x^n - ax^n]$ the same subtractive number remains; if we subtract the subtractive number from an empty power, $[0x^n - (-ax^n)]$ the same additive number remains. If we subtract an additive number from a subtractive number, the remainder is their subtractive sum; if we subtract a subtractive number from a greater subtractive number, the result is their subtractive difference; if the number from which one subtracts is smaller than the number subtracted, the result is their additive difference."[36]

Using these rules, al-Samaw'al could easily add and subtract polynomials by combining like terms. To multiply, of course, he needed the law of exponents.[37] Al-Karajī had, in essence, used this law with words for the powers, thus masking the numerical property of adding exponents. Al-Samaw'al decided that this law could best be expressed by using a table

[36] Quoted in Anbouba, 1975, p. 92.
[37] The material in this and the following three paragraphs is adapted from Berggren, 1986, pp. 113–118.

consisting of columns, each one of which represented a different power of either a number or an unknown. In fact, he also saw that he could deal with powers of $\frac{1}{x}$ as easily as with powers of x, and he clearly realized that the zeroth power of x was equal to 1 and was between the positive powers of x and the powers of $\frac{1}{x}$.

In his work, the columns are headed by the Arabic letters standing for the numerals, reading both ways from the central column labeled 0 (we will simply use the Arabic numerals themselves below). Each column then has the name of the particular power or reciprocal power. For example, the column headed by a 2 on the left is named "square," that headed by a 5 on the left is named "square-cube," that headed by a 3 on the right is named "part of cube," and so on (we just use powers of x). In his initial explanation of the rules, al-Samaw'al also put a particular number under the 1 on the left, such as 2, and then the various powers of 2 in the corresponding columns:

7	6	5	4	3	2	1	0	1	2	3	4	5	6	7
x^7	x^6	x^5	x^4	x^3	x^2	x	1	x^{-1}	x^{-2}	x^{-3}	x^{-4}	x^{-5}	x^{-6}	x^{-7}
128	64	32	16	8	4	2	1	$\frac{1}{2}$	$\frac{1}{4}$	$\frac{1}{8}$	$\frac{1}{16}$	$\frac{1}{32}$	$\frac{1}{64}$	$\frac{1}{128}$

Al-Samaw'al explained the law of exponents $x^n x^m = x^{n+m}$ using the chart in this way: "The distance of the order of the product of the two factors from the order of one of the two factors is equal to the distance of the order of the other factor from the unit. If the factors are in different directions then we count (the distance) from the order of the first factor towards the unit; but, if they are in the same direction, we count away from the unit."[38] So, for example, to multiply x^3 by x^4, count four orders to the left of column 3 and get the result as x^7. To multiply x^3 by x^{-2}, count two orders to the right from column 3 and get the answer x^1. Using these rules, al-Samaw'al could easily multiply polynomials in x and $\frac{1}{x}$ as well as divide such polynomials by monomials.

Al-Samaw'al also developed a method for dividing polynomials by polynomials. Given that such a division does not always result in a

---

[38] Quoted in Berggren, 1986, p. 114.

polynomial, he showed how to continue the process and express the answer in terms of powers of $\frac{1}{x}$ as well as x. He further understood that, although frequently the division process does not end, the partial results are better and better approximations of the "true" answer. Later in his career, he applied this idea to the division of whole numbers and thus explicitly recognized that fractions could be approximated more and more closely by calculating more and more decimal places. The work of al-Samaw'al—as well as that of al-Karajī—was thus extremely important in developing the notion that algebraic manipulations and numerical manipulations are parallel. Virtually any technique that applies to one can be adapted to apply to the other.

The Shining Book of Calculation also contained a version of the binomial theorem, evidently a result originally stated by al-Karajī. In modern notation, the binomial theorem states that

$$(a+b)^n = \sum_{k=0}^{n} C_k^n a^{n-k} b^k,$$

where n is a positive integer and where the values C_k^n—the so-called binomial coefficients—are the entries in what is now termed the Pascal triangle after the seventeenth-century philosopher and mathematician, Blaise Pascal (1623–1662). Since al-Samaw'al had no symbolism, he had to write this formula out in words in each individual instance. For example, in the case $n = 4$, he wrote, "For a number divided into two parts, its square-square [fourth power] is equal to the square-square of each part, four times the product of each by the cube of the other, and six times the product of the squares of each part."[39] He then provided a table of binomial coefficients (see figure 7.9) to show how to generalize this rule for greater values of n.

Al-Samaw'al's procedure for constructing this table is the one familiar today, namely, a given entry results from adding the entry to the left of it to the entry just above that one. The table then allowed him immediately to determine the expansion of any power up to the twelfth of "a number divided into two parts." In fact, al-Samaw'al gave detailed proofs of this result for small values of n. Thus, quoting the result for $n = 2$ as well

[39] Rashed, 1994a, p. 65. The quotations in the next two paragraphs are also found here. Al-Samaw'al's proof and related material have been translated and analyzed anew in Bajri, 2011.

x	x^2	x^3	x^4	x^5	x^6	x^7	x^8	x^9	x^{10}	x^{11}	x^{12}
1	1	1	1	1	1	1	1	1	1	1	1
1	2	3	4	5	6	7	8	9	10	11	12
	1	3	6	10	15	21	28	36	45	55	66
		1	4	10	20	35	56	84	120	165	220
			1	5	15	35	70	126	210	330	495
				1	6	21	56	126	252	462	792
					1	7	28	84	210	462	924
						1	8	36	120	330	792
							1	9	45	165	495
								1	10	55	220
									1	11	66
										1	12
											1

Figure 7.9. Al-Samaw'al's table of binomial coefficients.

known, he used that to prove the theorem for $n = 3$ and then continued to $n = 4$:

Let the number $\overline{AB}$ be divided into two parts, $\overline{AC}$ and $\overline{CB}$. I say that the square-square of $\overline{AB}$ is equal to the sum of the square-square of $\overline{AC}$, the square-square of $\overline{CB}$, four times the product of $\overline{AC}$ by the cube of $\overline{CB}$, four times the product of $\overline{CB}$ by the cube of $\overline{AC}$, and six times the product of the square of $\overline{AC}$ by that of $\overline{CB}$.

Al-Samaw'al's proof of this, all written out in words, begins with "the square-square of $\overline{AB}$ is the product of $\overline{AB}$ by its cube. But we showed [earlier] that the cube of $\overline{AB}$ is equal to the sum of the cube of $\overline{AC}$, of the cube of $\overline{CB}$, of three times the product of $\overline{AC}$ by the square of $\overline{CB}$, and three times the product of $\overline{CB}$ by the square of $\overline{AC}$." In symbols, if we let $c = \overline{AB}$, $a = \overline{AC}$, and $b = \overline{CB}$, then this sentence notes that $c^4 = cc^3$ and $c^3 = (a+b)^3 = a^3 + b^3 + 3ab^2 + 3ba^2$. We now continue the argument symbolically. The two preceding formulas imply that $(a+b)^4 = (a+b)(a+b)^3 = (a+b)(a^3 + b^3 + 3ab^2 + 3a^2b)$. Then, by repeatedly using the result $(r + s)t = rs + rt$, which al-Samaw'al quoted from Euclid's *Elements* II.1—once again turning a geometric result of Euclid into an algebraic result—he found that the latter quantity equals $(a + b)a^3 + (a + b)b^3 + (a + b)3ab^2 + (a + b)3a^2b = a^4 + a^3b + ab^3 + b^4 + 3a^2b^2 + 3ab^3 + 3a^3b + 3a^2b^2 = a^4 + b^4 + 4ab^3 + 4a^3b + 6a^2b^2$. Thus, not only is the result proved, but the proof also shows that the coefficients here are the appropriate ones from the table and that the new coefficients are formed from the old ones exactly as stated in the table construction. Al-Samaw'al next quoted the result for $n = 5$ and asserted his general result: "He who has understood what we have just said, can prove that for any number divided into two parts, its quadrato-cube [fifth power] is equal to the sum of the quadrato-cubes of each of its parts, five times the product of each of its parts by the square-square of the other, and ten times the product of the square of each of them by the cube of the other. And so on in ascending order."

There are two important ideas to note here. First, that al-Samaw'al had to write out everything in words prevented him from stating the binomial theorem in all generality. Even stating the case $n = 5$ was lengthy, and it is clear that a proof of that result, along the lines of his proof for $n = 4$, would have been very difficult for readers to follow. In fact, the manuscripts of the proof for $n = 4$ frequently leave out a phrase or add an extra phrase, probably because the copyist could not follow the argument. So, as in the case of Abū Kāmil discussed above, this makes manifest the real limitations of algebra expressed in words.

Second, though, al-Samaw'al's argument contained the two basic elements of an inductive proof. He began with a value for which the result is known, here $n = 2$, and used the result for a given integer to derive the result for the next. Although al-Samaw'al did not have any way of stating, and therefore of proving, the general binomial theorem, modern readers see his argument as only a short step from a full inductive proof. All that is needed is for the coefficients to be defined inductively in stating the theorem, exactly what al-Samaw'al did. And, presumably, understanding this construction is what al-Samaw'al meant when he stated that "he who has understood what we have just said" will be able to prove not only the $n = 5$ case of the binomial theorem but also the various further cases "in ascending order."

Another inductive-style proof of an important algebraic result—the formula for calculating the sum of the fourth powers of the integers—

was given by Abū 'Alī al-Ḥasan ibn al-Ḥasan ibn al-Haytham (965–1039) and is reminiscent of the Indian argument for the more general result discussed in chapter 6.[40] Ibn al-Haytham, who was born in Basra (now in Iraq), was called to Egypt by the caliph al-Ḥakim to work on controlling the Nile floods. Although he was unsuccessful as a flood-control engineer, he did spend the rest of his life doing scientific research, primarily in optics but also in mathematics. He wanted the formula under discussion to enable him to calculate the volume of a paraboloid formed by revolving a parabola around a line perpendicular to its axis, a volume that, in modern terms, requires the integral of a fourth-degree polynomial, which, in turn, requires the sum formula for fourth powers. Although his proof, like that of Jyeṣṭhadeva, used the result

$$(n+1)\sum_{i=1}^{n} i^k = \sum_{i=1}^{n} i^{k+1} + \sum_{p=1}^{n}\left(\sum_{i=1}^{p} i^k\right),$$

Ibn al-Haytham proved this not in general form but only for particular integers, namely, $n = 4$ and $k = 1, 2, 3$. Nevertheless, his proofs of both the double sum result and the formulas for the sum of powers of the integers are inductive in style and easily generalized to any values of n and k. He stopped at $k = 3$ presumably because that was all he needed for his volume calculation using the classical method of exhaustion. In any case, he did state the general result for fourth powers in words we can translate into the modern formula

$$\sum_{i=1}^{n} i^4 = \left(\frac{n}{5}+\frac{1}{5}\right) n \left(n+\frac{1}{2}\right)\left[(n+1)n-\frac{1}{3}\right].$$

THE SOLUTION OF CUBIC EQUATIONS

If arithmetization represented one strand in the development of algebra in the Islamic world, another was the application of conic sections to the solution of cubic equations. Recall from chapter 3 that Archimedes had used the intersections of two conic sections to solve an interesting

[40] The theorem and proof are given, in words, in Berggren, 2007, pp. 587–595.

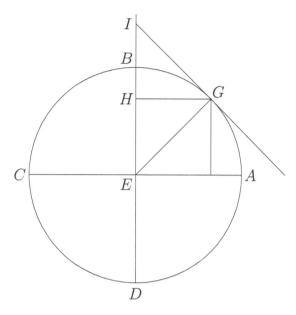

Figure 7.10.

geometric problem, one that we could translate into a cubic equation. Interestingly, several Islamic mathematicians by the eleventh century had converted Archimedes' geometric problem into an algebraic one and had attempted generalizations of the procedure to other types of cubic equations. In addition, Islamic mathematicians had continued their assault on the classical Greek problem of trisecting an angle and had found constructions—unattempted by the Greeks—of a regular hepta-gon. Of their numerous solutions to these problems, many involved the intersection of two conic sections.[41] It was, however, the mathematician and poet 'Umar ibn Ibrāhīm al-Khayyāmī (1048–1131) (usually known in the West as Omar Khayyam), who first systematically classified and then proceeded to solve all types of cubic equations by this method.

Al-Khayyāmī announced his project in a brief untitled treatise, in which he proposed to divide a quadrant AB of a circle at a point G such that, with perpendiculars drawn to two diameters, we have $AE : GH = EH : HB$ (figure 7.10).[42] Using the method of analysis, he assumed the

[41] See Hogendijk, 1984 for information on constructions of the regular heptagon.

[42] See Rashed and Vahabzadeh, 2000 for a translation of and commentary on this treatise and, indeed, on Omar's work on cubic equations in general as described below. They title this treatise,

problem solved and then constructed the tangent GI to the circle at G. After a few steps, he found that the right triangle EGI had the property that its hypotenuse EI equaled the sum of one of the sides EG and the perpendicular GH from the right angle to the hypotenuse. He concluded that if he could find such a right triangle, he could complete the synthesis of the problem.

To find the right triangle, however, al-Khayyāmī needed algebra. He considered a particular case, with $EH = 10$ and $GH = x$. Therefore, $GE^2 = x^2 + 100$. Since $GE^2 = EI \cdot EH$,

$$\frac{GE^2}{10} = \frac{x^2}{10} + 10 = \frac{EI \cdot EH}{10} = EI,$$

but since $EI = EG + GH$, $\frac{x^2}{10} + 10 = EG + GH = EG + x$. Therefore,

$$x^2 + 100 = EG^2 = \left(\frac{x^2}{10} + 10 - x\right)^2.$$

Simplifying this equation yielded a cubic equation in x: $x^3 + 200x = 20x^2 + 2000$. Noting that this could not be solved by "plane geometry," he proceeded to treat it using the intersection of a hyperbola and a semicircle. Given the solution, he could construct the right triangle and, finally, the point satisfying the original problem.

With this example as his point of departure, al-Khayyāmī went on to analyze all possible cubic equations, presenting the results in his algebra text, the *Risāla fi-l-barāhīn 'ala masā'il al-jabr wa'l-muqābala* (or *Treatise on Demonstrations of Problems of al-Jabr and al-Muqābala*). Although he suggested that his readers should be familiar with Euclid's *Elements* and *Data* as well as with the first two books of Apollonius's *Conics*, he addressed algebraic, not geometric, problems. He was really interested in solving equations.

Al-Khayyāmī was, in fact, disappointed that he could not find algebraic algorithms for solving cubic equations analogous to al-Khwārizmī's three algorithms for solving quadratics. As he put it, "when... the object of the problem is an absolute number, neither we, nor any of those who are concerned with algebra, have been able to solve this equation—perhaps

---

for convenience, *On the Division of a Quadrant of a Circle*. It was originally translated into English in Amir-Moéz, 1963.

others who follow us will be able to fill the gap."[43] It was not until the sixteenth century in Italy that his hope was realized (see chapter 9).

Al-Khayyāmī began his work, in the style of al-Khwārizmī, by giving a complete classification of equations of degree up through three. Since, for al-Khayyāmī as for his predecessors, all numbers were positive, he had to list separately the various types of cubic equations that might possess positive roots. Among these were fourteen types not reducible to quadratic or linear equations, including, of course, the form analyzed by Archimedes much earlier and the form needed in the quadrant problem. Al-Khayyāmī organized his types of equations into three groups depending on the number of terms: one binomial equation, $x^3 = d$; six trinomial equations, $x^3 + cx = d$, $x^3 + d = cx$, $x^3 = cx + d$, $x^3 + bx^2 = d$, $x^3 + d = bx^2$, and $x^3 = bx^2 + d$; and seven tetranomial equations, $x^3 + bx^2 + cx = d$, $x^3 + bx^2 + d = cx$, $x^3 + cx + d = bx^2$, $x^3 = bx^2 + cx + d$, $x^3 + bx^2 = cx + d$, $x^3 + cx = bx^2 + d$, and $x^3 + d = bx^2 + cx$. For each of these forms, he described the conic sections necessary for their solution, proved that his solution was correct, and discussed the conditions under which there may be no solutions or more than one solution.

That al-Khayyāmī gave this classification provides strong evidence of the major change in mathematical thinking that had occurred, primarily in Islam, in the more than thirteen hundred years since Archimedes.[44] Unlike his Greek predecessor, al-Khayyāmī was no longer concerned with just solving a specific geometric problem, even though such a problem may have sparked his interest. He wanted to find general methods for solving all sorts of problems that could be expressed in the form of equations. And, although he used only words and not our symbolic notation, there is no question that al-Khayyāmī was doing algebra, not geometry, because he was always finding an unknown length, given certain known quantities and relationships. Still, because he did not have an algorithmic and numerical method of solution, the only way he could express the equations was as equations between solids. For example, in his solution of $x^3 + cx = d$—or, as he put it, the case where "a cube and sides are equal to a number"[45]—since x represents a side of a cube,

---

[43] Kasir, 1931, p. 49 or, in a slightly different translation, Rashed and Vahabzadeh, 2000, p. 114.

[44] See Netz, 2004a for more details on this argument.

[45] Rashed and Vahabzadeh, 2000, p. 130. Al-Khayyāmī's discussion of the problem, summarized in the next two paragraphs, is found on pp. 130–132.

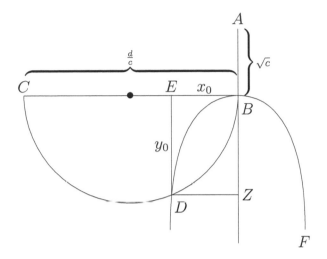

Figure 7.11.

c must represent an area (expressible as a square), so that cx is a solid, while d itself represents a solid. This said, the solution procedure did not manipulate solids but rather just used curves in the plane.

Al-Khayyāmī began the solution by setting AB equal in length to a side of the square c, or $AB = \sqrt{c}$ (figure 7.11). He then constructed BC perpendicular to AB so that $BC \cdot AB^2 = d$, or $BC = \frac{d}{c}$. Next, he extended AB in the direction of Z and constructed a parabola with vertex B, axis BZ, and parameter AB. In modern notation, this parabola has the equation $x^2 = \sqrt{c}\,y$. Similarly, he constructed a semicircle on the line BC. Its equation is

$$\left(x - \frac{d}{2c}\right)^2 + y^2 = \left(\frac{d}{2c}\right)^2 \qquad \text{or} \qquad x\left(\frac{d}{c} - x\right) = y^2.$$

The circle and the parabola intersect at a point D. It is the x coordinate of this point, here represented by the line segment BE, that provides the solution to the equation.

Al-Khayyāmī proved his solution correct using the basic properties of the parabola and the circle. If $BE = DZ = x_0$ and $BZ = ED = y_0$, then, first, $x_0^2 = \sqrt{c}\,y_0$ or $\frac{\sqrt{c}}{x_0} = \frac{x_0}{y_0}$, because D is on the parabola, and, second, $x_0(\frac{d}{c} - x_0) = y_0^2$ or $\frac{x_0}{y_0} = \frac{y_0}{\frac{d}{c} - x_0}$, because D is on the semicircle. It follows

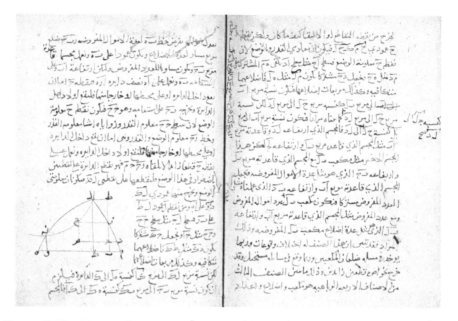

Figure 7.12. Manuscript page from a thirteenth-century manuscript of al-Khayyāmī's *Algebra*, copied in Lahore, containing a solution of one of the cases of a cubic equation, from the Smith Collection of the Rare Book and Manuscript Library, Columbia University.

that

$$\frac{c}{x_0^2} = \frac{x_0^2}{y_0^2} = \frac{y_0^2}{\left(\frac{d}{c} - x_0\right)^2} = \frac{y_0}{\frac{d}{c} - x_0}\frac{x_0}{y_0} = \frac{x_0}{\frac{d}{c} - x_0}$$

and then that $x_0^3 = d - cx_0$, so x_0 is the desired solution. Al-Khayyāmī noted here, without any indication of a proof, that this class of equations always has a single solution. In other words, the parabola and circle always intersect in one point other than the origin. The origin, though, does not provide a solution to the problem. Al-Khayyāmī's remark reflects the modern statement that the equation $x^3 + cx = d$ always has exactly one positive solution.

Al-Khayyāmī treated each of his fourteen cases similarly. In those cases in which a positive solution did not always exist, he noted that there were zero, one, or two solutions depending on whether the conic sections involved do not intersect or intersect at one or two points. His one failure in this analysis occurred in the case of the equation $x^3 + cx = bx^2 + d$, where he did not discover the possibility of three (positive) solutions. In general, however, he did not relate the existence of one or two solutions to conditions on the coefficients.

Al-Khayyāmī's methods were improved by Sharaf al-Dīn al-Ṭūsī (d. 1213), a mathematician born in Tus, Persia (and not to be confused with the more famous Naṣir al-Dīn al-Ṭūsī (1201–1274)).[46] Like his predecessor, al-Ṭūsı began by classifying the cubic equations, but because he was interested in determining conditions on the coefficients that dictate the number of solutions, his classification differed. Al-Ṭūsī's first group consisted of the equation $x^3 = d$ together with those equations that could be reduced to quadratics. The second group comprised the eight cubic equations that always have at least one (positive) solution. The third encompassed those types that may or may not have (positive) solutions, depending on the particular values of the coefficients. These were the forms $x^3 + d = bx^2$, $x^3 + d = cx$, $x^3 + bx^2 + d = cx$, $x^3 + cx + d = bx^2$, and $x^3 + d = bx^2 + cx$. Al-Ṭūsī solved the second group of equations as had al-Khayyāmī, but he went beyond his predecessor by always discussing carefully why the two conics, in fact, intersected. It was, however, relative to the third group that he made his most original contribution.

Consider his analysis of $x^3 + d = bx^2$, which was typical of how he handled the five equations in this group and very similar to Archimedes' treatment of the problem. Al-Ṭūsī began by putting the equation in the form $x^2(b - x) = d$.[47] He then noted that the question of whether the equation has a solution depends on whether the "function" $f(x) = x^2(b - x)$ reaches the value d or not. He therefore carefully proved that the value $x_0 = \frac{2b}{3}$ provides the maximum value for $f(x)$, that is, for any x between 0 and b, $x^2(b - x) \le (\frac{2b}{3})^2(\frac{b}{3}) = \frac{4b^3}{27}$. It follows that if $\frac{4b^3}{27}$ is less than the given d, the equation can have no solution. If $d = \frac{4b^3}{27}$, there is only one solution, namely, $x = \frac{2b}{3}$. Finally, if $\frac{4b^3}{27} > d$, there are two solutions, x_1 and x_2, where $0 < x_1 < \frac{2b}{3}$ and $\frac{2b}{3} < x_2 < b$.

Although he did not say how he arrived at this particular value of x_0, al-Ṭūsī probably found it through a close study of the Archimedean text. Still, al-Ṭūsī's work—like that of al-Khayyāmī—underscores how mathematics had changed since the time of the Greeks. Unlike Archimedes, who sought to solve a particular geometrical problem, al-Ṭūsī clearly saw

---

[46] Rashed, 1986 contains a translation of and commentary on the algebra of al-Ṭūsī.
[47] Rashed, 1986, pp. 9–16.

this as an algebraic problem and generalized it from one particular type of cubic to the entire group of five types, where the existence of a positive solution depends on the actual value of the coefficients. In addition, he showed that the solution of this type of equation was related to the solutions of other equations. In particular, he demonstrated that if one knows a solution X of the equation $x^3 + bx^2 = \frac{4b^3}{27} - d$, then the larger solution x_2 of the given equation is $x_2 = X + \frac{2b}{3}$. To find the remaining root x_1, he found the positive solution Y of the quadratic equation $x^2 + (b - x_2)x = x_2(b - x_2)$ and demonstrated that $x_1 = Y + b - x_2$. Hence, the root of the new polynomial is related to that of the old by this change of variable formula. Al-Ṭūsī thus clearly had a solid understanding of the nature of cubic equations and the relationship of their roots and coefficients, even if he was unable to determine the solutions of the cubic equation by an algebraic algorithm.

Al-Ṭūsī did solve cubic equations numerically by approximation techniques, however.[48] For example, he considered the equation $x^3 + 14{,}837{,}904 = 465x^2$. By the above method, he first calculated that $\frac{4b^3}{27} = 14{,}895{,}500$. Since this value is greater than $d = 14{,}837{,}904$, there are two solutions x_1, x_2, with $0 < x_1 < 310$ and $310 < x_2 < 465$. To find x_2, he needed to solve $x^3 + 465x^2 = 57{,}596$. He found that 11 is a solution and therefore that $x_2 = \frac{2b}{3} + 11 = 310 + 11 = 321$. To find x_1, he had to solve the quadratic equation $x^2 + 144x = 46{,}224$. The (positive) solution is an irrational number approximately equal to 154.73, a solution he found by a numerical method. The solution x_1 of the original equation is then 298.73.

****

By the fifteenth century, Islamic scientific civilization was in a state of decline, while mathematical activity had resumed in Europe.[49] A central factor in this revival was the work of the translators of the twelfth century who made available to Europeans a portion of the Islamic mathematical corpus, most notably the algebraic text of al-Khwārizmī. The work of Abū Kāmil also became available in Europe, chiefly through the inclusion of numerous problems from his work in Leonardo of Pisa's *Liber abbaci*

---

[48] Rashed, 1986, pp. 16–17.
[49] There is a lively debate in the literature as to the reasons for this decline. See, for example, Huff, 1993 and Saliba, 2007.

(1202) and the fifteenth-century translation of his book into Hebrew by Mordecai Finzi (1440–1475) in Italy. As far as is known, however, the more advanced algebraic materials of al-Karajī and al-Samaw'al did not reach Europe before or during the Renaissance, nor did the work on cubic equations of al-Khayyāmī or the maximization technique of Sharaf al-Dīn al-Ṭūsī. And, although a manuscript of ibn al-Haytham's work dealing with the sum formula for powers of integers exists in England, it was evidently only acquired in the nineteenth century and thus had no effect on similar developments in Europe in the sixteenth and seventeenth centuries.

Despite the general decline in Islamic science by the time of the European Renaissance, it is intriguing to consider how much further Islamic algebra might have gone had it not been constrained by its lack of symbolism and its heavy dependence on Greek geometry. It is clear from reading some of the Islamic texts, excerpts from which we considered above, that it would take detailed study and extreme concentration for anyone to follow the verbal algebraic arguments of Abu Kāmil, or al-Samaw'al, or al-Khayyāmī. In the absence of good notation, which so facilitates reading and understanding, only a few people could have mastered the material and developed it further. Moreover, since they based their work primarily on Greek geometry, Islamic algebraists could not see the generalizations possible by including negative numbers. Hints of nascent symbolism and of the use of negative numbers in some contexts may actually be found in some later Islamic work. These seeds, however, never had the chance to bear fruit, partly owing to the fact that Islamic rulers during this time were loath to allow many people to study mathematics. The same, however, was not true in medieval and early Renaissance Europe. Although the Europeans also had a heritage of Greek geometry and received Islamic algebra in its verbal form, mathematicians there, who flourished in the context of an evolving capitalist economy, broke free of those constraints and developed algebra in previously unforeseen directions.

8

Transmission, Transplantation, and Diffusion in the Latin West

When Hypatia died her violent death in 415 CE, the rise of Islam was more than two centuries in the offing, but the Roman Empire, of which her native Alexandria was a part, had long been in decline. One manifestation of that decline, to which her work in the Diophantine tradition served as a notable counterexample, was a waning interest in learning and in the preservation of classical Greek knowledge. By 500 CE, the Roman Empire had completely ceased to exist in the West, and the centralized government, established commercial networks, and urban stability that had characterized it had been replaced by the turmoil occasioned by invasions of Germanic tribes from the north and west as well as—during the course of the seventh and eighth centuries—by the march of Islam across North Africa and into the Iberian Peninsula.

The Latin West entered a five-hundred-year period—the so-called early Middle Ages, once termed the "Dark Ages"—marked both by a dwindling population that shifted demographically from the cities to the countryside and by a Catholic Church that evolved to represent the key cultural force unifying otherwise rival kingdoms. Relative to the history of science in general, David Lindberg put it well: "The contribution of the religious culture of the early Middle Ages to the scientific movement was ... one of preservation and transmission. The monasteries served as the transmitters of literacy and a thin version of the classical tradition (including science and natural philosophy) through a period when literacy and scholarship were severely threatened."[1] His remark holds equally well for the history of mathematics, and of algebra in particular.

---

[1] Lindberg, 1992, p. 157.

Figure 8.1. Map of medieval Europe.

From both a geopolitical and an educational point of view, the situation had begun to stabilize by 800. In that year, the pope named Charlemagne (ca. 742–814) emperor of a region that came to encompass modern-day France, Belgium, and the Netherlands as well as much of present-day Germany, part of Austria, and the northern half of Italy. In his effort to maintain control over this vast Carolingian Empire, Charlemagne endeavored to put into place the first centralized government since the fall of the Roman Empire and instituted educational reforms that mandated the establishment of monastery and cathedral schools throughout his domain. Although aimed simultaneously at strengthening the state and the clergy, these educational reforms may also be seen to have marked the beginning of a tradition of scholarship in the Latin West.[2]

The stability that the establishment of the Carolingian Empire promised was, however, short-lived. The last half of the ninth century witnessed the partitioning of the former empire into the kingdoms of France, Provence, Germany, Burgundy, and Italy as well as the coalescence of numerous smaller rivals—among them, the states of Aragon and Barcelona on the Iberian Peninsula, and the principalities of Capua-Benevento and Salerno in southern Italy—that vied for political and economic influence. One constant in all of this change was the Catholic Church and the schools that it continued to maintain as a result of the Carolingian reforms. Among the clergy who benefited from this institutionalization of learning was Gerbert of Aurillac (ca. 945–1003).[3]

Educated initially in the classical tradition at the Benedictine monastery in his hometown in south-central France, Gerbert next traveled to Catalonia in the northeast of Spain to study mathematics and music with the Bishop of Vic. Catalonia had, in the eighth century, come under the dominion of the Muslim Umayyad caliphate as it pushed its way across the Iberian Peninsula and into the part of Visigothic Gaul that is the western Mediterranean coastal region of modern-day France. By the turn of the ninth century, Charlemagne's forces had succeeded in taking back much of Catalonia and in forming the borderland known as the Spanish Marches that served as a buffer between the Christian and Islamic domains. Situated within the Marches, Vic was thus at the confluence of cultures. There, Gerbert undoubtedly came into contact with Islamic learning, and especially Islamic mathematics and astronomy. The evidence suggests that he brought that as well as his deep classical learning with him first as a student and then as headmaster at the cathedral school in Reims in northern France, briefly to northern Italy where he served as the abbot of the monastery in Bobbio, next back first to Reims and then to Italy in his posts as archbishop, and finally to Rome where he became Pope Sylvester II.

[2] Lindberg, 1992, p. 185.
[3] On Gerbert, his life and impact, see Brown, 2010; Darlington, 1947; Lindberg, 1978, p. 60–61; and Lindberg, 1992, pp. 188–190.

Gerbert did more than just transmit classical and Islamic learning, however. He was an active scholar who strove—through his innovative use of the abacus, celestial spheres, and armillary spheres of his own fabrication—to raise the level of learning of the mathematical sciences in the Latin West. Among his students was a generation of Catholic scholars who went on themselves to establish or to teach at cathedral schools and to influence educational reforms in royal courts throughout western Europe.

The two centuries following Gerbert's death in 1003 marked a period of transition characterized not only by a dramatic increase in population but also by a rcurbanization that conccntratcd rcsourccs back in thc cities and fostered the further development of schools and of intellectual culture more generally.[4] By the twelfth century, that intellectual culture had developed to such an extent that it sought out new texts and ideas. As the example of Gerbert and others had driven home, those fresh insights were abundantly available in manuscripts to be found in libraries on an Iberian Peninsula that had been progressively reclaimed from the Muslim caliphate by the Kingdoms of Portugal, Leon, Castile, and Aragon. Scholar-translators from as far away as the British Isles, Italy, Dalmatia, Hungary, and elsewhere began their self-appointed tasks of finding and translating from Arabic into Latin books in their own particular areas of interest.

Robert of Chester, for example, journeyed from England to Spain in the mid-twelfth century and produced Latin translations of both the Koran and al-Khwārizmī's *Al-kitāb al-mukhtaṣar fī ḥisāb al-jabr wa'l-muqābala* in the late 1140s before his appointment as archdeacon of Pamplona.[5] His contemporary, Gerard of Cremona (1114–1187), traveled from Italy to establish himself in the cathedral city of Toledo, where, over the course of his life, he not only served as canon of the cathedral but also executed over seventy major Latin translations, including al-Khwārizmī's text, Euclid's *Elements*, Ptolemy's astronomical treatise the *Almagest*, medical texts, and works of Aristotelian philosophy. Gerard's presence underscored the fact that by the mid-twelfth century, Toledo had become the main center for translation on the Iberian Peninsula thanks largely to the patronage of the church and to the linguistic expertise and

[4] Lindberg, 1992, pp. 183–213.
[5] Lindberg, 1978, p. 64.

intellectual curiosity of members of its clergy.[6] As such, it was also a key conduit of Islamic and, through the hands of Islamic scholars, classical Greek learning into the Latin West.[7]

THE TRANSPLANTATION OF ALGEBRAIC THOUGHT IN THE THIRTEENTH CENTURY

Another conduit was through direct contact with the Islamic world. By the beginning of the thirteenth century, not only had three religious Crusades been waged in the Holy Lands but a number of the Italian city-states, most notably Venice, Genoa, and Pisa, had well-established trade routes connecting the boot of Italy to the eastern Mediterranean rim and North Africa. Trade, and the commercial transactions it required, directly exposed at least one young man, Leonardo of Pisa (ca. 1170–ca. 1240 and also known in the literature as Leonardo Fibonacci), to the arithmetical practices that had developed in the Islamic and Mediterranean worlds as well as to the problem-solving techniques and mathematical ideas of al-Khwārizmī that he likely also read in Gerard's Latin translation.[8]

Son of an official assigned by the Republic of Pisa to oversee its customs house in the trading colony in Bugia (modern-day Bejaia, Algeria), Leonardo joined his father in North Africa sometime during the closing decade of the twelfth century and, at his father's instigation, studied mathematics in preparation for a mercantile career. Leonardo proved more than a willing student. As he put it, "there from a marvelous instruction in the art of the nine Indian figures [Hindu-Arabic numerals and the arithmetic associated with them], the introduction and knowledge of the art pleased me ... above all else, and I learnt from ... whoever was

---

[6] Burnett, 2001, p. 251–253. There was at least one other independent translator into Latin of al-Khwārizmī's text, William de Lunis, who flourished in the first half of the thirteenth century. See Hughes, 1982, pp. 35–36. For the complete list of extant Latin manuscripts of al-Khwārizmī's *Al-kitāb*, see Lejbowicz, 2012, pp. 31–32. Islamic material on the quadratic equation first made its way into Europe, however, via *The Treatise on Mensuration and Calculation* written in Hebrew by Abraham bar Hiyya of Barcelona (also known as Savasorda) sometime before his death in 1136. Plato of Tivoli translated Abraham bar Hiyya's manuscript into Latin in 1145, the same year that Robert of Chester translated al-Khwārizmī's text.

[7] The thirteenth century witnessed much direct translation from Greek into Latin in southern Italy and on Sicily, two areas in which Greek, Latin, and Arabic were all actively in use. See Lindberg, 1978, pp. 70–75.

[8] Rashed, 1994b, p. 148 and Høyrup, 2007, pp. 41–44.

learned in it, from nearby Egypt, Syria, Greece, Sicily, and Provence, . . . to which locations I travelled . . . for much study."[9] As a result of this exposure, he became convinced of the superiority of the Arabic and other indigenous techniques to which he was exposed and sought to introduce them "to the eager, and to the Italian people above all others, who up to now are found without a minimum" of instruction in these matters. To that end, Fibonacci composed his book, *Liber abbaci* (or *Book of Calculation*), in 1202.[10]

A massive compendium, the *Liber abbaci* opened with seven lengthy chapters that represented detailed tutorials on the basic mechanics of arithmetic using the base-ten, place-valued, Hindu-Arabic number system. Through numerous examples, Fibonacci instructed his readers how to add, multiply, subtract, and divide whole numbers as well as how to deal with fractions both simple and compound. He then put these techniques to use in a series of chapters devoted to particular types of problems encountered in trade such as monetary conversion, the conversion of different weights and measures, exchange rates, division of profits, bartering and trading, and the alloying of metals for coins. The problems collected in the first thirteen chapters, while often convoluted and sometimes fanciful, essentially boiled down to solving systems of linear equations in one or more unknowns, both determinate and indeterminate, while the fourteenth chapter treated square and cube roots and their manipulation.[11]

It was not until the fifteenth and final chapter that Fibonacci introduced what he termed "the method of algebra and almuchabala," namely, treatments—both algorithmic and geometric—of the quadratic equations inspired by and similar to al-Khwārizmī's.[12] To get the flavor of Fibonacci's text and in order to compare his exposition with his predecessor's, consider the Pisan's treatment of the case where "squares and numbers are equal to roots," that is, $ax^2 + c = bx$ (recall the previous chapter). Unlike al-Khwārizmī, who, as we saw, began with an example,

---

[9] Fibonacci, 1202/2002, pp. 15–16. For the quote that follows, see p. 16.

[10] Although modern editions of this work use the spelling *Liber abaci* for this text, Leonardo titled it *Liber abbaci*. We will use his spelling to help avoid the common confusion between his book's title and the Chinese counting device, the abacus.

[11] For a discussion of some of these many and varied problems in modern notation, see Sesiano, 1999, pp. 106–120.

[12] Fibonacci, 1202/2002, p. 554.

proceeded to a gloss that effectively translated into a version of our quadratic formula, and finished with a geometrical justification, Fibonacci opened with the solution algorithm, moved to a specific example, and ended with the geometrical verification. As he explained to his readers, "when it will occur that the census [that is, the square of the unknown] plus a number will equal a number of roots, then you know that you can operate whenever the number is equal to or less than the square of half the number of roots," in other words, the quadratic equation $x^2 + c = bx$ is solvable if $c \leq (\frac{b}{2})^2$ (we would say if the discriminant is positive), otherwise not.[13]

If it is indeed solvable, then Fibonacci distinguished three cases:

> if it is equal, then half of the number of roots is had for the root of the census, and if the number with which the census is equal to the number of roots is less than the square of the half of the number of roots, then you subtract the number from the square, and [the root of] that which will remain you subtract from half the number of roots; and if that which will remain will not be the root of the sought census, then you add that which you subtracted to the number from which you subtracted, and you will have the root of the sought census.

In modern terms, his algorithm, like that of al-Khwārizmī, corresponded to the following version of the quadratic formula:

$$x = \frac{b}{2} \pm \sqrt{\left(\frac{b}{2}\right)^2 - c}. \tag{8.1}$$

Fibonacci next illustrated this in the case of the particular example $x^2 + 40 = 14x$. As he explained, "indeed half of the number of roots is 7; from the square of this, namely 49, you subtract 40 leaving 9; the root of it, which is 3, you subtract from half the number of roots, namely 7; there will remain 4 for the root of the sought census." He then checked this solution—noting that "the census is 16 which added to 40 makes 56"— before moving on to determine a second solution. "Or," he continued, "you add the root of 9 to the 7; there will be 10 for the root of the sought

---

[13] Fibonacci, 1202/2002, p. 557. The quotes that follow in the next two paragraphs are also on this page.

census," and a quick check confirmed that "the census will be 100 which added to the 40 makes 140 that is 14 roots of 100, as the multiplication of the root of 100 by the 14 yields 140." Finally, he verified his algebraic procedure using, mutatis mutandis, the same geometrical, cut-and-paste construction in terms of line segments to be found in al-Khwārizmī's text.

Fibonacci concluded the lengthy fifteenth chapter with a series of some one hundred problems aimed at giving his readers practice in the "new" method of algebra and almuchabala that he had just introduced.[14] Drawn primarily from the works of a number of Arabic authors—among them, al-Khwārizmī, Abū Kāmil, and al-Karajī—these examples include many of the form "I divided 10 into two parts ..." (recall the discussion of the work of al-Khwārizmī and Abū Kāmil in the previous chapter) and, in particular, "I separated 10 into two parts, and I divided the greater by the lesser, and the lesser by the greater, and I added the quotients, and the sum was the root of 5 *denari*."[15] This translates, in modern notation, into the two equations

$$10 = x + y,$$

$$\frac{x}{y} + \frac{y}{x} = \sqrt{5},$$

which Fibonacci then proceeded to solve simultaneously by substituting $10 - x$ for y into the second equation.

Writing everything out in words (except for the expressions for the two unknowns, for which he used letters representing line segments), he performed what we would recognize as the various algebraic operations involved in simplifying the equations that resulted step by step. First, he noted that

$$xy \left(\frac{y}{x} + \frac{x}{y} \right) = x^2 + y^2.$$

---

[14] By 1240 when he wrote his *Practica geometriæ*, Fibonacci—like his Islamic predecessor Abū Bekr (fl. between the tenth and twelfth centuries CE) in the *Liber mensurationum*, his Hebrew predecessor Abraham bar Hiyya in the *Liber embadorum*, and his successor Johannes de Muris (ca. 1290–ca. 1351–1355 also known as Jean des Murs)—also showed how to use this new algebra in the solution of a variety of geometrical problems. For a comparative study of their work, see Moyon, 2012.

[15] Fibonacci, 1202/2002, p. 587.

Then, letting $y = 10 - x$, he found that

$$x^2 + y^2 = x^2 + (10 - x)^2 = 2x^2 + 100 - 20x$$

and

$$xy\left(\frac{y}{x} + \frac{x}{y}\right) = (10x - x^2)(\sqrt{5}) = \sqrt{500x^2} - \sqrt{5x^4}.$$

Setting the two right-hand sides of these equations equal, he had

$$\sqrt{500x^2} - \sqrt{5x^4} = 2x^2 + 100 - 20x,$$

which he modified by "restoring" the negative terms on both sides to get

$$2x^2 + 100 + \sqrt{5x^4} = \sqrt{500x^2} + 20x.$$

Combining like terms on the left-hand side yielded

$$(\sqrt{5} + 2)x^2 + 100 = 20x + \sqrt{500x^2},$$

so that multiplying both sides of the equation by $\sqrt{5} - 2$ finally resulted in the quadratic equation $x^2 + (\sqrt{50{,}000} - 200) = 10x$, a problem of the type "squares and numbers are equal to roots" treated above. Following through the algorithm for solving this type of quadratic, "there will remain 5 minus the root of the difference between 225 and the root of 50,000 ... and this is the one thing, namely one of the two parts of the 10; the other truly is the number ... 5 plus the root of the difference between 225 and the root of 50,000."[16] In other words, $x = 5 - \sqrt{225 - \sqrt{50{,}000}}$ and $y = 5 + \sqrt{225 - \sqrt{50{,}000}}$. Noting that these complicated square roots could be simplified, he used a formula that can also be found in al-Karajī's work (see chapter 7) to get that $\sqrt{225 - \sqrt{50{,}000}}$ is equal to $\sqrt{125} - 10$ and therefore that the two desired numbers were $x = 15 - \sqrt{125}$ and $y = \sqrt{125} - 5$. Fibonacci thus nicely showcased his expertise at algebraic manipulation.

---

[16] Fibonacci, 1202/2002, p. 588.

That expertise also manifested itself in a series of problems later in the fifteenth chapter in which he dealt with higher powers of the unknown expressed as a cube (x^3), census census (x^4), census census census or cube cube (x^6), and census census census census (x^8), the latter being the highest power of the unknown with which he dealt. All of these arose, however, in scenarios that ultimately reduced to quadratics solvable by his established techniques.

Consider, for example, the problem he posed in this way:[17] "There are three unequal quantities; if the least is multiplied by the greatest, then the product will be the same as the middle multiplied by itself; and if the greatest is multiplied by itself it will be as much as the sum of the smallest multiplied by itself and the middle multiplied by itself;" and if the smaller is multiplied by the middle you get 10. To put this in modern terms, we have $x < y < z$ where $xz = y^2$, $z^2 = x^2 + y^2$, $xy = 10$, and we want to solve for x, y, and z.

Fibonacci attacked the problem by first setting y equal to $\frac{10}{x}$ and then squaring to get $\frac{100}{x^2}$. Substituting that value into the first equation yielded

$$xz = \frac{100}{x^2} \text{ or } z = \frac{100}{x^3}.$$

Moving to the second equation, he replaced z there by $\frac{100}{x^3}$ to get

$$\frac{10,000}{x^6} = x^2 + \frac{100}{x^2}.$$

To simplify that, he multiplied both sides by x^6 resulting in

$$10,000 = x^8 + 100x^4.$$

As easy as that is for us to say in terms of our modern notation, Fibonacci's version of that one step reads with great prolixity as

You therefore multiply all that you have by the cube cube, and to multiply by the cube cube is to multiply by the census census census;

---

[17] Fibonacci, 1202/2002, pp. 601–602. The translation presented there is, however, faulty. The authors thank the second author's colleagues, especially Greg Hays and Tony Woodman in Classics and Elizabeth Meyer in History at the University of Virginia, for their help in clarifying the translation.

therefore if we multiply by the 10,000 divided by the cube cube, that is by the census census census, there results 10,000, and if we shall multiply the census, the square of the least quantity, by the census census census, we shall thence have census census census census, and if we multiply the square of the middle quantity, namely 100 divided by the census, by the census census census, then there results 100 census census; therefore census census census [census] plus 100 census census is equal to 10,000 *denari*.[18]

Finally, treating this as a quadratic in x^4, Fibonacci completed the problem by applying the algorithm he associated with the solution of quadratics of the form $y^2 + by = c$ to find that $x = \sqrt{\sqrt{\sqrt{12,500}} - 50}$. With this, and in a "radical" tour de force, he divided 10 by this quantity to get y as $\sqrt{\sqrt{\sqrt{12,500}} + 50}$ and z as

$$\sqrt{\left(\sqrt{\sqrt{\sqrt{12,500}} - 50}\right)^2 + \left(\sqrt{\sqrt{\sqrt{12,500}} + 50}\right)^2},$$

or more simply, $\sqrt{\sqrt{\sqrt{12,500}} - 50 + \sqrt{\sqrt{12,500}} + 50}$.

Around the time he finished the *Liber abbaci* in 1202, Fibonacci had returned to settle permanently in Pisa. There, he not only continued his mathematical work but also came, by the 1220s, to be associated with a group of scholars centered in Sicily at the court of Frederick II (1194–1250).[19] King of Germany before he was crowned Holy Roman Emperor in 1220, Frederick was known as "stupor mundi" or "the astonishment of the world," a moniker he earned not only because he was one of the most powerful medieval emperors of the Holy Roman Empire but also because he was an erudite ruler and patron especially of the sciences. In particular, Frederick fostered the active translation of texts from Greek and Arabic into Latin as well as the pursuit of mathematics, astrology, and philosophy, among other areas of inquiry. That Fibonacci interacted with members of Frederick's court is clear, since he referred to several of them by name in his works dating from that decade.

[18] Fibonacci, 1202/2002, p. 602.
[19] Fibonacci, 1225/1987, p. xvi.

Michael Scot (1175–ca. 1232) was one of the Catholic priests who, like Robert of Chester, journeyed from the British Isles to the Iberian Peninsula in search of Greek and Arabic texts. A noted translator of Aristotle while in Toledo, Michael moved to Sicily to join Frederick's court at some time before the end of the 1220s.[20] Fibonacci opened the dedication and prologue of the second edition (1228) of the *Liber abbaci* in recognition of Michael's influence. "You, my Master Michael Scot, most great philosopher," he acknowledged, "wrote to my Lord [Frederick II] about the book on numbers which some time ago I composed and transcribed to you; whence complying with your criticism, your more subtle examining and circumspection, to the honor of you and many others I with advantage corrected this work."[21] Several years before Fibonacci penned this dedication, in fact, Frederick and his court had been in Pisa, and Frederick had asked specifically to meet with the Pisan mathematician. At least two of Fibonacci's surviving texts—*Flos* (or *Flower*; 1225) and the *Liber quadratorum* (or *Book of Squares*; 1225)—stemmed from this visit.[22]

Fibonacci wrote *Flos* in direct response to questions posed by John of Palermo, a member of Frederick's court, during Fibonacci's audience with the Emperor. As he related it, John asked him to solve two problems: the cubic equation we would write as $x^3 + 2x^2 + 10x = 20$ and the Diophantine system we would express as

$$x^2 + 5 = y^2 \qquad \text{and} \qquad x^2 - 5 = z^2. \tag{8.2}$$

Relative to the cubic, Fibonacci argued that, based on his understanding of Euclid Book X, the root could be neither an integer nor a fraction nor one of the Euclidean irrationals. Thus, as he put it, "because I could not solve this question by some of the above written [quantities], I strived after an approximate solution. And I found the second approximation to be one and XXII minutes and VII seconds and XLII thirds and XXXIII fourths and IV fifths and XL sixths." Using the sexagesimal writing convention adopted in chapter 2, this solution, $x = 1; 22, 7, 42, 33, 4, 40$, is accurate to the fifth place and off in the sixth by only one and a

[20] Burnett, 2001, p. 253.
[21] Fibonacci, 1202/2002, p. 15.
[22] Vogel, 1971, p. 610.

half, although Fibonacci did not hint at how he arrived at his very close approximation.[23]

As for the Diophantine problem, in *Flos* Fibonacci merely presented—without explanation—a specific solution of it, the same specific solution to be found in al-Karajī's *Badī* (recall the previous chapter). To do his work on the question justice, however, he devoted an entire treatise, the *Liber quadratorum*, to its analysis. As he explained in that work's prologue addressed to Frederick II, the problem "pertain[ed] not less to geometry than to arithmetic," and "I saw, upon reflection, that this solution itself and many others have their origin in the squares and the numbers which fall between the squares."[24] Fibonacci's analysis hinged on the observation—known to the Pythagoreans—that "all square numbers ... arise out of the increasing sequence of odd numbers."[25] In other words,

$$1 + 3 + 5 + \cdots + (2n - 1) = n^2.$$

As he noted in the context of his first proposition, this fact instantly generates "two square numbers which sum to a square," that is, what we would call a Pythagorean triple, since

$$5^2 = 1 + 3 + 5 + 7 + 9 = (1 + 3 + 5 + 7) + 9 = 4^2 + 3^2.$$

Exploiting this and other facts he proved along the way, Fibonacci was able finally to address John of Palermo's problem in his seventeenth proposition. In his eleventh and twelfth propositions, Fibonacci had proven the general number-theoretic fact that the equations $x^2 + n = y^2$ and $x^2 - n = z^2$ can be solved for integers x, y, and z only if n is what he termed a congruous number, that is, only if n is of the form

[23] See Franci and Toti Rigatelli, 1985, p. 27 (for the quotation) and Vogel, 1971, p. 610. Note Fibonacci's use of Roman, instead of Hindu-Arabic, numerals. Vogel conjectured that Fibonacci may have used the Horner method, a technique known to both Arab and Chinese mathematicians. This same problem, in the form "a root and two in number and ten parts of a root are equal to twenty parts of a square," appeared in al-Khayyāmī's *Treatise on Algebra* as part of his discussion of equations containing the inverse of the unknown. After converting it to an ordinary cubic equation, he solved it geometrically by means of the intersection of a circle and a hyperbola. On this problem, see Rashed and Vahabzadeh, 2000, p. 159. For an exposition of Fibonacci's argument, see Bashmakova and Smirnova, 2000, p. 57.
[24] Fibonacci, 1225/1987, p. 3.
[25] Fibonacci, 1225/1987, p. 4. For the example that follows, see pp. 5–6.

$ab(a+b)(a-b)$ when $a+b$ is even and $4ab(a+b)(a-b)$ when $a+b$ is odd. Moreover, he proved that congruous numbers are always divisible by 24.[26] Since the 5 in John of Palermo's query is clearly not divisible by 24, Fibonacci had demonstrated that (8.2) is not solvable in integers. As the Pisan mathematician showed in his seventeenth proposition, however, $720 = 12^2 \cdot 5$ is congruous (for $a = 5$ and $b = 4$), and since $41^2 + 720 = 49^2$ and $41^2 - 720 = 31^2$, if both equations are divided by 12^2, then $x = \frac{41}{12}$, $y = \frac{49}{12}$, $z = \frac{31}{12}$ is a *rational* solution of (8.2).[27] Described as a "first-rate scientific achievement [that] shows Leonardo as a major number theorist,"[28] the *Liber quadratorum* represents, at least from an historical point of view, an important advance beyond the *Arithmetica*, despite the fact that it seemingly had no influence on contemporaneous mathematical developments and remained largely unknown until its rediscovery in the mid-nineteenth century.

In the originality of his work, Fibonacci was rivaled by at least one known contemporary, Jordanus of Nemore (fl. 1220). Although little about his life is known with certainty, "it appears to be a fair assumption," according to historian of mathematics Jens Høyrup and based on the extant evidence, "that Jordanus taught in Paris in the earlier part of the thirteenth century, and that until the mid-thirteenth century, a 'Jordanian circle' stayed active around the same center under fairly direct influence from the master."[29] Regardless of the particulars of his life, Jordanus nevertheless survives through some dozen manuscripts that treat topics in both mathematics and mechanics and was perhaps best known for his work in the latter.[30]

Relative to mathematics in general and to algebra in particular, however, Jordanus tilled some of the same ground as Fibonacci. His work, *De numeris datis* (or *On Given Numbers*), for instance, focused on solving linear and quadratic equations, although he developed there a

---

[26] Fibonacci, 1225/1987, pp. 44–52. For more on Fibonacci's work in the *Liber quadratorum*, see Weil, 1984, pp. 10–14.

[27] Fibonacci also treated, albeit in different and original ways, problems that may be found in Diophantus's *Arithmetica*. See, for example, Fibonacci, 1225/1987, pp. 93–105, which, while identical to Diophantus's II.19 discussed in chapter 4, is solved in quite a unique manner. According to Rashed, 1994b, pp. 156–160, Fibonacci likely drew from a tenth-century Arabic Diophantine tradition.

[28] Vogel, 1971, p. 610.

[29] Høyrup, 1988, p. 351.

[30] See Brown, 1978, pp. 190–197 for an overview of Jordanus's work on the theory of statics.

presentational style and approach significantly different from those of his Pisan counterpart. Whereas Fibonacci's treatment of algebra—influenced by al-Khwārizmī's methods—may be considered elementary, Jordanus's was advanced, thereby building on those more elementary introductions.[31] What's more, whereas Fibonacci's text dealt with algebra only in its fifteenth and final chapter, Jordanus's was devoted solely to the topic.

A work in four books, *De numeris datis* has been described as "not only a work on algebra, but also a work written in hidden dialogue with the Islamic algebraic tradition."[32] Jordanus seems to do for algebra what Euclid had done for geometry in the *Data*, namely, *demonstrate* that the problem-solving techniques that had developed and that had long been cast in the context of particular problems were, in fact, general and correct. He sought to do this, moreover, based on the firm foundation that he had built for arithmetic and its practice in his text, *Arithmetica*. There, Jordanus had crafted a Euclidean-style presentation of arithmetic in terms of definitions, axioms, and postulates that was carefully ordered in ten books and that aimed "to elevate this subject ... to the level of Euclidean order."[33] If algebra was firmly grounded in arithmetic and arithmetic was solidly and rigorously established, then algebra was solidly and rigorously established as well. Through *De numeris datis*, Jordanus thus sought to transform algebra from what Høyrup has termed a "mathematically dubious" practice "into a genuine piece of *mathematical theory*."[34]

In his text, Jordanus first stated a general problem, then provided an algorithm for solving it, and finally illustrated it in terms of a specific numerical example. To get a flavor for his work, consider his proposition I.5: "If a number is separated into two parts whose difference is known and whose product is also known, then the number can be found."[35] In modern notation, this problem considers the two equations $x - y = b$ and $xy = d$, where b and d are given, and asks that $x + y$ be found. Jordanus's solution hinges on the identity we would write as $(x - y)^2 + 4xy = (x + y)^2$.

---

[31] Hughes, 1981, pp. 3–4. Mahoney, 1978, pp. 159–162 provides a brief comparison of the work of Fibonacci and Jordanus.
[32] Høyrup, 1988, p. 332–333.
[33] Høyrup, 1988, p. 327.
[34] Høyrup, 1988, p. 336 (his emphasis).
[35] Hughes, 1981, p. 129.

To set up the problem, Jordanus used the same single-letter notation that he employed in his *Arithmetica* rather than the two-letter notation of line segments found in the texts of al-Khwārizmī or Fibonacci. For Jordanus, however, the single letters, while representative and general, were not symbols that were in any sense manipulated. Letters were labels and were assigned to the components of a problem in a roughly alphabetical way from start to finish. Using them, he first described his algorithm and then illustrated it in the context of a particular example. As he put it,

> As before, [let the given number abc be divided into ab and c.] Then b, the difference of the parts is given as is d, their product, whose double is e. To the double of e is added h, which is the square of the difference, and the sum is f, which will be the square of abc. So abc is given.[36]

To put this in more modern notation, Jordanus's algorithm sets $x - y = b$ and $xy = d$. Then $e = 2d = 2xy$ and $f = 2e + b^2 = 4xy + (x - y)^2$ [= $(x + y)^2$] so that $abc = x + y$ is the number to be found.

By way of example, Jordanus let the difference of the parts be 6 and their product 16. Twice this is 32, and twice again is 64. To this he added 36, the square of 6, to make 100. The root of this is 10, the number that was separated into 8 and 2.[37] In other words, letting $x - y = b = 6$ and $xy = d = 16$, we have $4xy = 64$ and $(x - y)^2 = 36$ so that $4xy + (x - y)^2 = 64 + 36 = 100$. Thus, since $4xy + (x - y)^2 = (x + y)^2$, it is the case that $x + y = \sqrt{100} = 10$. At this point, Jordanus decided also to determine x and y, for which he used proposition I.1. In the present context, the algorithm provided there translates as, find x and y given $x + y = 10$ and $x - y = 6$. Since $y + (x - y) = x$, it follows that $y + y + (x - y) = x + y$. Hence, $10 - 6 = 2y$ and $y = \frac{10-6}{2} = 2$ so that $x = 8$.[38]

As he used I.1 in his solution of I.5, so Jordanus used I.5 in I.7, his treatment of what we would write as the second-degree equation

---

[36] See Hughes, 1981, p. 59 for the Latin; this is our translation. Jordanus's use of multiple letters such as abc and ab to denote certain components of this problem may be understood geometrically, with each component representing a line segment. Thus, abc represents the line composed of the adjacent, consecutive line segments a, b, and c. For this interpretation of Jordanus's notation, see Puig, 2011, pp. 36–39.

[37] Hughes, 1981, p. 129.

[38] Compare Hughes, 1981, p. 127.

$x^2 + bx = c$. Again, unlike al-Khwārizmī or Fibonacci, whose algorithmic solutions of this particular case translate in modern terms as

$$x = \sqrt{\left(\frac{b}{2}\right)^2 + c} - \frac{b}{2},$$

Jordanus employed a different—arithmetic as opposed to geometric—algorithm. Recognizing $x^2 + bx$ as the product of $x + b$ and x, he acknowledged that the difference b and the product c of these two numbers were thus given, precisely as in I.5. Using the algorithm developed there yielded $x + (x + b) = \sqrt{4c + b^2}$, which is equivalent, rather, to

$$x = \frac{1}{2}\left(\sqrt{4c + b^2} - b\right).$$

Jordanus, in rigorizing and arithmetizing algebra, so to speak, was striking out on a path different from both his Islamic predecessors and his Pisan contemporary. "His replacement of traditional Latin mathematics with works oriented toward Ancient standards was an acceptance of the 'new learning' and a rejection of scholarly traditionalism" and reflected his engagement, again in contrast to Fibonacci, in the professional environment of the incipient universities.[39]

THE DIFFUSION OF ALGEBRAIC THOUGHT ON THE ITALIAN PENINSULA AND ITS ENVIRONS FROM THE THIRTEENTH THROUGH THE FIFTEENTH CENTURIES

Despite the fact that he was likely in a university setting—or actually perhaps because of it—Jordanus and his work ultimately had no influence. The nascent medieval universities, dominated as they were by the traditional Latin quadrivium, were unlikely seedbeds for the new kind of approach Jordanus was sowing. Moreover, the small mid-thirteenth-century "Jordanian circle" in Paris "accepted his theorems, the letter symbolism for numbers, his rigor" but "did not share his care for

---

[39] Høyrup, 1988, p. 334 (for the example) and p. 353 (for the quotation).

Ancient purity and standards."[40] Høyrup put it aptly when he wrote that Jordanus's "standards were defined by those Ancient mathematical works to which he had access; therefore they reflected the needs and the structure of a social community long since deceased: the circle of corresponding mathematicians connected with the Alexandria Museum and schools ....Jordanus stands out as a Don Quixote, fighting a battle in which no one really needed his victory."[41]

If Jordanus, the academician, was, in some sense, both an anachronism and a trailblazer, Fibonacci, the merchant, sought to share what he viewed as the highly superior Islamic "method of algebra and almuchabala" with his fellow Italians.[42] That goal was embraced by his successors in the fourteenth and fifteenth centuries. Their textbooks, the *libri d'abbaco* of the *botteghe d'abbaco* or "abbacus schools," aimed to convey to those engaged in mercantilistic pursuits—and in the vernacular as opposed to in the learned language of Latin—at least the high points of the kind of mathematics to be found in Fibonacci's *Liber abbaci* and elsewhere.[43]

The earliest known work of this type containing a treatment of algebra dates from 1307 and was written in Italian in the French city of Montpellier by one Jacopo of Florence. Entitled *Tractatus algorismi* (or *Treatise on Algorism*), it presented "the algebraic 'cases' in a different order than al-Khwārizmī, Abū Kāmil, and Leonardo Fibonacci" and gave examples following those cases that "also differ from those of the same predecessors."[44] The evidence is compelling that Jacopo's text stemmed not directly from Fibonacci's *Liber abbaci* but rather emerged— like Fibonacci's work and the entire Italian abbacus tradition—from within the broader mathematical context of mathematical practices

---

[40] Høyrup, 1988, p. 342. This group consisted of one Richard of Fournival, who had Jordanus's works copied and collected in his personal library, the mathematician and natural philosopher Campanus of Novara (ca. 1220–1296), and possibly Roger Bacon (ca. 1214–1294), who is believed to have taught at the University of Paris between 1241 and 1246.

[41] Høyrup, 1988, p. 343.

[42] Fibonacci, 1202/2002, p. 554.

[43] Høyrup, 2007, pp. 27–44. The terminology *libro d'abbaco* and *botteghe d'abbaco* is reminiscent of the title of Fibonacci's book and, as noted above, has no connection to the calculating device known as the abacus.

[44] Høyrup, 2007, p. 3. Here, the English translation "algorism" of the title of Jacopo's book refers to the introduction of computation using Hindu-Arabic numerals. It should not be confused with the term "algorithm," which derives from the Latinization of al-Khwārizmī's name.

indigenous to Provence and Catalonia.[45] A plurality of algebras, subtly different in their implementations and ultimate goals, were in practice in Renaissance Europe.[46]

The *Tractatus*, Høyrup has argued convincingly, served as the model for a number of subsequent thirteenth-century *libri d'abbaco*.[47] In particular, another Florentine, Paolo Gerardi, who lived and worked in Montpellier some two decades after Jacopo penned his text, composed the *Libro di ragioni* (or *Book of Problems*) in the Italian vernacular in 1328. Gerardi's text comprised some 193 examples, the first 178 of which, like the problems in the first fourteen chapters of Fibonacci's *Liber abbaci*, aimed at illustrating calculational techniques for solving practical, commercially oriented problems cast in terms of Hindu-Arabic numerals.[48] The last fifteen problems, however, dealt with solving algebraic equations.

Gerardi opened this part of his text with a treatment of the six types of linear and quadratic equations. For example, he gave an algorithm for solving a quadratic of the type $ax^2 + bx = c$ and illustrated it in the context of this quasi-real-world problem concerning interest: "A man loaned 20 lire to another for two years at compound interest. When the end of 2 years came he gave him 30 lire. I ask you at what rate [were the lire] loaned [at *denari* per lira] per month?"[49] Knowing that 1 lira equaled 20 *soldi* and 1 *soldo* equaled 12 *denari*, this problem boiled down, as Gerardi argued, to solving the equation $x^2 + 40x = 2000$. As he expressed it, the answer was $x = \sqrt{600} - 20$, a rather unusual irrational value for the number of *denari* per lira per month!

Gerardi's treatment of the equations of degree at least two presented no types of solutions that were not already to be found in the texts of Fibonacci or Jacopo of Florence. It moved beyond the work of his predecessors, however, by also treating nine types of cubics, five of which are not reducible to lower degrees. The language here is carefully chosen. Gerardi may have *treated* cubic equations, but that does not mean that he treated them *correctly*.

---

[45] Høyrup, 2007, pp. 41–44.

[46] On this notion of a "plurality of algebras," see the essays in Rommevaux, Spiesser, and Massa Esteve, ed., 2012.

[47] Høyrup, 2007, pp. 159–182. For a survey of fourteenth-century Italian algebra manuscripts, see Franci, 2002.

[48] Van Egmond, 1978, pp. 156–162.

[49] Van Egmond, 1978, p. 176.

In modern notation, Gerardi considered these nine cubic equations:

$$ax^3 = d, \quad ax^3 = \sqrt{d}, \quad ax^3 = cx, \quad ax^3 = bx^2, \quad ax^3 = bx^2 + cx,$$

$$ax^3 = cx + d, \quad ax^3 = bx^2 + d, \quad ax^3 = bx^2 + cx + d,$$

$$ax^3 + bx^2 = cx.$$

Of them, he solved the first two by taking cube roots; the next three as well as the last one reduce to equations of degrees lower than the third and so are solvable using the algorithms that may be found in Fibonacci's and Jacopo's texts and that Gerardi laid out as well; only the sixth, seventh, and eighth (middle row above) are irreducible cubics. In introducing examples such as the latter, Gerardi moved beyond both of his predecessors; neither had dealt with *irreducible* cubics.

Consider Gerardi's treatment of $ax^3 = cx + d$. "When the *cubi* [that is, the cubes or the cube term in the unknown] are equal to the *cose* [that is, the "things" or the linear term in the unknown] and to the number," he explained, "you must divide by the *cubi* [that is, the coefficient of the cube term] and halve the *cose* [that is, the coefficient of the linear term] and multiply it by itself and add it onto the number, and the root of this plus half the *cosa* will equal the *cosa*."[50] This algorithm, for which he gave no justification, amounted to

$$x = \sqrt{\frac{d}{a} + \left(\frac{c}{2a}\right)^2} + \frac{c}{2a}.$$

If this seems familiar, that is because it is precisely the algorithm for solving a *quadratic* of the form $ax^2 = cx + d$. Gerardi gave exactly the same algorithmic solution (replacing c by b) for the case of the cubic $ax^3 = bx^2 + d$.

His "solution" of the final irreducible cubic $ax^3 = bx^2 + cx + d$ equally reflected his lack of understanding. In that case, the algorithm he provided reduced to

$$x = \sqrt{\frac{c + d}{a} + \left(\frac{b}{2a}\right)^2} + \frac{b}{2a},$$

[50] Van Egmond, 1978, p. 180. For the quote that follows, see p. 181.

namely, his solution of the previous two cases *except* that he added the "extra" coefficient c to the numerator of the first fraction under the square root. Had Gerardi checked his work, much less provided any sort of a justification for his algorithms, he would presumably have realized that he had not generated valid solutions in these three cases. This and other facts have led at least one historian to conclude "that the author of this text was a relatively ill-educated calculator, a person who was interested in solving numerical problems, but little more."[51]

While that may have been the case, Gerardi nevertheless extended in his *Libro di ragioni* a tradition anchored in Jacopo's *Tractatus* of treating cubic equations—and sometimes even the very same ones—that persisted through the end of the fifteenth century in works such as the anonymous *Trattato dell'alcibra amuchabile* (or *Treatise on Algebra and Almuchabala*) (ca. 1365) and the *Questioni d'algebra* (or *Algebraic Problems*) written by one Gilio of Siena in 1384. Others—like Gerardi's Florentine contemporary, Biagio the Elder (d. ca. 1340) about whom we know only through Benedetto's work more than a hundred years later, but, more importantly, Master Dardi of Pisa in his *Aliabraa argibra* (or *Algebra*) of 1344—extended the algebraic analysis to various cases of equations of the fourth degree.[52] There was thus a centuries-long Western tradition, in Italy at least, of working with equations of degree higher than two before the celebrated sixteenth-century work of Niccolò Tartaglia, Girolamo Cardano, and others (see the next chapter).[53]

Dardi's text is of particular interest here, given that it was the first book to have come down to us *written in the vernacular* and devoted *solely to algebra* as opposed to commercial problem solving with a nod toward more theoretical concerns.[54] After defining the terms *cosa, censo, cubo,* and *censo di censo*—that is, the first four powers of the unknown—and laying out the basic rules for algebraic calculation such as the multiplication of binomials, Dardi proceeded to enumerate and solve 194 problems. All but 17 of them amount to equations with only two or three terms made up

---

[51] Van Egmond, 1978, p. 162.

[52] Jacopo had also given rules for dealing with various reducible cases of the quartic. See Høyrup, 2007, pp. 321–323.

[53] Van Egmond, 1978, p. 163; Franci and Toti Rigatelli, 1983 and 1985; Gilio of Siena, 1384/1983, pp. xiv–xv; Van Egmond, 1983; and Dardi of Pisa, 1344/2001, pp. 21–33.

[54] Franci and Toti Rigatelli, 1985, p. 36. The analysis of Dardi's work presented here follows that on pp. 36–39. Compare Van Egmond, 1983 and van der Waerden, 1985, pp. 47–52. For the full edition, see Dardi of Pisa, 1344/2001, pp. 37–297.

of the first four powers of x as well as the square and cube roots of those powers. Dardi typically arrived at his solutions by making substitutions such as $x^3 = y^2$ or $x^2 = y$ or by raising to appropriate powers in order to clear out roots.

Four additional problems, however, merit special consideration. Inserted near the end of the text between his treatments of problems 182 and 183, their solutions are not, as Dardi acknowledged, as generally applicable as the solutions he had presented elsewhere. "Here below," he explained, "some chapters are written the equations of which are valid only for their problems ....Therefore we have put them aside from the chapters we have said before which are perfectly regulated to their equations whatever are the problems from which the equations come."[55]

Consider the first of these interpolated problems. Like that considered above from Gerardi's *Libro di ragioni*, it involves the calculation of an interest rate, but with a "cubic" twist: "A man lent 100 lire to another and after 3 years he gives him 150 lire for principal and interest at yearly compound interest. I ask you at what rate was the lira lent per month?" Setting x *denari* equal to the interest that one lira earns in one month (where, recall, 1 lira = 20 *soldi* and 1 *soldo* = 12 *denari*), 1 lira bears $\frac{1}{20}x$ lire (= x *soldi* = $12x$ *denari*) in interest in one year. Thus, Dardi's scenario yields the equation $100 + 15x + \frac{3}{4}x^2 + \frac{1}{80}x^3 = 150$ or $x^3 + 60x^2 + 1200x = 4000$. Thinking of the general cubic of this form, namely, $ax^3 + bx^2 + cx = d$, the solution algorithm that Dardi gave for this problem may be expressed as

$$x = \sqrt[3]{\left(\frac{c}{b}\right)^3 + \frac{d}{a}} - \frac{c}{b},$$

which, while not a solution for this general type of cubic, *is* a solution for any of the cubic equations arising in the calculation of interest over a three-year period, since any such problem results in a cubic equation, like this one, where the left-hand side can be completed to a cube.

[55] Franci and Toti Rigatelli, 1985, p. 38. The quote that follows in the next paragraph is also on this page.

Thus, since

$$x^3 + 60x^2 + 1200x = 4000,$$

$$x^3 + 60x^2 + 1200x + 8000 = 4000 + 8000,$$

$$(x + 20)^3 = 12{,}000,$$

it is easy to see why Dardi's "formula" makes sense.

The second, third, and fourth of Dardi's interpolated problems involve quartic equations, and they, too, are solved using algorithms that yield—in the special numerical cases considered—correct solutions.[56] In modern notation, they translate as

$$ax^4 + bx^3 + cx^2 + dx = e, \qquad x = \sqrt{\left(\frac{d}{b}\right)^2 + \frac{e}{a}} - \sqrt{\frac{d}{b}},$$

$$ax^4 + dx = bx^3 + cx^2 + e, \qquad x = \sqrt[4]{\left(\frac{c}{2a}\right)^2 + \frac{e}{a} + \frac{b}{4a}} - \sqrt{\frac{d}{2b}},$$

$$ax^4 + cx^2 + dx = bx^3 + e, \qquad x = \sqrt[4]{\left(\frac{c}{2a}\right)^2 + \frac{e}{a} + \frac{b}{4a}} - \sqrt{\frac{d}{2b}}.$$

The four interpolated problems presented in Dardi's *Aliabraa argibra* clearly piqued the interest of subsequent mathematical authors, since they reappeared, often verbatim, in manuscripts throughout the fourteenth and fifteenth centuries. However, the caveat that Dardi was careful to articulate—namely, that the solutions were not general and applied only to the particular problems at hand—was lost in transmission.[57] While this may have reflected a certain lack of comprehension of the algebraic art on the part of Dardi's successors as well as an uncritical acceptance of work viewed as authoritative, the relatively large number of copies of his text—either in whole or in part that date into and through the fifteenth century—suggests sustained awareness of more than just his

[56] For transcriptions of all four problems, see Dardi of Pisa, 1344/2001, pp. 269–274 as well as Van Egmond, 1983, p. 416
[57] Franci and Toti Rigatelli, 1985, p. 39.

four interpolated problems.[58] Fourteenth- and fifteenth-century Italian abbacists, like their Islamic predecessors, persistently engaged with the problem that we would recognize as solving equations of degrees three and higher, even if they only succeeded in obtaining solutions in various special cases.

The work of at least one other noted fourteenth-century mathematician, the Florentine Antonio de'Mazzinghi (1353–1383), has come down to us because of its interest to at least one later mathematician. In 1463, Maestro Benedetto also of Florence wrote a *Trattato di praticha d'arismetrica* (or *Treatise on the Practice of Arithmetic*) in sixteen books that may be viewed as a compendium of the mathematical knowledge that had thus far issued from the abbacus schools of the fifteenth century.[59] Like Fibonacci's *Liber abbaci*, Benedetto's *Trattato* opened with numerous lengthy chapters— ten in all—on the basic principles of arithmetic cast in terms of the then usual types of mercantile problems. Next followed two chapters on issues, like working with roots, needed in the study of algebra per se that then occupied chapters thirteen through fifteen. The treatise closed with a treatment of Fibonacci's results on squares. In particular, Benedetto devoted the three parts of the fifteenth chapter to the work of three of his esteemed predecessors: Fibonacci, one Giovanni di Bartolo, and his own teacher, Antonio de'Mazzinghi, who ran an abbacus school in Florence.[60] The third part contained forty-four problems from Antonio's book, entitled *Fioretti* (or *Little Flowers*), eight of which were commercial problems while the remaining thirty-six were theoretical. Among the latter, Antonio considered basically three types of scenario: (1) divide a given number into two or three parts such that ..., (2) find two numbers such that ..., and (3) find three, four, or five numbers in continuous proportion such that ....[61]

To get a feel for Antonio's algebraic virtuosity as well as for his willingness to break with tradition, consider his treatment of this problem (cast in modern notation):[62] Divide 10 into two parts u and v such that

---

[58] On the different editions of Dardi's text, compare Van Egmond, 1983, pp. 419–420 and Dardi of Pisa, 1344/2001, p. 25. Dardi's text was even translated into Hebrew by the Jewish mathematician, Mordecai Finzi, in Mantua in 1473.

[59] Franci and Toti Rigatelli, 1983, p. 299.

[60] Franci and Toti Rigatelli, 1983, p. 309 and 1985, p. 40.

[61] Franci and Toti Rigatelli, 1983, p. 312.

[62] The treatment of this example follows Franci and Toti Rigatelli, 1985, pp. 43–44.

$\sqrt{97 - u^2} + \sqrt{100 - v^2} = 17$. Antonio set $u = 5 + x$ and $v = 5 - x$ to get

$$\sqrt{72 - 10x - x^2} + \sqrt{75 + 10x - x^2} = 17.$$

He then exploited the following fact: $\sqrt{a} + \sqrt{b} = \sqrt{a + b + 2\sqrt{ab}}$. In his setting, he thus had not only

$$2\sqrt{ab} = 2\sqrt{(72 - 10x - x^2)(75 + 10x - x^2)}$$
$$= \sqrt{21{,}600 + 4x^4 - 988x^2 + 120x}$$

and

$$a + b = (72 - 10x - x^2) + (75 + 10x - x^2) = 147 - 2x^2,$$

but also

$$\sqrt{a} + \sqrt{b} = \sqrt{97 - u^2} + \sqrt{100 - v^2} = 17,$$

so that $\left(\sqrt{a} + \sqrt{b}\right)^2 = 289$. Therefore,

$$\sqrt{21{,}600 + 4x^4 - 988x^2 + 120x} = 289 - 147 + 2x^2.$$

Squaring both sides and simplifying yields $1556x^2 + 120x = 1436$. Finally, the standard algorithm for this type of quadratic equation produces the root $x = \frac{359}{389}$, yet Antonio not only recognized that $x = -1$ is another root but also used that *negative* root—despite the general avoidance in his day of negative numbers much less negative roots—to conclude that $u = 4$ and $v = 6$.[63]

In another radical departure from his predecessors and contemporaries, Antonio, in a number of his problems, also actually used *two distinct* unknowns which he termed *cosa* and *quantità*. Consider, for example, his approach to this problem: "Find two numbers such that multiplying the one into the other makes 6 and their squares are 13, that is adding

---

[63] On the appearance of negative solutions in medieval mathematics, see Sesiano, 1985.

together the squares of each of them we obtain 13. We ask [what] are those numbers."[64] In modern notation, he sought two numbers s and t that satisfy

$$st = 6,$$
$$s^2 + t^2 = 13.$$

To find them, Antonio let "the first number be one *cosa* minus the root of some *quantità*, [and] the other be one *cosa* plus the root of some *quantità*." In other words, he set

$$s = x - \sqrt{y},$$
$$t = x + \sqrt{y}.$$

He then used the second of the initial conditions to get

$$13 = s^2 + t^2 = (x - \sqrt{y})^2 + (x + \sqrt{y})^2 = 2x^2 + 2y$$

or $y = \frac{13}{2} - x^2$. This then yielded

$$s = x - \sqrt{\frac{13}{2} - x^2},$$

$$t = x + \sqrt{\frac{13}{2} - x^2},$$

which, in light of the first initial condition gave

$$6 = \left(x - \sqrt{\frac{13}{2} - x^2} \right) \left(x + \sqrt{\frac{13}{2} - x^2} \right) = x^2 - \frac{13}{2} + x^2 = 2x^2 - \frac{13}{2}$$

[64] Franci and Toti Rigatelli, 1985, p. 44. The following quote and a discussion of Antonio's solution may also be found on this page. See also Heeffer, 2010a, b for more on this algebraic innovation. As Heeffer argues, "the importance of the use of multiple unknowns in the process leading to the concept of an equation cannot be overstated." See Heeffer, 2012, p. 130.

or $x^2 = \frac{25}{4}$. Hence,

$$s = \frac{5}{2} - \sqrt{\frac{13}{2} - \frac{25}{4}} = 2,$$

$$t = \frac{5}{2} + \sqrt{\frac{13}{2} - \frac{25}{4}} = 3.$$

Benedetto's *Trattato*, in which many of Antonio de'Mazzinghi's problems were preserved, represents only one example of a fifteenth-century mathematical compendium. Perhaps the best known, the *Summa de arithmetica, geometria, proportioni, e proportionalita* (or *Compendium on Arithmetic, Geometry, Proportions, and Proportionality*) (1494), was written by Luca Pacioli (ca. 1445–1517) over the course of the closing decades of the century. Born into a family of modest means in the Tuscan town of Borgo San Sepolcro, Pacioli came under the influence, at different moments in his life, of his fellow townsmen, the noted painter Piero della Francesca (ca. 1415–1492) (who allegedly painted Pacioli at least twice) and the humanist scholar, artist, and architect Leon Battista Alberti (1404–1472).[65] While Piero may have given him early instruction in mathematics, Alberti helped him to establish himself first in Venice and then in Rome in the 1460s and 1470s. In Venice, in particular, Pacioli served the wealthy merchant, Antonio Rompiansi (also spelled Rompiasi), as business assistant and tutor to the three Rompiansi sons, while availing himself of the educational opportunities Venice had to offer. With its commercial and seafaring focus, Venice had been the first of the Italian city-states to support chairs in the mathematics of navigation and to fund public lectures in algebra. Moreover, "as early as 1446 public schools were founded [there] for the teaching of the humanities, and public readers were subsidized to expound the scriptures and translate them into the vulgar tongue."[66] Venice was thus ideally suited to the needs of a young man talented in mathematics, anxious to educate himself further, and desirous of earning a living compatible with his interests. Pacioli's first mathematical text, in fact, was an arithmetic written in 1470 and dedicated to the young Rompiansis.

---

[65] On Pacioli's life, see Taylor, 1942. Comments sprinkled throughout Pacioli's mathematical writings provided many of the details presented there. It is interesting to note that Piero wrote an undated and untitled *trattato d'abbaco* at some point in the latter half of his life.

[66] Taylor, 1942, p. 53.

Figure 8.2. Painting of Luca Pacioli teaching by Jacopo di Barbari (early sixteenth century), from the Capodimonte Museum of Naples, courtesy Wikimedia.

After some six years in the City of Bridges, Pacioli followed Alberti to Rome. When that association came abruptly to an end with Alberti's death in 1472, Pacioli decided to enter the priesthood. At some point in the 1470s, he joined the Franciscan order and began the life of an itinerant scholar and lecturer, teaching at the University of Perugia for stretches in the 1470s and 1480s, in Florence and Rome at various times in the 1480s, and in Naples in the 1480s and 1490s. This itinerancy, as well as Pacioli's focus on his scholarly writing as opposed to his monastic duties, got him into trouble with his order in the early 1490s, at precisely the time that he was finishing his *Summa* and preparing it for a Venetian press.

A massive, 600-page compendium, the *Summa* covered the theoretical and practical arithmetic and algebraic material to be found in Fibonacci's *Liber abbaci* as well as in the texts of the subsequent *maestri d'abbaco*. It also contained accounts of weights and measures and of double-entry

Figure 8.3. Title page of Pacioli's *Summa*, from the Smith Collection of the Rare Book and Manuscript Library, Columbia University.

bookkeeping as well as a summary of Euclidean geometry.[67] Its treatment of algebra, in particular, was important. "By reformulating algebraic derivations of abbacus masters as theorems of algebra, and using Euclid's

---

[67] For a brief overview of the mathematical contents of the *Summa*, see Speziali, 1973, pp. 95–100. For an analysis of Pacioli's sources, and especially his reliance on texts of the *maestri d'abbaco*, see Heeffer, 2010a.

theorems for algebraic quantities," historian of mathematics Albrecht Heeffer has convincingly argued, Pacioli "introduce[d] a new style of argumentative reasoning which was absent from abbacus algebra."[68]

Pacioli's book, moreover, had the distinction of being both one of the few mathematical texts among the roughly ten percent of *scientific* incunabula—or books printed in or before 1500—and the earliest printed work of a general mathematical nature.[69] Although hardly original—Pacioli himself acknowledged his indebtedness to Euclid, Fibonacci, and many, many others—the very fact that the *Summa* was *printed* made it highly influential. In the words of historian of science George Sarton, the printing press had made it "possible not simply to publish a text, but to standardize it" by "transmitting *identical* knowledge" and thereby allowing scholars "to *accumulate* that knowledge under the same conditions" and enter into competition with others "as it were, without handicap."[70] In the case of Pacioli's text, that standardization involved, among other things, the very way in which mathematics was conveyed. Its actual printing, moreover, allowed mathematics to become more democratic, that is, accessible to a more broadly literate readership.

Although the *Summa* is certainly not short on text, Pacioli employed a kind of shorthand in places for mathematical expressions that a Fibonacci would have written out completely in words. Although the details of the problem behind it need not concern us here, consider, for example, the passage on the left-hand side of figure 8.4 from the *Summa*. Knowing that co. and ce. are abbreviations for the Italian words "cosa" or "thing" and "censo" or "wealth," respectively, that m̄. and p̄. abbreviate the Italian words "meno" or "less" and "più" or "more," respectively, that R̪ₓ abbreviates the Latin word "radix" or "root," and that v indicates "all that follows," this passage translates into modern notation as the right-hand side of figure 8.4. This kind of abbreviated style was adopted and adapted by Pacioli's successors throughout the sixteenth century[71] as was the sense

---

[68] See Heeffer, 2010a, p. 19 for the quote and pp. 16–18 for a specific example of Pacioli's kind of reinterpretation of the work of the *maestri d'abbaco*.

[69] See Sarton, 1938, p. 62 and Bühler, 1948, p. 164. Of the roughly 30,000 incunabula, some 3100 are scientific.

[70] Sarton, 1938, p. 60 (his emphases).

[71] The use of p̄, m̄, and R̪ₓ was, in fact, already becoming common in fifteenth-century Italian manuscripts before Pacioli's work was actually printed. See Flegg, Hay, and Moss, ed., 1985, 334. Heeffer, 2010a, pp. 5–6 points out interesting differences between the printed version of the *Summa*, which employs abbreviations, and an earlier manuscript of a text on arithmetic

p^a.	1.co.m̄.R̲ₓv.1.ce.	m̄.36	1st.	$x - \sqrt{(x^2 - 36)}$	
2^a.	6		2nd.	6	
3^a.	1.co.p̄.R̲ₓv.1.ce.	m̄.36	3rd.	$x + \sqrt{(x^2 - 36)}$	
	2.co.p̄.6	216		$2x + 6$	216
	2.co.	210		$2x$	210
	valor rei	105.		value of x	105.

Figure 8.4. A sample passage from Pacioli's *Summa* and its translation into modern notation. (See Cajori, 1928, p. 110 and Parshall, 1988a, pp. 140–141.)

that he conveyed of key problems that remained to be solved. In fact, the latter resulted, among Pacioli's sixteenth-century Italian successors, in an example of precisely the kind of competition that Sarton's analysis implied.

THE DIFFUSION OF ALGEBRAIC THOUGHT AND THE DEVELOPMENT OF ALGEBRAIC NOTATION OUTSIDE OF ITALY

Before considering their work, however, it is important to realize that the Italian Peninsula was not the only site of the diffusion and development of algebraic thought in fifteenth- and sixteenth-century Europe. Algebraic ideas circulated quite probably from Italy to France, the Iberian Peninsula, Germany, and England, where they both became part of shared local algebraic literatures—that, as in Italy, often served the needs of developing economies—and inspired a variety of algebras that depended in their particulars on the individuals practicing them.[72] In the commercial city of Lyon in France, for example, Nicolas Chuquet (d. ca. 1487) composed a three-part treatise aptly entitled *Triparty* (1484) that aimed, more like the Italian *libri d'abbaco* than like the work of his thirteenth-century, Paris-based predecessor Jordanus de Nemore, to

---

and algebra written by Pacioli while he was teaching in Perugia, which employs a consistent symbolism.

[72] For a more detailed overview of the algebraic work done elsewhere in Europe at this time, see Katz, 2009, pp. 389–399. On the notion of a plurality of algebras as reflected in early printed European texts, see Stedall, 2012.

provide instruction in basic arithmetic and algebraic techniques.[73] That work informed the algebraic ideas that the younger Lyonnais, Étienne de la Roche (1470–1530), published in 1520 in his text, *L'Aristmetique*, the first *printed* book on algebra written in French.[74] In England, Oxford-trained Robert Recorde (1510–1558), through his *Whetstone of Witte* (1557), brought the algebraic thought and notational conventions that had been developing in Germany to the British Isles, also introducing the modern symbol for equality: "To avoid the tedious repetition of these words—is equal to—I will set as I do often in work use, a pair of parallels, or gemow [twin] lines of one length, thus $=$, because no two things can be more equal."[75] A few years earlier, and drawing directly from the work of Pacioli and the *maestri d'abbaco*, the Portuguese mathematician Pedro Nuñes (1502–1578), began work on the manuscript he would eventually publish as the *Libro de algebra* (or *Book on Algebra*) (1567). Both as Royal Cosmographer under King John III (the Pius) of Portugal and as professor of mathematics at the University of Coimbra, Nuñes gave instruction in the art of navigation to those associated with Portugal's merchant and naval fleets. His *Libro de algebra* provided the mathematical underpinnings of that instruction—and much more—adopting Pacioli's abbreviated notational style and treating the solution not only of linear and quadratic equations but also that of a cubic equation of the type $x^3 + cx = d$ following the spectacular mid-sixteenth-century work of the Italians Niccolò Tartaglia and Girolamo Cardano (see the next chapter).[76]

In Germany, Christoff Rudolff (ca. 1500–ca. 1550) published *Coss* around 1525, one of the first works in German on algebra. Taking his title from the Italian word for "thing" that the *maestri d'abbaco* (and, as we have just seen, Luca Pacioli and Pedro Nuñes as well) used for

[73] Large portions of Chuquet's *Triparty* may be found in English translation in Flegg, Hay, and Moss, 1985. For an analysis of his work in the context of French commercial arithmetic, see Spiesser, 2006.

[74] On de la Roche's algebraic ideas, see Heeffer, 2012.

[75] Recorde, 1557, f. Ff1^r and Katz, 2009, p. 396.

[76] Nuñes's *Libro de algebra* is available online in the Google Book archive at http://www.archive.org/details/librodealgebrae00nunegoog. For his account of the cubic case, see Nuñes, 1567, pp. 333–341. For recent analyses of Nuñes's work—in particular his sources, his use of a second unknown, and his sense of algebra as both a science superior to geometry and as an art for solving mathematical problems—see Labarthe, 2012 and Spiesser, 2012. For more on the development of algebra in Spain, see Massa Esteve, 2012.

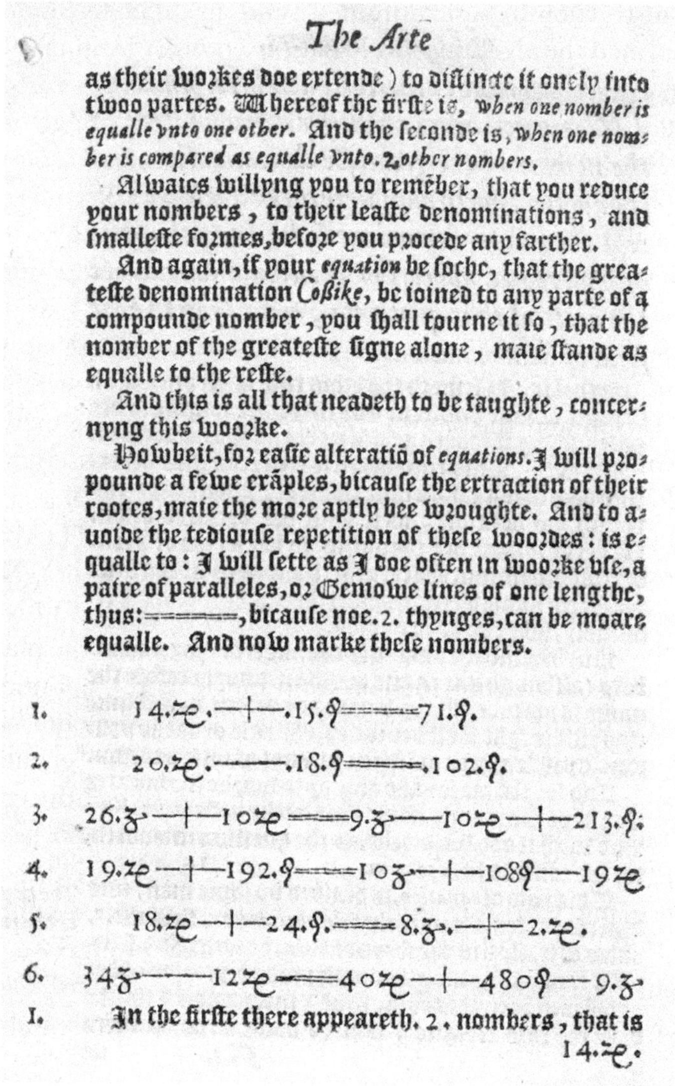

The Arte

as their workes doe extende) to distincte it onely into twoo partes. Whereof the firste is, *when one nomber is equalle vnto one other*. And the seconde is, *when one nomber is compared as equalle vnto.2.other nombers.*

Alwaies willyng you to remēber, that you reduce your nombers, to their leaste denominations, and smalleste formes, before you procede any farther.

And again, if your *equation* be soche, that the greateste denomination *Coßike*, be ioined to any parte of a compounde nomber, you shall tourne it so, that the nomber of the greateste signe alone, maie stande as equalle to the reste.

And this is all that neadeth to be taughte, concernyng this woorke.

Howbeit, for easie alteratiō of *equations*. I will propounde a fewe exāples, bicause the extraction of their rootes, maie the more aptly bee wroughte. And to auoide the tediouse repetition of these woordes: is equalle to: I will sette as I doe often in woorke vse, a paire of paralleles, or Gemowe lines of one lengthe, thus:=====, bicause noe.2. thynges, can be moare equalle. And now marke these nombers.

1. 14.ze.—+—.15.ϕ=====71.ϕ.
2. 20.ze.——.18.ϕ=====.102.ϕ.
3. 26.ʒ—+—10ze=====9.ʒ——10ze—+—213.ϕ.
4. 19.ze—+—192.ϕ=====10ʒ—+—108ϕ——19ze
5. 18.ze—+—24.ϕ.=====8.ʒ.—+—2.ze.
6. 34ʒ——12ze=====40ze—+—480ϕ——9.ʒ

1. In the firste there appeareth.2. nombers, that is 14.ze.

Figure 8.5. Page from *The Whetstone of Witte* by Robert Recorde, giving the first use of the equals sign, from the Plimpton Collection of the Rare Book and Manuscript Library, Columbia University.

the unknown, Rudolff presented to his German reading audience much of the background material needed to carry out algebraic calculations. He also included a brief discussion of three cubic problems and several problems in which he, like Antonio de'Mazzinghi, introduced a second

dragma	φ	number
radix	x	x
zensus	$\mathit{3}$	x^2
cubus	$\mathit{æ}$	x^3
zens de zens	$\mathit{33}$	x^4
sursolidum	β	x^5
zensicubus	$\mathit{3æ}$	x^6
bisursolidum	$b\beta$	x^7
zenszensdezens	$\mathit{333}$	x^8
cubus de cubo	$\mathit{ææ}$	x^9

Figure 8.6. Rudolff's system for powers of the unknown.

distinct unknown.[77] In particular, Rudolff employed a well-developed notational system for the first nine powers of the unknown, as shown in figure 8.6.

To help the reader understand these terms, Rudolff gave as examples the powers of various numbers. He then showed how to add, subtract, multiply, and divide expressions formed from his new symbols. Because it is not obvious how they multiply, he presented an explicit multiplication table, which showed, for example, that x times $\mathit{3}$ was $\mathit{æ}$.[78] He made things even more transparent by labeling *radix* 1, *zensus* 2, *cubus* 3, etc. and noting that to multiply expressions one simply added the corresponding numbers to find the correct symbol. Using these notational conventions, Rudolff treated binomials—that is, terms connected by an operation sign—and used, for the first time in an algebra text, the symbols $+$ and $-$ not to indicate excess and deficiency (as they had originally been used in a 1489 text by Johann Widman; ca. 1460–after 1498) but actually to represent addition and subtraction. Finally, he sometimes used a period as a symbol for "equals," as in $1\mathit{x}.2$ for our $x = 2$, but at other times he simply used the German word "*gleich*" meaning "equals."[79]

[77] On Rudolff's sources, see Heeffer, 2012.

[78] Rudolff, 1525, f. Dvi^r. For a detailed discussion of Rudolff's work, see Kaunzner, 1996, and compare Katz, 2009, pp. 392–394.

[79] For more on Rudolff's notational conventions, see Katz, 2009, p. 392.

Rudolff devoted the second half of *Coss* to solving algebraic equations. After presenting, in words, the rule for the solution of each type of equation, he gave several hundred illustrative examples, many of which are commercial in nature, dealing with buying and selling, exchange, wills, and money, and some of which are recreational, including a version of the classic 100-birds-for-100-coins problem. In general, the practical problems involved linear equations, while those needing a version of the quadratic formula were artificial like the oft-found "divide 10 into two parts such that ...."

Rudolff was, however, quite ingenious in inventing "quasi-real" problems such as one of the earliest travel problems on record:

There are 140 miles between two cities. One traveler goes 6 miles each day, while a second in the opposite direction goes 1 mile the first day, 2 the second day, and so on. When will they meet?[80]

To solve this, note that in x days, the first traveler goes $6x$ miles, while the second goes $\frac{1}{2}x + \frac{1}{2}x^2$. This yields the equation $\frac{1}{2}x^2 + 6\frac{1}{2}x = 140$, or $x^2 + 13x = 280$. Interestingly, the solution is irrational, since $x = \sqrt{322\frac{1}{4}} - 6\frac{1}{2}$ or approximately 11 days, 10 hours, and 48 minutes. Like al-Karajī, who presented a similar problem 500 years earlier, Rudolff did not indicate how this answer met the original conditions, given that the formula for summing an arithmetic sequence only applies when x is an integer.

Rudolff also posed a profit and loss problem:

A man bought a horse for a sum of money and sold it for 27. He lost per hundred $\frac{1}{3}$ as much as the horse cost (i.e., he lost $\frac{1}{3}x\%$ on the transaction). How much did he pay?[81]

He solved the problem this way. Since 100 florins result in a loss of $\frac{1}{3}x$, x florins result in a loss of $\frac{1}{300}x^2$. So $\frac{1}{300}x^2 = x - 27$, or $\frac{1}{300}x^2 + 27 = x$, and $x = 30$. Curiously, at the end of his text, Rudolff presented three irreducible cubic equations with answers but without a method of solution. He simply noted that his successors would continue to pursue the algebraic art and show how to deal with these.

---

[80] In both Rudolff, 1525 and Stifel, 1553, this is the example ("another example") following Rudolff's thirtieth example of the kind "squares and roots equal to numbers."

[81] In both Rudolff, 1525 and Stifel, 1553, this is the fourteenth example of the kind "roots and numbers are equal to squares."

Michael Stifel (1487–1567) brought out a new edition of Rudolff's text in 1553, nine years after he had published his own work, the *Arithmetica integra* (or *Complete Arithmetic*). In the latter work, Stifel had used the same symbols as Rudolff for the powers of the unknowns, but he more consistently employed the correspondence between letters and integral "exponents." He surpassed Rudolff by writing out a table of powers of 2, along with their exponents, which included the negative values -1, -2, -3 corresponding to $\frac{1}{2}$, $\frac{1}{4}$, and $\frac{1}{8}$, respectively. More importantly, however, he introduced two innovations when dealing with a problem in multiple unknowns. First, although keeping the cossic symbol $æ$ for the first unknown, he represented the second and third unknowns by $1A$ and $1B$, respectively. Since he thought of $1A$ as shorthand for $1Aæ$, he wrote the square of that unknown as $1A\mathfrak{z}$. Second, he also used juxtaposition to represent multiplication. Thus, as he wrote, "I wish to multiply $3A$ into $9B$; this is $27AB$, which is $27A$ multiplied into $1B$."[82] Yet, since the first unknown and its powers were still represented by the standard cossic symbols, there were problems with Stifel's notation. For example, he wrote the product of $3\mathfrak{z}$ by $4A$ as $12\mathfrak{z}A$, while 12 times the square of A was written as $12A\mathfrak{z}$. Similarly, he would have written the product of the square of A with the square of the first unknown as $\mathfrak{z}A\mathfrak{z}$, even though the notation $A\mathfrak{z}\mathfrak{z}$ would have meant A raised to the fourth power. Since the multiplication of cossic terms was always commutative, in either case the expressions should designate the same idea.

By the time Stifel brought out his revision of Rudolff's text, however, he had modified the notation to deal with this problem. Namely, he now wrote $1AA$ for the square of his second unknown, so that $12A\mathfrak{z}$ and $12\mathfrak{z}A$ always represented $12A$ times the square of the first unknown. Also, $12\mathfrak{z}$ multiplied by the square of A would be written as $12AA\mathfrak{z}$ or as $12\mathfrak{z}AA$, without ambiguity. On the other hand, he still had no notation for the square of a binomial; to write the square of $1æ+1A$, he had to multiply everything out.

Stifel also went further than any previous author in operating explicitly on equations. In single equations in one unknown, the basic idea was simply to manipulate the terms so as to put the equation into a standard form to which a particular rule applied. For two equations in two unknowns, however, the goal was usually to eliminate one of the

[82] Stifel, 1544, f. 252^r.

DE ARITH. HIER. CARDANI   313

Item hoc exemplum, &c.

Quærantur duo numeri, quorum multiplicatio inter se, re motis à producto ipsis numeris, faciat 14. Et quadrata eorum addita ad se, cum radicibus suis superadditis faciant 62.

Figura exempli huius sic stat.

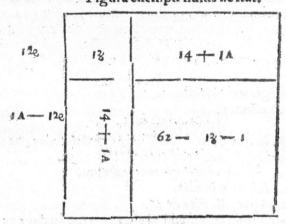

Exemplum aliud Hiero. Car. nisi quod numeri sunt mutati.

QVærantur duo numeri, quorum differentia multiplicata in differentiam quadratorum suorum, faciat 792. aggre gatum'q numerorum illorum duorum, multiplicatum in ag= gregatum quadratorum eorum, faciat. 5720.

Maior numerus. 1A +— 12e
Minor numerus. 1A — 12e
Differentia eorum est. 2 2e
Aggregatum eorum est. 2 A.
Quadratum maioris. 1 A ȝ +— 1 ȝ +— 2 2e A.
Quadratum minoris. 1 A ȝ +— 1 ȝ — 2 2e A.
Differentia quadratorum. 4 2e A.
Aggregatum quadratorum. 2 A ȝ +— 2 ȝ.

Itaq

Figure 8.7. Page from *Arithmetica integra* by Michael Stifel, illustrating his use of symbolism in solving equations in two unknowns, from the Plimpton Collection of the Rare Book and Manuscript Library, Columbia University.

unknowns by hook or by crook. Stifel introduced two new techniques: the multiplication of an entire equation by a number and the addition of two equations to form a new one. We see these techniques, as well as Stifel's notation, in use in the following problem: "Find two numbers so that the sum of both multiplied by the difference of their squares equals 675. But if one multiplies the difference of the two numbers by the sum of their squares, one gets 351. What are these numbers?"[83] Letting the two numbers be $1\!\!\imath$ and $1A$, whose squares are $1\mathfrak{z}$ and $1AA$ and whose cubes are $1\mathit{ce}$ and $1AAA$, Stifel multiplied $1\!\!\imath\!+\!1A$ by $1\mathfrak{z}-1AA$ to get

$$1\mathit{ce} +1\mathfrak{z}A - 1\!\!\imath AA - 1AAA \text{ gleich } 675.$$

Similarly, the second condition results in the equation

$$1\mathit{ce}-1\mathfrak{z}A + 1\!\!\imath AA - 1AAA \text{ gleich } 351.$$

Noting that $13 \cdot 675$ is equal to $25 \cdot 351$, Stifel multiplied the first equation by 13, the second by 25, and equated the results. Reducing the resulting equation by addition and subtraction produced the new equation:

$$38\mathfrak{z}A- 38\!\!\imath AA \text{ gleich } 12\mathit{ce}-12AAA.$$

"So I now divide each side by $12\!\!\imath -12A$. So

$$3\tfrac{1}{6} \!\!\imath A \text{ gleich } 1\mathfrak{z}+ 1\!\!\imath A + 1AA. \quad (*)$$

Then I add to each side $1\!\!\imath\ A$ so that I can extract the square roots." In other words, since he had equated the expressions $4\tfrac{1}{6}\!\!\imath A$ and $1\mathfrak{z}+2\!\!\imath A + 1AA$, he could take the square root of each side, giving the new equation $\sqrt{4\tfrac{1}{6}\!\!\imath A}$ gleich $1\!\!\imath\!+\!1A$. Stifel next subtracted $3\!\!\imath A$ from equation (*) and again took square roots to get $\sqrt{\tfrac{1}{6}\!\!\imath A}$ gleich $1\!\!\imath -1A$. "So make from these two equations a single equation by adding. This will be $2\!\!\imath$ gleich $\sqrt{6\!\!\imath A}$." Finally, he easily completed the solution by squaring each side of this

[83] Stifel, 1553, ff. $470^r - 471^v$. The quotations that follow in the next paragraph are also on this page.

equation and dividing to get that A was equal to $\frac{2}{3}x$ and so the two numbers were 6 and 9.

What distinguished Stifel's approach to algebra from that of his predecessors, then, was, in the words of historian of mathematics Jackie Stedall, "not so much in the range of problems [he] was prepared to tackle as in his conceptual understanding of what algebra was. For Stifel, algebra was no longer to be seen as a collection of specific techniques but as a general method encapsulated in a single rule."[84]

Stifel's example of explicitly using a symbolic equation was followed in mid-sixteenth-century France by Jacques Peletier (1517–1582) and Jean Borrel (ca. 1492–ca. 1568 also known as Johannes Buteo, the Hellenized version of his name), both working in a mid-century liberal arts context characterized by an emphasis on rhetoric and the honing of reasoning skills and both focused mathematically on solving systems of linear equations. Peletier, in his *Algèbre* (or *Algebra*) of 1554, adopted Stifel's notation $1A$ and $1B$ for the second and third unknowns, although he used $1R$ for the first (standing for "radix") and reverted to using $p.$ and $m.$ for plus and minus. Still, he treated a system of three equations as a single object, identified the equations by number as he worked on them, and explicitly added and subtracted equations (and their multiples) together in order to eliminate two of the unknowns. Similarly, Borrel, in his *Logistica* (or *Logistic*) of 1559, used A, B, and C for his three unknowns (although he used a comma to represent plus and a left square bracket for equals) as he systematically manipulated the equations in a system of linear equations to eliminate first one and then a second unknown.[85]

These works by Peletier and Borrel—both published in Paris—were only two manifestations of algebraic interest in France. The decade of the 1550s also witnessed the publication in Paris of texts with significant algebraic content by the Tübingen professor of mathematics, Johann Scheybl (1494–1570); the mathematics instructor and from 1560 to 1573 royal reader in mathematics, Pierre Forcadel (fl. mid-sixteenth century); and the noted professor, humanist, and educational reformer, Pierre de la Ramée (1515–1572 also known by his Anglicized name, Peter

---

[84] Stedall, 2012, p. 234.
[85] See Heeffer, 2010b, for more discussion of these two authors as well as the use of algebraic symbolism in the sixteenth century. For much more on Peletier and his place in the history of sixteenth-century French algebra, in particular, see Cifoletti, 1992.

Ramus).[86] These mid-sixteenth-century French developments, taking place largely within a university setting as opposed to within that of the abbacus schools, marked the beginning of a significant shift in algebra from practical tool to academic discipline that would continue into the seventeenth century.[87]

****

By the middle of the sixteenth century, algebra as a subject had diffused to and developed in much of Europe. Yet, while it constituted an increasingly shared body of knowledge internationally and was in some sense "a single subject with a variety of names," it was also "a single name with evolving meanings."[88] Even as late as 1608, the German Jesuit and prominent educational reformer, Christoph Clavius (1538–1612), would open his text *Algèbre* pondering the multiplicity of names by which the subject went as well as the widely varying accounts of its origins and development.[89] A plurality of intimately related yet subtly different algebras had emerged.

This plurality was also reflected in the notations employed by different authors. Although notation had developed significantly and although symbolism had begun to appear in the work of various authors, there was still little consistency.[90] Algebra owed its most substantive growth beyond the bounds of the *Liber abbaci* and the mathematics of the abbacus schools to the work of several sixteenth-century mathematicians active in both Italy and France, to which we now turn.

---

[86] For an analysis of all of these works, see Loget, 2012.
[87] For more on this theme, see Cifoletti, 1996.
[88] Stedall, 2012, p. 219.
[89] For more on Clavius's ideas on algebra, see Rommevaux, 2012.
[90] On the question of mathematical notation and its development, see the classic Cajori, 1928 as well as the more targeted Serafati, 2005.

9

The Growth of Algebraic Thought in Sixteenth-Century Europe

In his compendium *Summa de arithmetica, geometria, proportioni, e proportionalita*, Luca Pacioli, like his fifteenth-century contemporaries and successors, devoted most of his algebraic efforts to solutions of equations of the first and second degree but could not resist treating those higher-degree equations for which solutions were known.[1] He noted, for example, that given any one of the six types of linear and quadratic equations, multiplication by any given power of the unknown generated a solvable higher-order equation. More interestingly, he considered eight cases of fourth-degree equations, which he wrote as

(1) "censo de censo equale a numero" or $ax^4 = e$,

(2) "censo de censo equale a cosa" or $ax^4 = dx$,

(3) "censo de censo equale a censo" or $ax^4 = dx^2$,

(4) "censo de censo e censo equale a cosa" or $ax^4 + cx^2 = dx$,

(5) "censo de censo e cosa equale a censo" or $ax^4 + dx = cx^2$,

(6) "censo de censo e numero equale a censo" or $ax^4 + e = cx^2$,

(7) "censo de censo e censo equale a numero" or $ax^4 + cx^2 = e$, and

(8) "censo de censo equale a numero e censo" or $ax^4 = e + cx^2$.

As he remarked, the fourth and fifth cases were "*impossibile*"—they reduce to cubics involving the terms x^3, x, and a constant—but Pacioli

---

[1] See the discussion in Franci and Toti Rigatelli, 1985, pp. 64–65. The quotations that follow in this paragraph may also be found there.

elaborated further:

> But of number, *cosa* and *cubo* how ever they are compound[ed], or
> of number, *censo* and *cubo*, or of number, *cubo* and *censo de censo*
> nobody until now has formed general rules, because they are not
> proportional among them....And therefore, until now, for their
> equations, one cannot give general rules, except that, sometimes, by
> trial (as I have said above) in some particular cases, . . . and therefore
> when in your equations you find terms with different intervals
> without proportion, you shall say that the art, until now, has not
> given the solution to this case....Even if the case may be possible.

"Until now." Pacioli had signaled—and *in print*—major open problems in
what he had earlier termed "the great art"[2] of algebra, and they involved
precisely those long-enticing cubics and quartics.

Equations of the third and fourth degrees sparked quite a few algebraic
fireworks in the first half of the sixteenth century. Their solutions marked
the first major European advances beyond the algebra contained in
Fibonacci's thirteenth-century *Liber abbaci*. By the end of the century,
algebraic thought—through work on the solutions of the cubics and quartics but, more especially, through work aimed at better contextualizing
and at unifying those earlier sixteenth-century advances—had grown
significantly beyond the body of knowledge codified in Pacioli's *Summa*.
Algebra was evolving in interesting ways.

SOLUTIONS OF GENERAL CUBICS AND QUARTICS

Scipione del Ferro (1465–1526), who had become lecturer in mathematics
at his hometown University of Bologna in 1496, made the first key
breakthrough. Sometime during the course of the first two decades of
the sixteenth century, he had found a general solution of cubic equations
of the form $ax^3 + dx = e$ or, more simply, of the form

$$x^3 + px = q, \tag{9.1}$$

---

[2] Franci and Toti Rigatelli, 1985, p. 62.

exactly the kind of equation that Pacioli had singled out at the opening of the passage quoted above. While the *Summa* had been published in 1494, and while Pacioli himself had lectured at Bologna in 1501–1502, it remains unclear whether Pacioli spurred del Ferro to consider this type of equation. Regardless of how he came to the problem, however, del Ferro recorded his solution of it in a notebook that passed to his son-in-law, Annibale della Nave (ca. 1500–1558), at the time of del Ferro's death. That solution algorithm translates into modern notation as

$$x = \sqrt[3]{\sqrt{\left(\frac{q}{2}\right)^2 + \left(\frac{p}{3}\right)^3} + \frac{q}{2}} - \sqrt[3]{\sqrt{\left(\frac{q}{2}\right)^2 + \left(\frac{p}{3}\right)^3} - \frac{q}{2}},$$

but he gave no indication as to how he might have arrived at it. It is, however, easy to see with a bit of algebraic manipulation[3] that he could have arrived at it by comparing (9.1) and the identity $(u - v)^3 + 3uv(u - v) = u^3 - v^3$. Setting $x = u - v$ would have necessitated finding u and v simultaneously satisfying the equations $u^3 - v^3 = q$ and $uv = \frac{p}{3}$, a straightforward task (at least given our modern notation) in light of the identity

$$\left(\frac{u^3 + v^3}{2}\right)^2 = \left(\frac{u^3 - v^3}{2}\right)^2 + u^3 v^3.$$

Girolamo Cardano (1501–1576) ultimately succeeded in clothing all of this geometrically in 1545 in his book, *Ars magna, sive de regulis algebraicis* (or *The Great Art, or the Rules of Algebra*).

Unlike the publish-or-perish ethos that seems to have characterized contemporary academe for at least half a century, new discoveries like del Ferro's represented prized, closely guarded intellectual property in early sixteenth-century Italy. Since university posts were not permanent and depended on favorable reviews and renewals by a university's senate, one way to impress the senators was to perform well at renewal time in a public challenge against a potential rival. In the case of a mathematics position, each man (and they were always men) would pose a set of problems to the other. After an agreed upon length of time, the two would come back together and publicly present their respective solutions. They spoke a common mathematical language; mathematics became

---

[3] Compare the mathematical discussion in Sesiano, 1999, p. 135.

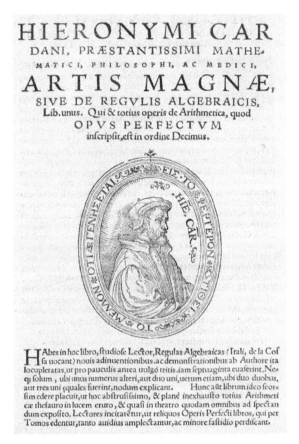

Figure 9.1. Title page of the *Ars magna* by Girolamo Cardano, from the Smith Collection of the Rare Book and Manuscript Library, Columbia University.

increasingly democratic as its questions, terminology, and notation became increasingly shared. Since the mathematician who successfully solved the most problems won the contest, possessing a solution unknown to the mathematical community represented real job security. The notebook that del Ferro bequeathed to his son-in-law and successor at Bologna and that contained del Ferro's solution algorithm for (9.1) was thus extremely valuable property.

Late in his life, del Ferro had also apparently shared his solution with his student, Antonio Maria Fior.[4] In an effort to establish himself

---

[4] The story that follows of the entanglements of Fior, Tartaglia, and Cardano has been told many times. See, for example, Ore, 1953, 62–65; Parshall, 1988a, 142–143; Sesiano, 1999, 135–140; or Katz, 2009, 399–401.

in Venice in 1535, Fior challenged Niccolò Fontana (ca. 1500–1557), a teacher of mathematics there, to a public contest. Confident that only he knew how to solve cubics like (9.1), Fior posed thirty questions, all of which boiled down to precisely that type of equation. Fior, however, had met his match. Fontana, also known as Tartaglia or the "stammerer," recorded with surprise his success in solving (9.1) on the night of 12 February 1535. He was thus able to handle all of Fior's problems, while Fior faltered in dealing with the thematically widely ranging set of problems Tartaglia had posed. Not only had the better mathematician won, but Tartaglia, too, had found the general solution of (9.1). The secret that cubics of the form $x^3 + px = q$ were solvable was out of the bag.

Cardano, a trained physician who would later be renowned as an astrologer, doctor, mathematician, and natural philosopher, was, in the 1530s, lecturing on mathematics at the Piatti Foundation in Milan and trying to gain admission to the College of Physicians there.[5] In 1539, the same year that the College of Physicians finally accepted him into their fold (his reputation for irascibility and his illegitimate birth had been at issue), Cardano published both a textbook, *Practica arithmetice et mensurandi singularis* (or *Practice of Mathematics and Individual Measurements*), on the basics of arithmetic and algebra up to and including equations of degree two and hosted Tartaglia in Milan. Envisioning the publication of a more comprehensive and advanced algebra than his *Practica arithmetice*, Cardano had already asked Tartaglia once to share his solution of (9.1) so that it could be included, with all due credit, in his proposed work. It is perhaps not surprising that Tartaglia demurred, preferring to publish his solution in a work by his own hand in due course. As Cardano's guest in 1539, however, Tartaglia found it harder to refuse his host's renewed entreaties and gave him, under an oath of secrecy, the solution algorithm in the form of a poem.[6] Thus began one of the most famous priority controversies in the history of mathematics.

With knowledge of Tartaglia's solution, Cardano fairly quickly managed to solve the remaining cases of the cubic, while his student and secretary, Ludovico Ferrari (1522–1565), succeeded in 1540 in solving

---

[5] For more on Cardano's colorful life, see his autobiography Cardano, 1575/2002 and Fierz, 1983, among other sources.

[6] For the poem, see Sesiano, 1999, 137–138.

the quartics as well. They were making stunning progress on the next book project, but Cardano's sworn oath to Tartaglia thwarted them in their goal of fully laying out "the great art."[7] That changed in 1543 when they traveled to Bologna and met with Annibale della Nave, who showed them his deceased father-in-law's notebooks. There, they found del Ferro's solution to (9.1), a solution that had long preceded Tartaglia's. Since del Ferro, not Tartaglia, had priority of discovery, Cardano considered himself released from any promise he may have made to the latter.

When he ultimately published the *Ars magna* in 1545, Cardano was thus very careful to lay out the historical sequence of events. "In our own days," he explained, "Scipione del Ferro of Bologna has solved the case of the cube and first power equal to a constant, a very elegant and admirable accomplishment....In emulation of him, my friend Niccolò Tartaglia of Brescia, wanting not to be outdone, solved the same case when he got into a contest with his [Scipione's] pupil, Antonio Maria Fior, and, moved by my many entreaties, gave it to me."[8] Cardano was, however, conspicuously silent on the matter of the promise. Feeling supremely betrayed, Tartaglia entered into a long and nasty correspondence with Ferrari that lasted until 1548, when the two men engaged in a public mathematical contest in Milan. After the first day of the challenge, it was clear that this time Tartaglia, although the more experienced mathematician, had met *his* match. He left Milan, the contest unresolved, and thereby secured Ferrari's reputation.

The published solution that had caused all of this drama appeared in the eleventh chapter of Cardano's mathematical magnum opus, the *Ars magna*. A fascinating text that both broke from and sought, by and large, to uphold the geometrical tradition in algebra perpetuated by Fibonacci,[9] the *Ars magna* opened with a discussion of first- and second-degree equations, moved on to treat all of the cubics, and closed with solutions of the quartics. Cardano acknowledged both his embeddedness in the mathematical environment of his day and his willingness to move

---

[7] For a finely grained analysis of Cardano's mathematical writing and the evolution of his thought between the publication of the *Practica arithmetice* in 1539 and the *Ars magna* in 1545, see Gavagna, 2012.

[8] Cardano, 1545/1968, p. 8.

[9] For a discussion and interpretation of Cardano's pivotal place in the evolution of algebraic thought, see Parshall, 1988a.

beyond it in his book's opening chapter. There, he explained that "for as *positio* [the first power] refers to a line, *quadratum* [the square] to a surface, and *cubum* [the cube] to a solid body, it would be very foolish for us to go beyond this point. Nature does not permit it. Thus, it will be seen, all those matters up to and including the cubic are fully demonstrated, but the others which we will add, either by necessity or out of curiosity, we do not go beyond merely setting out."[10] Like al-Khwārizmī, Abu Kāmil, Fibonacci, and others before him, Cardano accepted the geometrical standards of justification that went back at least as far as Euclid. He showed how successfully to extend them to the third dimension in his *justifications* of the solutions of the cubics—thereby surpassing his thirteenth-, fourteenth-, and fifteenth-century predecessors as well as his sixteenth-century contemporaries. Ferrari's solutions of the quartics, however, although solutions they were, simply could not be justified by geometrical constructions of the complete-the-square or complete-the-cube varieties. "Nature does not permit it," but Cardano did not let that stop him. To get a feel first for his extension to three dimensions of the cut-and-paste constructions we last encountered in Fibonacci's work, consider Cardano's discussion of (9.1).

His chapter "On the Cube and First Power Equal to the Number" opened with yet another acknowledgment of Tartaglia's work, but one that carefully distinguished between the rule or algorithm for the solution and its actual geometrical justification. "Scipione Ferro of Bologna well-nigh thirty years ago discovered this rule ... [Tartaglia] gave it to me ... though withholding the demonstration. Armed with this assistance, I sought out its demonstration in [various] forms."[11] In the case at hand, Cardano asked his readers to consider a particular example: "let GH^3 plus six times its side GH equal 20, and let AE and CL be cubes the difference between which is 20 and such that the product of AC, the side [of one], and CK, the side [of the other], is 2, namely one-third the coefficient of x. Marking off BC equal to CK [see figures 9.2a and 9.2b],

---

[10] Cardano, 1545/1968, p. 9. Unlike Pacioli, Cardano used the abbreviations po. for the unknown, quad. for x^2, cu. for x^3, and quadr. quad. for x^4. See Cajori, 1928, 1:117 for an overview of Cardano's notation. Cardano had used Pacioli's co., ce., and cu. for x, x^2, and x^3, respectively, in his *Practica arithmetice* just six years earlier in 1539. The $\tilde{p}$ and $\tilde{m}$ for "plus" and "minus," respectively, that appeared in Pacioli's printed *Summa* had become fairly standard and were adopted by Cardano as well.

[11] Cardano, 1545/1968, p. 96. The following quotation is also from this page.

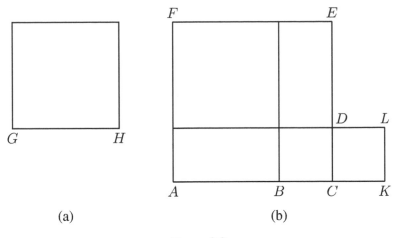

Figure 9.2.

I say that, if this is done, the remaining line AB is equal to GH and is, therefore, the value of x, for GH has already been given as [equal to x]."[12]

With this overview of the construction, Cardano next provided the details of effecting it and the justification that it yielded the desired result. Essentially, he completed the cubes and the other six solids defined by considering figure 9.2 in three dimensions (figure 9.3), and he used various identities that he had proven earlier in his text to rearrange and reinterpret the pieces.[13] In algebraic dress, his geometrical construction boiled down to this set of steps. If, in equation (9.1), $x = u - v$, where $u^3 - v^3 = q$ and $u^3 v^3 = (\frac{p}{3})^3$, then

$$
\begin{aligned}
x^3 + px &= (u - v)^3 + p(u - v) \\
&= u^3 - 3u^2 v + 3uv^2 - v^3 + 3uv(u - v) \\
&= u^3 - v^3 - 3uv(u - v) + 3uv(u - v) \\
&= q.
\end{aligned}
$$

Having justified his solution algorithm geometrically, he next stated it rhetorically and used it to find numerical solutions to three different

---

[12] Here and below, the translator has used x, where Cardano had the Latin word "res" or "thing."

[13] For a fuller discussion of his geometrical argument, see Parshall, 1988a, pp. 144–146, and compare Cardano, 1545/1968, pp. 96–98.

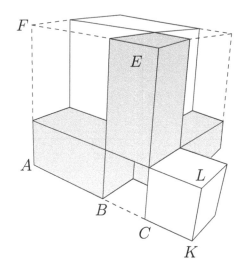

Figure 9.3.

examples. Interestingly, the first of these,

$$x^3 + 6x = 20,$$

which appeared in the 1545 printed version of the *Ars magna* in a Latinized abbreviated form as

$$cub^9 \, p : 6 \, reb^9 \; \text{æ}\bar{q}lis \; 20,$$

clearly has, by inspection, $x = 2$ as a solution, yet the algorithm yields the complicated-looking solution $x = \sqrt[3]{\sqrt{108} + 10} - \sqrt[3]{\sqrt{108} - 10}$. As Cardano noted at the end of the chapter, this expression is actually equivalent to 2, although he gave no further explanation other than to tell his readers that this "was perfectly clear if it is tried out."[14]

A series of chapters devoted to the solutions of the other forms of the cubic equation as well as to other algebraic techniques and manipulations followed before Cardano tackled the case of the quartic in chapter thirty-nine. As we have seen, quartic—like cubic—problems had long been part

[14] Cardano, 1545/1968, pp. 99–100. For a facsimile of the page from Cardano's 1545 text, see Struik, 1986, p. 66. We have rendered printers' marks here as [9], although these should in no way be construed as exponents. Here, "reb" abbreviates the ablative plural of "res."

of the mathematical discourse of the *maestri d'abbaco*, even if they had solutions that were only valid in certain restricted settings. Thanks to the work of his student Ferrari, Cardano was able not only to extend that earlier work but also to provide solutions that applied to more general cases. As he explained, "there is another rule, more noble than the preceding. It is Lodovico Ferrari's, who gave it to me on my request. Through it we have all the solutions for equations of the fourth power, square, first power, and number, or of the fourth power, cube, square, and number....In all these cases [he had listed twenty], therefore, which are indeed [only] the most general as there are sixty-seven others, it is convenient to reduce those involving the cube to equations in which x is present."[15] The basic technique, then, for those equations involving a cube was first to make—as can always be done—a linear substitution in order to eliminate the x^3 term. The resulting equation then involved only the x^4, x^2, x, and constant terms and could be manipulated so that both sides of it became perfect squares. Depending on the type of example, that manipulation could result in a cubic (today called the resolvent) that would then be solved using one of the algorithms established earlier. Since "nature did not permit" geometrical constructions of the quartics analogous to those Cardano had given in the cases of the cubics, he and Ferrari contented themselves with a geometrical, complete-the-square-like justification of an identity they often needed in the course of their quartic manipulations, namely,[16]

$$(x^2 + a + b)^2 = (x^2 + a)^2 + 2x^2 b + 2ab + b^2. \tag{9.2}$$

To see Ferrari's algorithm in action, consider the following problem posed by Cardano: "Find a number the fourth power of which plus four times itself plus 8 is equal to ten times its square," that is, $x^4 + 4x + 8 = 10x^2$.[17] Since this example did not involve an x^3 term, Cardano dived right in by reorganizing the terms so as to ready the equation for the completion of squares on both sides. He first shifted the x term to the right-hand side of the equation and then added $-2x^2$ to both sides to get $x^4 - 2x^2 + 8 = 8x^2 - 4x$. Subtracting 7 from both sides results in a perfect

---

[15] Cardano, 1545/1968, pp. 237–238.
[16] See Cardano, 1545/1968, pp. 238–239, and compare van der Waerden, 1985, p. 57.
[17] Cardano, 1545/1968, p. 248. For the solution that follows, compare pp. 248–249.

square on the left-hand side but a nonsquare on the right-hand side:

$$(x^2 - 1)^2 = x^4 - 2x^2 + 1 = 8x^2 - 4x - 7.$$

In order to keep the left-hand side a perfect square and to transform the right-hand side into one (with the identity (9.2) in mind), Cardano added $-2bx^2 + b^2 + 2b$ to both sides to get

$$[x^2 - (b + 1)]^2 = (8 - 2b)x^2 - 4x + (b^2 + 2b - 7). \tag{9.3}$$

Now, the right-hand side of (9.3) is a perfect square provided its discriminant is zero,[18] in other words, provided

$$(8 - 2b)(b^2 + 2b - 7) = 4,$$
$$-2b^3 + 4b^2 + 30b - 56 = 4,$$
$$b^3 + 30 = 2b^2 + 15b.$$

This, however, is a cubic that, by inspection, has $b = 2$ as a solution "whether working according to the rule or by one's own sense alone."[19] Substituting $b = 2$ back into (9.3) then gives

$$(x^2 - 3)^2 = (2x - 1)^2 \tag{9.4}$$

or $x^2 - 3 = 2x - 1$, which is solvable using the quadratic algorithm known to al-Khwārizmī and Fibonacci and discussed earlier in the *Ars magna*. Thus, taking the positive solution, $x = \sqrt{3} + 1$ is a root of the original quartic, as Cardano then verified with a simple check. Interestingly, Cardano further noted that since (9.4) could equally well be interpreted as $(x^2 - 3)^2 = (1 - 2x)^2$ so that the quadratic equation to solve was $x^2 - 3 = 1 - 2x$, then $x = \sqrt{5} - 1$ is another root. In neither instance, however, did he recognize the corresponding negative roots.

Indeed, the insistence on justification by geometrical construction excluded negative roots from the justifiable. Physically, squares and cubes

---

[18] If $ax^2 + bx + c = 0$, the discriminant is $b^2 - 4ac$. Note that $b^2 - 4ac = 0$ is the same as $ac = \left(\frac{b}{2}\right)^2$.

[19] Cardano, 1545/1968, p. 249.

cannot have negative sides. Again, "nature does not permit it," but, that said, Cardano could not deny that something was afoot. Already in chapter one, for example, he had noted that "if the square of a square is equal to a number and a square, there is always one *true* solution and another *fictitious* solution equal to it. Thus, for $x^4 = 2x^2 + 8$, then x equals 2 or -2."[20] Here, and elsewhere, he acknowledged the existence of a negative root, but it was "fictitious" in that it was outside the realm of the geometrical.

Even worse, whole classes of problems could be illegitimate. As Cardano observed, "for an odd power, there is only one true solution and no fictitious one when it is equated to a number alone. Thus, ... if $2x^3 = 16$, x equals 2," and "it is always presumed in this case, of course, that the number to which the power is equated is true and not fictitious," for "to doubt this would be as silly as to doubt the fundamental rule itself." In other words, since an equation like $2x^3 = -16$ represents a geometrical absurdity from the start, it makes no sense even to consider the question of whether it has roots, true or fictitious.

Cardano encountered yet another type of "false" problem in chapter five during the course of his discussion of quadratics of the type "the first power is equal to the square and number."[21] Among the examples he considered was the quadratic $10x = x^2 + 6$, and he noted that there are two "true" solutions: $5 + \sqrt{19}$ and $5 - \sqrt{19}$. However, "if the number cannot be subtracted from the square of one-half the coefficient of the first power, the problem itself is a false one and that which has been proposed cannot be. It must always be observed as a general rule throughout this treatise that, when those things which have been directed cannot be carried out, that which is proposed is not and cannot be."

Negatives under square roots clearly represented one such no-no, that is, until Cardano came to expound "on the rule for postulating a negative" in chapter thirty-seven. There, he posed the following problem:

> If it should be said, Divide 10 into two parts the product of which is 30 or 40, *it is clear that this case is impossible. Nevertheless* we will work thus: We divide 10 into two equal parts, making each 5. These we square, making 25. Subtract 40, if you will, from the 25 thus

---

[20] Cardano, 1545/1968, p.11 (our emphases). The quotation in the next paragraph is also on this page.

[21] Cardano, 1545/1968, p. 38. For the quote that follows in this paragraph, see p. 39.

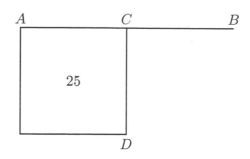

Figure 9.4.

produced ... leaving a remainder of -15, the square root of which added to or subtracted from 5 gives parts the product of which is 40. These will be $5 + \sqrt{-15}$ and $5 - \sqrt{-15}$. [22]

This was precisely one of the "I divided 10 into two parts..." problems that Fibonacci and his Arabic predecessors had treated using their quadratic techniques of algebra and almuchabala, so it was natural for Cardano to consider it in that light as well, by trying to provide a complete-the-square construction to verify that the solution algorithm he had given was correct.

The problem boiled down to one of the form $x^2 + c = bx$ (recall Fibonacci's treatment of this in equation (8.1) in the previous chapter), and so it was the construction for that type of quadratic that Cardano evoked. As he explained,

> In order that a true understanding of this rule may appear, let AB be a line which we will say is 10 and which is divided in two parts [see figure 9.4], the rectangle based on which must be 40. Forty, however, is four times 10; wherefore we wish to quadruple the whole of AB. Now let AD be the square of AC, one-half of AB, and from AD subtract $4AB$, ignoring the number. The square root of the remainder, then—if anything remains—added to or subtracted from AC shows the parts. *But since such a number is negative, you will have to imagine* $\sqrt{-15}$—that is, the difference between AD and $4AB$—which you add to or subtract from AC, and you will have that which you seek, namely $5 + \sqrt{25 - 40}$ and $5 - \sqrt{25 - 40}$, or

---

[22] Cardano, 1545/1968, p. 219 (our emphasis). For the quote that follows in the next paragraph, see pp. 219–220 (our emphases).

$5 + \sqrt{-15}$ and $5 - \sqrt{-15}$. *Putting aside the mental tortures involved,* multiply $5 + \sqrt{-15}$ by $5 - \sqrt{-15}$, making $25 - (-15)$ which is $+15$. Hence this product is 40. Yet the nature of AD is not the same as that of 40 or of AB, since a surface is far from the nature of a number and from that of a line, though somewhat closer to the latter. *This is truly sophisticated,* since with it one cannot carry out the operations one can in the case of a pure negative and other [numbers].

Thus, as figure 9.4 attests, the complete-the-square construction ultimately could not be realized, even though the algebraic manipulation was effected easily enough from a mechanically algebraic—if not perhaps from a philosophical—point of view.

At the same time that Cardano tried to uphold the geometrical standard of rigor adopted by al-Khwārizmī, Fibonacci, and others, he broke significantly from this standard in his acknowledgment of "fictitious" or negative solutions, his willingness to treat what we recognize as complex numbers, and his more general considerations of fourth-degree equations. Historian of mathematics Roy Wagner put it well when he said that "to understand Renaissance algebra one must recognise the impact of both ... conserving distinctions and undermining them."[23] Cardano's *Ars magna* represents a key case in point. A work embedded in past traditions, it began noticeably to move algebra beyond prior boundaries.[24]

TOWARD ALGEBRA AS A GENERAL PROBLEM-SOLVING TECHNIQUE

While, as we have just seen, Cardano struggled to work within what he understood to be the geometrical parameters of algebraic discourse as defined by Fibonacci and before him by such Arabic scholars as al-Khwārizmī, a "new" model of algebraic thinking had come to light in Europe as early as 1463. The Renaissance humanist, Johannes Müller of Königsberg (1436–1476) (or Regiomontanus after the Latinization of his place of birth in Bavaria), reported that he had found in Venice a Greek text of "the fine thirteen Books of Diophantus, in which the very flower

[23] Wagner, 2010, p. 492.
[24] Compare the argument in Parshall, 1988a, p. 149.

of the whole of Arithmetic lies hid, the *ars rei et census*, which today they call by the Arabic name of Algebra."[25] Actually, the text that he found—and that he had hoped to bring out in Latin translation—contained the six Greek books of Diophantus's *Arithmetica* that we analyzed in chapter 4. While it is, of course, unclear how the course of the history of algebra in the West might have been affected had Regiomontanus realized his desired translation in the fifteenth century, by the 1560s at least two other mathematical adepts, Rafael Bombelli (1526–ca. 1572), an engineer-architect involved in the draining of the Val di Chiana in central Italy,[26] and his friend Antonio Maria Pazzi, a reader in mathematics at the University of Rome, had not only pored over a manuscript copy of the *Arithmetica* in the Vatican Archives in Rome but also completed a translation of the first five (of the six Greek) books.[27] Bombelli's reading of Diophantus proved significantly to affect his interpretation of algebra as it had developed through the 1560s.

Sometime in the 1550s, Bombelli had begun work on a mathematical manuscript of his own aimed at rendering Cardano's *Ars magna* more readily comprehensible to an interested mathematical readership. As he put it, "no one has explored the secrets of algebra except Cardano of Milan, who in his *Ars magna* dealt with the subject at length, but was not clear in his exposition."[28] "In order to remove finally all obstacles before the speculative theoreticians and the practitioners of this science [algebra] ...," he continued, "I was taken by a desire to bring it to perfect order."[29] Building, as did Cardano, on the work of the *maestri d'abbaco* before him, Bombelli sought to produce a text, which, unlike Cardano's, was fully self-contained. To realize that goal, Bombelli felt the need first to lay a solid arithmetic foundation by providing appropriate definitions of basic concepts such as number, power, and root and by showing how properly to execute the basic operations of addition, subtraction, multiplication, division, and the extraction of roots.[30] Then, with that as his basis, he could move on to algebra per se, by detailing

---

[25] Heath, 1964, p. 20.

[26] For the painstakingly researched details known of Bombelli's life, see Jayawardene, 1963, 1965, and 1973.

[27] Jayawardene, 1963, p. 392 and Parshall, 1988a, p. 153.

[28] Jayawardene, 1973, p. 513 (his translation).

[29] Wagner, 2010, p. 490 (his translation).

[30] On changing techniques for calculating, specifically, square and cube roots from Fibonacci to Bombelli, see Rivoli and Simi, 1998.

algebraic notation, indicating how arithmetic operations "translated" into algebraic settings, laying out rules for solving algebraic equations of the first through the fourth degrees, and illustrating those rules through a series of examples.

The manuscript, entitled *L'Algebra*, that Bombelli initially produced, accomplished these two tasks in Books I and II, respectively, and provided further instruction in Book III by means of over 150 "daily life" and recreational exercises characteristic of and taken largely from prior sources such as Fibonacci's *Liber abbaci*, Pacioli's *Summa*, and various of the *libri d'abbaco*, among others.[31] Initially, Bombelli cast his manuscript in what had become the traditional Italian terms of "cosa" for the unknown and "censo" for the square of the unknown. By the time the first three books had appeared in print in 1572, however, the "practical" problems in Book III had been replaced by over 140 more abstract and indeterminate problems drawn directly from Diophantus and cast in terms of "tanto" and "potenza," Bombelli's Italian equivalents of the Diophantine "number (arithmos)" and "power (dynamis)" (recall the discussion in chapter 4).[32]

As Bombelli announced in his introduction to his revised Book III, he was no longer satisfied with casting problems "in the guise of human actions (buying, selling, barter, exchange, interest, defalcation, coinage, alloys, weights, partnership, profit and loss, games and other numerous transactions and operations relating to daily living)."[33] He aimed, in fine humanistic style, to return to what he viewed as the more pristine knowledge of the ancient Greeks, in his case to Diophantus, rather than to persist in a line of thought that stemmed from the medieval Islamic mathematicians through the *maestri d'abbaco*.[34] Yet, although Diophantus superficially supplanted the *maestri d'abbaco* and their Islamic predecessors in the revised version of Book III, Bombelli's text, even in its printed form, represented, like Cardano's *Ars magna*, a complex mélange of old and new, a text steeped in tradition but fundamentally

[31] For a complete analysis of the sources of Bombelli's exercises in the unpublished, manuscript version of Book III, see Jayawardene, 1973, pp. 513–522.

[32] For a comparison of the problems to be found in the works of Diophantus, Cardano, Bombelli, and Viète, see Reich, 1968.

[33] Bombelli as quoted in Jayawardene, 1973, p. 511.

[34] On the role of Renaissance humanism in the history of mathematics in general, see Rose, 1975. On its role in the history of algebra in particular, see Cifoletti, 1996.

influenced by "new" ways of thinking.[35] To get at least a sense of how Bombelli refined and "perfected" Cardano's exposition, consider both his treatment of what we would term complex numbers in Book I and the notation he adopted in Book II for his algebraic discussions.

As we saw at the close of the previous section, Cardano had not only encountered complex numbers in the context of solving certain types of quadratic equations but also manipulated them despite "the mental tortures involved." Still, he did not treat them more comprehensively, and this was precisely one of the lacunae that Bombelli had recognized in Cardano's text. To fill it, Bombelli rather nonchalantly and without any particular fanfare included a discussion of them in his general treatment of arithmetic in Book I. There, he recognized two new "numbers": $+\sqrt{-1}$, which he called "più di meno" or "plus from minus" and which he recognized as "neither positive nor negative" and $-\sqrt{-1}$, which he called "meno di meno" or "minus from minus."[36] He then presented a multiplication table that governed their behavior:

"più di meno via più di meno fa meno" $[(+\sqrt{-1})(+\sqrt{-1}) = -1]$
"più di meno via meno di meno fa più" $[(+\sqrt{-1})(-\sqrt{-1}) = +1]$
"meno di meno via più di meno fa più" $[(-\sqrt{-1})(+\sqrt{-1}) = +1]$
"meno di meno via meno di meno fa meno" $[(-\sqrt{-1})(-\sqrt{-1}) = -1]$

and proceeded to show how to perform various arithmetic operations on the new numbers.

Bombelli's treatment of the complex numbers, in fact, reflected innovations that were both purely notational and structurally algebraic. Whereas Pacioli, Cardano, and others had used abbreviations for aggregation such as $v.$ signifying "all that follows," Bombelli introduced the pairing $\lfloor$ and $\rfloor$ to indicate inclusion unambiguously just like our modern-day parentheses.[37] Thus, using $R.c.$ to denote $\sqrt[3]{}$ and $R.q.$ to denote $\sqrt{}$,

---

[35] For a detailed and finely grained discussion of this point, see Wagner, 2010. It should be noted here that Bombelli intended L'Algebra to be a text in five books, with books IV and V presenting applications of geometry to algebra as well as of algebra to geometry. Unpublished drafts of books IV and V were only uncovered and published early in the twentieth century by the Italian historian of mathematics, Ettore Bortolotti. For the texts, see Bombelli, 1966.

[36] For these and the following quote, see Bombelli, 1966, pp. 133–134, and compare Bashmakova and Smirnova, 2000, p. 73.

[37] Already in his 1484 manuscript, *Triparty*, Nicolas Chuquet had used underlining in a similar spirit. See Cajori, 1928, 1:384–385.

he expressed

$$\left(\sqrt[3]{2+\sqrt{-3}}\right)\left(\sqrt[3]{2-\sqrt{-3}}\right)$$

as

Moltiplichisi, $R.c.\lfloor 2$ più di meno $R.q.3\rfloor$ per $R.c.\lfloor 2$ meno di meno $R.q.3\rfloor$.[38]

From an algebraic point of view, moreover, Bombelli sometimes acknowledged—although he sometimes also failed to acknowledge—complex roots of the equations he considered,[39] while from a geometrical point of view, he showed little concern for the fact that a complex root could not properly correspond, say, to the side of a square. To get a feel for Bombelli's manipulation and understanding of complex roots, consider his treatment of the irreducible cubic equation

$$x^3 = 15x + 4, \tag{9.5}$$

an example that also illustrated well his self-appointed task of creating a text more self-contained than Cardano's *Ars magna*.[40]

In this kind of a case, namely, one "when the cube of one-third the coefficient of x is greater than the square of one-half the constant of the equation," Cardano had, indeed, acknowledged difficulties.[41] The problem was that the solution algorithm he had elaborated in chapter twelve of the *Ars magna* boiled down to

$$x = \sqrt[3]{\frac{q}{2} + \sqrt{\left(\frac{q}{2}\right)^2 - \left(\frac{p}{3}\right)^3}} + \sqrt[3]{\frac{q}{2} - \sqrt{\left(\frac{q}{2}\right)^2 - \left(\frac{p}{3}\right)^3}},$$

so that problems arose under the square roots when $\left(\frac{p}{3}\right)^3 > \left(\frac{q}{2}\right)^2$. Rather than pressing forward to deal with the resulting imaginary numbers as

---

[38] Bashmakova and Smirnova, 2000, p. 73.

[39] Wagner, 2010, p. 513.

[40] For the discussion that follows, compare Bashmakova and Smirnova, 2000, pp. 73–74 and van der Waerden, 1985, 60–61. For the printed Italian version, see Bombelli, 1966, pp. 225–226.

[41] Cardano, 1545/1968, p. 103.

he would later in the "impossible" quadratic case (recall the discussion above), Cardano referred his readers to chapter twenty-five where he laid out various algorithmic "fixes" that resulted in real roots.

In *L'Algebra*, however, Bombelli tackled the problem head on. As he acknowledged, applying Cardano's algorithm to equation (9.5) yielded the solution $x = \sqrt[3]{2+\sqrt{-121}} + \sqrt[3]{2-\sqrt{-121}}$, while clearly $x = 4$ is a root by inspection. Bombelli proceeded actually to show that these two roots are equivalent, that is, that $x = \sqrt[3]{2+\sqrt{-121}} + \sqrt[3]{2-\sqrt{-121}}$ equals 4 algebraically.[42] His approach to this was clever; he sought to equate $\sqrt[3]{2+\sqrt{-121}}$ to a "simple" complex number $a+\sqrt{-b}$. This was possible provided he could solve

$$\sqrt[3]{2+\sqrt{-121}} = a+\sqrt{-b},$$
$$2+\sqrt{-121} = (a^3 - 3ab) + (3a^2 - b)\sqrt{-b},$$

for a and b. Clearly, this equation, as well as its "conjugate" $\sqrt[3]{2-\sqrt{-121}} = a - \sqrt{-b}$, holds if the following two conditions are satisfied: $2 = a^3 - 3ab$ and $\sqrt{-121} = (3a^2 - b)\sqrt{-b}$. Multiplying the two "conjugate" equations together then yielded $\sqrt[3]{125} = a^2 + b$ or

$$b = 5 - a^2, \tag{9.6}$$

and considering this in light of the equation for the first condition gave $4a^3 = 15a + 2$, a solution of which is clearly $a = 2$ so that $b = 1$. Finally, then,

$$\sqrt[3]{2+\sqrt{-121}} = 2+\sqrt{-1},$$
$$\sqrt[3]{2-\sqrt{-121}} = 2-\sqrt{-1},$$

and

$$x = \sqrt[3]{2+\sqrt{-121}} + \sqrt[3]{2-\sqrt{-121}} = (2+\sqrt{-1}) + (2-\sqrt{-1}) = 4.$$

---

[42] For the discussion that follows, compare van der Waerden, 1985, pp. 60–61.

Whereas Cardano had avoided confronting the problem that his solution algorithm posed, Bombelli meaningfully explored the algebraic equivalences between these new kinds of numbers—which were neither "positive nor negative"—and those numbers with which his readers were much more familiar. In so doing, he brought the complex numbers explicitly into the algebraic realm even if he was still less than perfectly comfortable with them himself. As he put it, "at first, the thing seemed to me based more on sophism than on truth, but I searched until I found the proof."[43]

Bombelli's notational conventions for and algebraic manipulations of the unknown and its powers are equally noteworthy. Unlike Pacioli, Cardano, and others, who used abbreviations for the powers of x, Bombelli adopted a notation that, at least to our modern eyes, looks remarkably like our power notation. He denoted the first power of the unknown by $\underset{\smile}{1}$, the second power of the unknown by $\underset{\smile}{2}$, etc. To multiply powers of the unknown, moreover, he added the corresponding numbers. Thus, he understood the product of the second and third powers of the unknown, namely, $\underset{\smile}{2}\,\underset{\smile}{3}$, to be $\underset{\smile}{5}$ since $2 + 3 = 5$.[44] Using this notational convention, equation (9.5) would have appeared more or less as

$$1\overset{3}{\smile}. \text{ eguale a } 15\overset{1}{\smile}.\tilde{p}.4,$$

although the "superscripts" indicating the appropriate power of the unknown would have actually appeared directly over the associated coefficient.[45]

In addition to perfecting and extending Cardano's exposition in the *Ars magna* and to developing further the notation in which it was cast, Bombelli helped to bring Diophantus's ideas back to light by treating a host of Diophantine problems in Book III of *L'Algebra*. For example, in problem thirty-eight of that book, Bombelli asked his readers to "find

---

[43] Bombelli as quoted in van der Waerden, 1985, p. 61.

[44] In his 1484 manuscript, *Triparty*, Nicolas Chuquet had already adopted a notation similar to this—without the semicircle—to denote the powers of x. See Cajori, 1928, 1:102–103 and 124–125. Bombelli's younger contemporary, the Dutch mathematician Simon Stevin (1548–1620), also devised a notational variation on this theme in his book entitled *L'Arithmétique* (1585). He denoted successive powers of the unknown by encircling the integers 1, 2, 3, etc.

[45] Compare Bombelli, 1966, p. 225, which has adopted a somewhat more modernized typesetting convention, and see Cajori, 1928, 1:125.

two squares, which when added together, make a square."[46] If this sounds familiar, that is because it reads remarkably like II.8 in Diophantus's *Arithmetica* that we discussed in chapter 4, namely, "to divide a given square number into two squares." As he treated the problem, however, Bombelli was really considering a special case of Diophantus's II.10, that is, "to find two square numbers having a given difference" (recall chapter 4, note 14). To see why, consider Bombelli's text. He asked his readers to "take a square at your pleasure (let's say 9)" and to "find two squares which subtracted one from the other leave 9."[47] To do this, Bombelli let one of the squares be x^2 and the other $(x+2)^2$, which he wrote as $\underset{\smile}{2}$ and $\underset{\smile}{2}$ $p.$ 4 $\underset{\smile}{1}$ $p.$ 4, respectively.[48] As he then explained, "subtract[ing] one from the other leaves $4x + 4$ and this equals 9." Solving this equation for x yields $1\frac{1}{4}$. Now, "its square is $1\frac{9}{16}$, and this is one of the required numbers, and the other is 9, which added to $1\frac{9}{16}$ equals $10\frac{9}{16}$, which is a square, the side of which is $3\frac{1}{4}$." In other words, $9 + \frac{25}{16} = \frac{144+25}{16} = \frac{169}{16}$. "But if," he continued, "you want the squares without fractions, you have to multiply them by 16, and [you] come up with 25 and 144, which when added together make 169."[49]

Bombelli's technique here was exactly that used by Diophantus in the more general setting (where the difference is not necessarily taken to be a square) of II.10. Taking the difference to be 60, Diophantus tersely said, "Side of one number x, side of the other x plus *any number the square of which is not greater than* 60, say 3."[50] In other words, solving the equation that we would write as $(x + 3)^2 - x^2 = 60$, Diophantus found that $x = 8\frac{1}{2}$ so that the two squares were $72\frac{1}{4}$ and $132\frac{1}{4}$. In his text, however, Bombelli went beyond Diophantus in explicitly stating "the following rule":

> If one has to find two squares, which added together make a square, take two squares different from each other, and unequal to each other. Subtract the lesser from the greater and what remains you

---

[46] Bombelli, 1579, p. 439. We thank Cristina Della Coletta, the second author's colleague in the Department of Spanish, Italian, and Portuguese at the University of Virginia, for her help with our translation of this problem.

[47] Bombelli, 1579, p. 439. The quotes that follow in this paragraph may also be found here.

[48] Note the difference in typesetting between the 1572 and the 1579 printed editions relative to the powers of the unknown.

[49] Here, the printed text has 69 instead of 169.

[50] Heath, 1964, p. 146 (our emphasis). Compare Bombelli's treatment below.

divide by the double of the side of the lesser [square], and the square of the result will be one of the squares, and the greater of the two taken will be the other.[51]

In other words, given a square $\mathbb{D}$, we want to find two squares $\mathbb{X}$ and $\mathbb{Y}$ ($\mathbb{Y} < \mathbb{X}$) such that $\mathbb{D} = \mathbb{X}^2 - \mathbb{Y}^2$. Letting $\mathbb{Y} = x$ and $\mathbb{X} = x + h$, $\mathbb{D} = \mathbb{X}^2 - \mathbb{Y}^2 = (x + h)^2 - x^2$, so that $x = \mathbb{Y} = \frac{\mathbb{D} - h^2}{2h}$. Note, however, that whereas Diophantus specified that what we have denoted here by h must be *a number the square of which is not greater than* $\mathbb{D}$, Bombelli was silent on that point.

As this brief discussion of Bombelli's work should now make clear, the 1572 published version of *L'Algebra* went farther than Cardano's *Ars magna* notationally, in its willingness to engage with a whole class of numbers—the complex numbers—that could not be understood in constructive, Euclidean geometric terms, and in its incorporation of the "new" types of indeterminate problems found in Diophantus's *Arithmetica*. At the same time that he sought to perfect and to render more transparent Cardano's mathematical ideas to potential readers, Bombelli produced a text that introduced those same readers to problems very different in spirit from what they would have found in the *libri d'abbaco*. Still, he approached algebra in essentially the same way as had his predecessors in that tradition, namely, as a series of cases dependent on underlying distinctions between types of quantities—integers, rational numbers, roots, etc.[52] While he may thus be seen as the last proponent of abbacist algebra, Bombelli was also the first mathematician to begin the humanistic integration of newly rediscovered Diophantine thought into the algebraic milieu and thereby the first mathematician fundamentally to alter it.

Another mathematician who engaged in this humanistic agenda and who had read Diophantus, although in Xylander's 1575 Latin translation, was the French mathematician, Guillaume Gosselin (d. ca. 1590). An avid student of algebra, Gosselin had read and absorbed the work of the Italians, Pacioli, Cardano, and Tartaglia (whose *Arithmetic* he translated into French), the Portuguese Nuñes, and the Germans, Rudolff, Stifel, and Schebyl in addition to that of his French predecessors, de la Roche,

---

[51] Heath, 1964, p. 440.
[52] Compare the discussion in Wagner, 2010, pp. 516–519.

Forcadel, Borrel, and Ramus.[53] It was against this rich intellectual backdrop that he incorporated his reading of Diophantus into his own four-book treatment, *De arte magna* (or *On the Great Art*) of 1577, unaware of Bombelli's work of just five years earlier. Like Bombelli, however, Gosselin was deeply impressed by the "newness" and seeming abstraction of Diophantus's approach and devoted two of the thirteen chapters of his book three to Diophantine problems cast in purely algebraic terms.[54] Of equal importance to his inclusion of Diophantine problems, however, was his sense of the purpose and scope of algebra. In his view, "whatever [algebra] may be, it is the science which, of absolutely all, occupies the first rank of dignity, because it teaches how to conduct all calculations, and if something remains unclear to us, it is the fault of the artisan and not of the art."[55] This sense of the immense potential and promise of algebra was not lost on Gosselin's younger French contemporary, François Viète (1540–1603), who both studied his works and self-consciously promoted his humanistic agenda even more aggressively.[56]

Well connected within the successive governments of Charles IX, Henry III, and Henry IV of France, Viète officially served his kings during the course of the religious wars that deeply divided the country in the second half of the sixteenth century, at the same time that he unofficially pursued his mathematical and astronomical interests. That serious sideline had resulted by 1591 in the publication in Tours of *In artem analyticem isagoge* (or *The Introduction to the Analytic Art*), a text in which Viète proposed nothing short of a complete refashioning of algebra as it was then understood. In the dedication of that work to his former student, the French humanist and Protestant activist, Catherine of Pathenay (1554–1631), Viète made his position clear. "Behold," he wrote,

> the art which I present is new, but in truth so old, so spoiled and defiled by the barbarians, that I considered it necessary, in order to introduce an entirely new form into it, to think out and publish a new vocabulary, having gotten rid of all its pseudo-technical terms lest it should retain its filth and continue to stink in the old way,

---

[53] On Gosselin's sources, see Loget, 2012, pp. 75–76 and Kouteynikoff, 2012, pp. 153–154.

[54] Kouteynikoff, 2012, pp. 154–156.

[55] Quoted in French from Gosselin's *De arte magna* in Kouteynikoff, 2012, p. 156 (our translation).

[56] On the particular flavor of humanism that developed in the French legal circles within which Gosselin and Viète moved, see Cifoletti, 1996, especially pp. 137–140.

but since till now ears have been little accustomed to it, it will be hardly avoidable that many will be offended and frightened away at the very threshold. And yet underneath the Algebra or Almucabala which they lauded and called "the great art," all Mathematicians recognized that incomparable gold lay hidden, though they used to find very little.[57]

In contrast to his predecessors, moreover, "who vowed hecatombs and made sacrifices to the Muses and to Apollo if any one would solve *some one problem* or other," Viète was able to solve problems "*freely by the score, since our art is the surest finder of all things mathematical.*" Rather than viewing algebra merely as the search for solutions of particular equations, he understood it as the analysis of an actual theory of equations.

The analytic art that Viète devised, while fully informed by the work of Cardano and his predecessors in the tradition of the *maestri d'abbaco*, also drew from the humanist and mathematical texts of Peter Ramus and Gosselin as well as from works such as Diophantus's *Arithmetica* and the 1588 Latin translation of Pappus of Alexandria's *Mathematical Collection* made by Federico Commandino (1509–1575). The latter two texts, in particular, represented more ancient knowledge—more ancient, that is, than the work of the medieval Islamic scholars—and so by humanist lights purer and more pristine knowledge *un*spoiled and *un*defiled "by the barbarians." Pappus and his successor, Theon, for example, had codified the distinction, going back at least to Plato, between analysis and synthesis. For Viète, that distinction—in light of Diophantus's kind of indeterminate analysis—proved especially suggestive.[58] As Viète explained, "analysis [was]. . . defined. . . as assuming that which is sought as if it were admitted [and working] through the consequences [of that assumption] to what is admittedly true, as opposed to synthesis, which is assuming what is [already] admitted [and working] through the consequences [of that assumption] to arrive at and to understand what is sought."[59] (Recall the discussion in chapter 4.) Although the Greeks recognized two flavors of analysis—zetetic and poristic—Viète added a third which he called rhetic

---

[57] Viète, 1591/1968, pp. 318–319. For the quotes that follow in this paragraph, see p. 319 (our emphases).

[58] On Viète's sources, see Reich, 1968 and Mahoney, 1973, pp. 26–48, and compare Parshall, 1988a, pp. 153–157.

[59] Viète, 1983, p. 11. For the next two quotations, see p. 12.

or exegetic. He distinguished between the three in this way: "It is properly zetetics by which one sets up an equation or proportion between a term that is to be found and the given terms, poristics by which the truth of a stated theorem is tested by means of an equation or proportion, and exegetics by which the value of the unknown term in a given equation or proportion is determined." In his view, then, "the whole analytical art, assuming this three-fold function for itself, may be called the science of correct discovery in mathematics."

If Viète's capstone notion of the exegetic art "encompassed nothing less than the theory of equations addressed to their solutions,"[60] his conceptualization of the zetetic art had equally important consequences for the development of algebra. As he explained, "zetetics... has its own method of proceeding. It no longer limits its reasoning to numbers, a shortcoming of the old analysts, but works with a newly discovered *symbolic* logistic which is far more fruitful and powerful than *numerical* logistic for comparing magnitudes with one another."[61] That "symbolic logistic," unlike the "numerical logistic" of Diophantus or Viète's predecessors from Fibonacci through Bombelli, "employs symbols [*species*] or signs for things as say, the letters of the alphabet." And those "things" may explicitly be *either* unknowns—for which expressions like Pacioli's co. and ce. and Bombelli's $\underline{1}$, etc. had been evolving throughout the fifteenth and sixteenth centuries—or indeterminate magnitudes. As Viète expressed it, "in order to assist this work by another device, given terms are distinguished from unknown by constant, general and easily recognized symbols, as (say) by designating unknown magnitudes by the letter A and the other vowels E, I, O, U and Y and given terms by the letters B, G, D and the other consonants."

If this seems reminiscent in principle of our modern notation of x, y, and z for unknowns and a, b, c, etc. for indeterminate magnitudes, a convention which we owe to René Descartes in the seventeenth century (see the next chapter), it is important to recognize that Viète's symbols or "species," unlike ours, carried explicit geometrical meaning. They had dimension, and only expressions of the same dimension were commensurate. For Viète,

---

[60] Mahoney, 1973, p. 34.
[61] Viète, 1983, p. 13 (our emphases). For the next two quotations, see pp. 17 and 24, respectively.

The prime and perpetual law of equations or proportions..., since it deals with their homogeneity, is called the law of homogeneous terms....

Homogeneous terms must be compared with homogeneous terms, for... it is impossible to understand how heterogeneous terms [can] affect each other. Thus,

> If one magnitude is added to another, the latter is homogeneous with the former.
> If one magnitude is subtracted from another, the latter is homogeneous with the former.
> If one magnitude is multiplied by another, the product is heterogeneous to [both] the former and the latter.
> If one magnitude is divided by another [the quotient] is heterogeneous to the former [that is, to the dividend].

Much of the fogginess and obscurity of the old analysis is due to their not having been attentive to these [rules].[62]

To Viète's way of thinking, then, the addition of a one-dimensional unknown A to a one-dimensional indeterminate B was denoted simply by $A + B$ (we would write $x + b$), but in two dimensions, the sum appeared as A square $+ B$ plane (or our $x^2 + b$) and in three dimensions as A cube $+ B$ solid (or our $x^3 + b$).

To get a sense of what he had in mind here, consider the following example:

> Let B in square $+$ D plane in A equal Z solid.
> I say that by reduction

$$A \text{ square} + \frac{D \text{ plane in } A}{B} \text{ equals } \frac{Z \text{ solid}}{B}.^{63}$$

Here, each of the terms on each side of the original equation has the same "degree of homogeneity," namely, three. Division by B, a term of

---

[62] Viète, 1983, pp. 15–16.

[63] Viète, 1983, pp. 26–27. We have preserved (in English translation) the original Viètian notation here, whereas the translator of Viète, 1983 has (unfortunately) adopted a shorthand, exponential notation. It is important to note, too, that various of the editions and translations of Viète's works utilize various notational conventions. While Viète did not employ our =, he did use $+$ and $-$.

degree one, thus yields a new equation, each term of which has degree two. Homogeneity is preserved, as required, in performing the algebraic operation.

At the same time, then, that Viète sought to purify and extend the algebra of his Islamic and even his ancient Greek predecessors, he insisted on grounding his algebra philosophically in geometrical strictures that had persisted since at least the time of Euclid, geometrical strictures that, viewed through our modern, post-Cartesian eyes, clearly mark Viète as an algebraist very much rooted in the sixteenth century. This insistence on geometric homogeneity—a convention that would effectively be abandoned within a generation—in no way prevented Viète from making key advances in developing both a general problem-solving technique and an actual theory of equations. To get a sense of his accomplishments, consider just a few examples from his works.

Responding directly to Diophantus's *Arithmetica*, Viète's *Zeteticorum libri quinque* (or *Five Books of Zetetics*), published in the same year as *The Introduction to the Analytic Art*, showcased that new analytic art by providing general treatments of many of the problems that Diophantus had considered. It opened, for example, with Viète's interpretation of the solution of Diophantus's I.1, namely, "given the difference between two roots and their sum, to find the roots" (recall the discussion in chapter 4).[64] Whereas Diophantus had merely solved a particular case in which the given difference was 40 and the given sum was 100—even though his solution technique was certainly generalizable to cases involving different specific numbers—Viète set up and solved the problem in complete generality. Letting B be the difference of the two roots and D be their sum, Viète let A be the smaller of the two roots, which made the larger $A + B$ and so the sum $A2 + B$ equals D.[65] Thus, $A2$ equals $D - B$, and A equals $D\frac{1}{2} - B\frac{1}{2}$. Similarly, if E denotes the greater root, the smaller will be $E - B$, and an analogous argument yields that E equals $D\frac{1}{2} + B\frac{1}{2}$. Viète then followed these algebraic manipulations with a purely rhetorical statement of the solution, namely, "given... the difference between two roots and their sum, the roots can be found, for half the sum of the roots

---

[64] Viète, 1983, p. 83. For the quotation that follows, see p. 84. Recall from the previous chapter that Jordanus had also treated this problem generally, but he expressed his answer completely in words.

[65] Note that Viète wrote coefficients after the variables to which they were associated.

minus half the difference is equal to the smaller root, and [half the sum of the roots] plus [half their difference is equal] to the greater." He then closed with the specific example he had encountered in the *Arithmetica*, letting $B = 40$ and $D = 100$ to solve for $A = 30$ and $E = 70$.

He extended his theory of equations to quadratics, cubics, and beyond in his posthumously published *De Æquationum recognitione et emendatione tractatus duo* (or *Two Treatises on the Understanding and Amendment of Equations*), works that Viète apparently intended to follow his *Five Books of Zetetics*. In particular, he gave what might be considered the first statement of the now-so-familiar quadratic formula in addition to providing a more truly algebraic treatment of the solution of the cubics. He approached the quadratic this way.[66] Considering A square $+ B2$ in A equals Z plane, he let $A + B$ equal E and noted that then E square equals Z plane $+ B$ square. In our notation, this simply says that if $x^2 + 2bx = c$ and $x + b = y$, then $y^2 = c + b^2$. As an immediate corollary, he concluded that "hence A, the original unknown, is

$$\ell.\overline{Z \text{ plane} + B \text{ square}} - B"$$

(here ℓ stands for the Latin "latus" or "side" and indicates taking a square root), that is, in modern notation, he gave the following formula for the solution of this particular quadratic:

$$x = \sqrt{c + b^2} - b, \tag{9.7}$$

and illustrated it in the particular example $x^2 + 2x = 20$ to get $x = \sqrt{21} - 1$.

Viète also put his brand of analysis to work on the various types of cubics and, in particular, on the case of a "cube and first power equal to the number." Recall that Cardano had looked at this same case in the context of the cubic $x^3 + 6x = 20$ and in terms of a construction that effectively completed a cube. For his part, however, Viète considered, in modern notation,

$$x^3 + 3cx = 2d, \tag{9.8}$$

and provided a sequence of more purely algebraic steps in order to effect a solution.[67] Geometrical considerations were not, however, completely absent in Viète's work. For him, the coefficient c in (9.8) was "plane" or two-dimensional (since x is linear) and the constant term d was "solid" or three-dimensional, owing to the law of homogeneity.

To solve this kind of a cubic—one, as he put it, "with a positive linear affection"—Viète effectively transformed it into "a square based on a solid root that is similarly affected" and solved the resulting quadratic equation.[68] To do this, he first introduced a new variable y such that

$$y^2 + xy = c \tag{9.9}$$

or $x = \frac{c-y^2}{y}$. Substituting this into (9.8), multiplying through both sides of the resulting equation by y^3 "and arranging everything artfully" then yields

$$(y^3)^2 + 2dy^3 = c^3,$$

the desired quadratic in y^3. Equation (9.7) now gives the solution

$$y^3 = \sqrt{c^3 + d^2} - d,$$

or

$$y = \sqrt[3]{\sqrt{c^3 + d^2} - d}. \tag{9.10}$$

Substituting this back into (9.9), Viète could then solve for x as

$$\frac{c - \left(\sqrt[3]{\sqrt{c^3 + d^2} - d}\right)^2}{\sqrt[3]{\sqrt{c^3 + d^2} - d}}.$$

By way of example, he considered $x^3 + 81x = 702$. Since $c = 27$ and $d = 351$, a bit of arithmetic perseverance yields $x = \frac{27 - \left(\sqrt[3]{27}\right)^2}{\sqrt[3]{27}}$ or $x = 6$.[69]

---

[67] See Viète, 1983, pp. 287–288 for the argument that follows, and compare van der Waerden, 1985, pp. 66–67.

[68] Viète, 1983, p. 287. The quotations that follow in this paragraph are also on this page.

[69] Viète, 1983, pp. 287–288.

As he realized, however, he could equally well have made the initial substitution[70]

$$y^2 - xy = c \tag{9.11}$$

to get $x = \frac{y^2 - c}{y}$. Substituting this into (9.8) and performing the same algebraic manipulations as before results in

$$(y^3)^2 - 2dy^3 = c^3,$$

another quadratic in y^3, the solution of which, by (9.7), is

$$y^3 = \sqrt{c^3 + d^2} + d,$$

or

$$y = \sqrt[3]{\sqrt{c^3 + d^2} + d}. \tag{9.12}$$

Substituting this back into (9.11), Viète could then also solve for x as

$$\frac{\left(\sqrt[3]{\sqrt{c^3 + d^2} + d}\right)^2 - c}{\sqrt[3]{\sqrt{c^3 + d^2} + d}}.$$

Finally, he concluded that, in general, the unknown sought in (9.8) is the difference between (9.12) and (9.10) or

$$\sqrt[3]{\sqrt{c^3 + d^2} + d} - \sqrt[3]{\sqrt{c^3 + d^2} - d},$$

a result in complete accord with Cardano's geometrical result in the *Ars magna* but expressed in Viète's notation as

$$\ell.c.\ell.\overline{B \text{ planeplaneplane} + Z\text{solidsolid} + Z\text{solid}}$$

$$- \ell.c.\ell.\overline{B \text{ planeplaneplane} + Z\text{solidsolid} - Z\text{solid}}.$$

[70] Viète, 1983, p. 288.

Viète also considered the case in which the cubic had a "negative linear affection," namely,[71]

$$x^3 - 3cx = 2d.$$ (9.13)

Once again, he introduced a new variable y, but this time such that $xy - y^2 = c$ or $x = \frac{c+y^2}{y}$. Substituting this into (9.13) yielded, after suitable manipulation, $2d(y^3) - (y^3)^2 = c^3$, a quadratic in y^3 for which he had already determined the solution, namely, mutatis mutandis,

$$y^3 = d \pm \sqrt{d^2 - c^3}.$$

As Viète concluded, however, "it is apparent...from the properties of the reduced equation that the square of [d] must be greater than the cube of [c]." In other words, whereas Bombelli had forged ahead and dealt with the "complex" implications that arose when $d^2 < c^3$, Viète insisted on maintaining the geometrically imposed restrictions on square roots.

Even though Viète may not have gone as far as Bombelli down this particular mathematical path, he was nevertheless formulating an actual *theory* of equations as further exemplified by his analysis of relationships between equations and their (positive) roots. (Again, for largely geometrical reasons, Viète did not consider negative roots.) In considering (in modern notation) the quadratic equation

$$bx - x^2 = c,$$ (9.14)

he recognized the relationship between its two positive roots and its coefficients.[72] To do this, he considered the two roots x_1 and x_2. They thus satisfy the two equations $bx_1 - x_1^2 = c$ and $bx_2 - x_2^2 = c$. Equating these yielded the expression

$$bx_1 - x_1^2 = bx_2 - x_2^2,$$

[71] See Viète, 1983, p. 289 for this and the quote that follows.
[72] See Viète, 1983, pp. 209–210.

on which Viète then performed the algebraic manipulations

$$x_2{}^2 - x_1{}^2 = bx_2 - bx_1$$

and

$$\frac{x_2{}^2 - x_1{}^2}{x_2 - x_1} = \frac{bx_2 - bx_1}{x_2 - x_1}$$

to get

$$x_2 + x_1 = b.$$

As he recognized, this meant that the coefficient of the linear term is the same as the sum of the roots. Moreover, substituting $x_2 + x_1 = b$ into $bx_1 - x_1^2 = c$ gave

$$(x_2 + x_1)x_1 - x_1^2 = x_2 x_1 = c,$$

or the fact that the constant term is the same as the product of the roots. Similarly, he realized that if x_1, x_2, and x_3 are three (positive) roots of a cubic $x^3 + bx^2 + cx = d$, then $b = -x_1 - x_2 - x_3$, $c = x_1 x_2 + x_1 x_3 + x_2 x_3$, and $d = x_1 x_2 x_3$. Analogously, he also gave results for fourth- and fifth-degree monic polynomials. In his eyes, these observations were so "beautiful" that they well served as "the end and crown" of his *Two Treatises on the Understanding and Amendment of Equations*.[73]

As this discussion of the work of Viète suggests, by the end of the sixteenth century, the plurality of algebras that had existed earlier in the century had begun not only to take on a whole new look but also to coalesce—as Peletier had advocated and foreseen—into a general problem-solving technique, the objective of which was "to solve every problem."[74] The work of Cardano, Tartaglia, and Ferrari had provided general justifications for the solutions of whole categories of cubic and

---

[73] See Viète, 1983, p. 310.

[74] Viète, 1983, p. 32. As Cifoletti has argued, Peletier was "at the origin of the French algebraic tradition," within which Gosselin, Viète, and Descartes subsequently matured as mathematicians. "Algebra became with them," in her view, "an important Latin discipline, whereas according to them, it had been nothing but an obscure technical calculation in Arabic and vernacular languages." See Cifoletti, 2004, p. 131.

quartic equations, whereas only solutions of particular special cases had been effected in prior centuries by the *maestri d'abbaco*. Moreover, while their proofs of the solutions of the cubics continued to rely on geometrical constructions, those of the quartics had largely had to leave the geometrical standard of rigor behind as did Bombelli's exploration of complex numbers as roots of equations. Viète may have continued to preserve vestiges of those earlier standards in his insistence on the law of homogeneity, but modern eyes unquestionably recognize his late sixteenth-century demonstrations as *algebraic* and not *geometric* and his approach as one geared more toward general theory building than toward individual problem solving. As the work of these mathematicians reflects, the sixteenth century represents a fascinating stage in the evolution of an algebra in the West increasingly distinct from both arithmetic and geometry, an algebra on the verge of becoming a new mathematical discipline pursued by mathematicians internationally who increasingly shared a common mathematical language.

10

From Analytic Geometry to the Fundamental Theorem of Algebra

Although Viète's work represented an advance on earlier algebraic work in the generality of its treatment of equations, his continued use of words and abbreviations rather than true symbols made it difficult for his successors actually to realize his aim to "solve every problem." In particular, the problems of motion on earth, as studied by Galileo Galilei (1564–1642), and motion in the heavens, as explored by Johannes Kepler (1571–1630), required new algebraic ideas in order for their solutions to be understood and for further questions about motion to be resolved. It was one thing to solve a cubic or quartic equation, which only had a limited number of unknown solutions, but Galileo wanted to know how a projectile moved, and Kepler sought to understand the orbit of Mars. The solution of such problems was not a *number*; it was an entire *curve*.

Galileo claimed famously in his 1623 text, "The Assayer," that nature "is written in this grand book, the universe, which stands continually open to our eyes. But the book cannot be understood unless one first learns to comprehend the language and read the letters in which it is composed. It is written in the language of mathematics, and its characters are triangles, circles, and other geometric figures without which it is humanly impossible to understand a single word of it; without these, one wanders about in a dark labyrinth."[1] For Galileo, then, to understand nature meant to understand mathematics, but he had apparently read neither Viète nor any other writers on algebra. For him, mathematics was geometry. His work on projectiles thus hearkened back to the classical Greeks and, especially, to Apollonius, who had *geometrically* defined the parabolic curve that Galileo needed to describe projectile motion.

---

[1] Drake, 1957, pp. 237–238.

Similarly, Kepler built his analysis of the elliptical orbit of Mars and the other planets on the basis of a deep understanding of Apollonius's work, again with no mention of algebra. As a result of their reliance on geometry, mathematical explanations in the texts of both men were prolix and difficult to penetrate. Neither natural philosopher realized that the algebra already being developed in his lifetime would allow for his arguments to be expressed effectively in a fraction of the space and to be rendered comprehensible ultimately to a much greater audience.

Viète's algebra, however, was not up to the task of solving the problems Kepler and Galileo posed. Ironically, two of Viète's followers—Thomas Harriot (1560–1621) and Pierre de Fermat (1601?–1665)—helped to transform that algebra into the problem-solving tool he had envisioned, but their work, although circulated in manuscript, remained unpublished and so limited in impact and influence. It was René Descartes (1596–1650), in his text *La Géométrie* published in 1637, who truly realized the grand ambitions for algebra that Viète shared with his French predecessors Peletier and Gosselin. At the same time, Descartes helped to circulate widely Fermat's ideas of identifying a geometric curve with an algebraic equation in two variables. Interestingly, because Descartes's French text ultimately reached few readers (only 500 copies were printed and not all of those had been sold or distributed by the time of Descartes's death) and because he used very complicated examples to illustrate his ideas, it was not until the *Géométrie* appeared in Latin editions in 1649 and again a decade later—each time with commentaries explaining and extending the Frenchman's ideas—that it began to exert real influence.

THOMAS HARRIOT AND THE STRUCTURE OF EQUATIONS

Little is known about the early life of Thomas Harriot, other than the fact that he entered the University of Oxford in 1577.[2] By 1585, he had joined an expedition to Virginia as an expert in cartography, ultimately writing an influential report on the colony and its native inhabitants. During the last decade of the sixteenth century and following his return from the New World, Harriot established a reputation as a mathematician and

---

[2] The best source for biographical information on Harriot is Fox, 2000, a selection of essays on various aspects of Harriot's life and work.

soon found a patron, Sir Henry Percy, who supported him in his scientific research for the remainder of his life. Among Harriot's friends was Richard Torporley (1564–1632), another Oxford graduate interested in mathematics, who, sometime around 1600, met Viète in Paris and became his amanuensis. Torporley kept Harriot informed about Viète's work, and soon Harriot became one of the few people in England familiar with the "new" algebra. A convert to Viète's ideas, the Englishman extended them by making algebraic arguments even more symbolic.

Harriot recorded many of his innovations in his *Treatise on Equations*, probably written during the first decade of the seventeenth century. The potentially groundbreaking ideas this work contained did not become more widely known until 1631, when Harriot's executors published some of them in the *Artis analyticae praxis* (or *Practice of the Analytic Art*). There is good evidence, however, that mathematicians in England, and perhaps even on the Continent, actually read Harriot's work in circulated manuscript prior to its partial publication. In his *A Treatise on Algebra* of 1685, John Wallis claimed credit for his fellow countryman Harriot for many of the discoveries generally attributed to Descartes, as did Charles Hutton in the entry on "Algebra" in his *Mathematical and Philosophical Dictionary* of 1795–1796. In particular, Hutton claimed that Harriot "shewed the universal generation of all the compound or affected equations, by the continual multiplication of so many simple ones... thereby plainly exhibiting to the eye the whole circumstances of the nature, mystery and number of the roots of equations."[3] That "exhibition" involved Harriot's modifications of Viète's notation.

Harriot took over from Viète the idea of using vowels for unknowns and consonants for knowns, although he used lower case instead of Viète's capital letters. He also consistently employed juxtaposition for multiplication, an idea that had manifested itself earlier to a limited context in the work of Michael Stifel, among others. Thus, Harriot wrote *ba* where Viète wrote "B in A," and *aaaa* in place of Viète's "A square-square." He also denoted "equals" in a way that differed from our modern sign only by virtue of two short vertical strokes between the longer parallel horizontal ones. In addition, he used the now standard < and > for "less than" and "greater than," respectively, and our usual signs for square and cube roots. Harriot was thus able to simplify Viète's rules considerably.

---

[3] Hutton, 1795–96, 1:96.

For example, in Harriot's hands, Viète's discussion of reduction (recall the previous chapter), namely,[4]

Let B in A square $+ D$ plane in A equal Z solid.
  I say that by reduction

$$A \text{ square} + \frac{D \text{ plane in } A}{B} \text{ equals } \frac{Z \text{ solid}}{B}.$$

became simply[5]

Suppose $baa + dca = zcd$, I say that $aa + \frac{dca}{b} = \frac{zcd}{b}$.

This looks quite modern, but note that where Viète used the expressions "D plane" and "Z solid," Harriot, although he used symbols, still felt constrained by the notion of homogeneity and replaced these by dc and zcd rather than by single letters. And, as noted above, he did not use exponential notation, but just repeated a letter an appropriate number of times to indicate a power.

As Hutton argued, though, perhaps Harriot's most important innovation was the idea that equations could be generated from their (positive) roots $b, c, d, \ldots$ by multiplying together expressions of the form $a - b$, $a - c, a - d, \ldots$. Thus, not only was he convinced that an equation had as many roots as its degree, but he was also led to the basic relationship between the roots and the coefficients of the equation. For example, Harriot multiplied together $a - b$, $a - c$, and $a - d$ to get the equation $aaa - baa - caa - daa + bca + bda + cda - bcd = 0$, from which he noted that the only roots are b, c, and d. The sum of these, moreover, is the negative of the coefficient of the square term, and their product is the constant term. As we saw in the previous chapter, Viète had also recognized these relationships between the roots and the coefficients of an equation, but his method of discovery was considerably more complicated than Harriot's, since he had to assume that different solutions $x_1, x_2, x_3, \ldots$ of a polynomial equation $p(x) = c$ all produced the same constant c. He then had to manipulate cleverly the resulting equations $p(x_i) = p(x_j)$ to produce the relationships. (Recall chapter 9 for the degree 2 case.)

---

[4] Viète, 1983, pp. 26–27.
[5] Harriot, 2003, p. 43.

Harriot made good use of these relationships in deriving the solutions of cubic and quartic equations. Although "omitted from these derivations are linear and quadratic equations, as being sufficiently known,"[6] it is easy enough to determine Harriot's methods for quadratics. He had to solve those in order to deal with higher-degree equations.

Consider his solution of the cubic equation he wrote as $2ccc = 3bba + aaa$.[7] He began by recalling Cardano's method and setting the unknown a equal to the difference $q - r$, with $qqq - rrr = 2ccc$ and $qr = bb$. He then set $e = q$, solved the latter equation for r and substituted into the former to get $eee - \frac{bbbbbb}{eee} = 2ccc$, which reduces to a quadratic equation in eee:

$$eeeeee - 2ccceee = bbbbbb.$$

To solve this, he added $cccccc$ to both sides to get $eeeeee - 2ccceee + cccccc = bbbbbb + cccccc$. Since the left-hand side is a perfect square, he took both square roots to get

$$eee - ccc = \sqrt{bbbbbb + cccccc}$$

and

$$ccc - eee = \sqrt{bbbbbb + cccccc}.$$

The first square root gave the solution $eee = ccc + \sqrt{bbbbbb + cccccc}$. The second square root, however, gave a negative value for eee. Since Harriot generally rejected negative solutions, he went back to the beginning, set $e = r$, and found a second positive value for eee: $\sqrt{bbbbbb + cccccc} - ccc$. It followed that the solution of the original cubic equation was

$$a = q - r = \sqrt[3]{\sqrt{bbbbbb + cccccc} + ccc} - \sqrt[3]{\sqrt{bbbbbb + cccccc} - ccc}.$$

Recall that Cardano had numerous procedures—one for each type—for solving cubic equations. Because Harriot always used substitution to

[6] Harriot, 2003, p. 204.
[7] Harriot, 2003, pp. 187–188.

reduce cubics with square terms to those without, he only needed three different formulas. Similarly, he solved quartic equations using the method of Ferrari but also showed, in that case, how to eliminate the cubic term from any quartic by a suitable substitution.

It was in connection with a quartic equation that Harriot actually dealt with negative and complex solutions and, in fact, demonstrated his understanding that the relationships between the roots and coefficients of an equation held even for such solutions. In considering the equation $12 = 8a - 13aa + 8aaa - aaaa$, for example, he saw first that two of the roots are 2 and 6. He next noted that "this equation has no other real roots besides 2 and 6, because their sum is equal to the coefficient of the cube term. If there were more [positive roots], the sum of all of them would be greater, which is against the canon for four roots."[8] He realized, however, that there ought to be four solutions, so he used the substitution $a = 2 - e$ to remove the cubic term. The new equation in e was $-20e + 11ee - eeee = 0$, the real roots of which are 0 and -4. After reducing this to a cubic, $eee - 11e = -20$, with one root equal to -4, he concluded that the sum of the remaining roots must be 4 and their product 5. Earlier, however, he had given a general formula showing how to solve this kind of problem,[9] namely, if the sum of the roots is x and the product is $xx - df$, then the roots are, in fact,

$$e = \frac{x}{2} \pm \sqrt{df - \frac{3xx}{4}}.$$

In this case, since $x = 4$, we must have $df = 11$, so the two desired values are $e = 2 + \sqrt{-1}$ and $e = 2 - \sqrt{-1}$. Therefore, the complex roots of the original equation are $a = 2 - (2 + \sqrt{-1}) = -\sqrt{-1}$ and $a = 2 - (2 - \sqrt{-1}) = +\sqrt{-1}$.

Harriot's notational innovations made Viète's ideas an even more useful tool for manipulating and solving equations. As we shall see below, Fermat, although keeping his mentor's notation, fashioned these ideas into a tool for dealing algebraically with *curves* in the plane.

---

[8] Harriot, 2003, p. 268.
[9] Harriot, 2003, p. 202.

PIERRE DE FERMAT AND THE *INTRODUCTION TO PLANE AND SOLID LOCI*

Pierre de Fermat began his study of mathematics at the University of Toulouse in the context of a curriculum that covered little more than an introduction to Euclid's *Elements*.[10] After completing his baccalaureate degree and before beginning his legal education, however, he spent several years in Bordeaux studying mathematics with former students of Viète, who, in the late 1620s, were engaged in editing and publishing their teacher's work. Fermat thus became familiar with Viète's new ideas for denoting algebraic concepts and for algebraic analysis in general. He attempted to use Viète's formulation in his own project of reconstructing Apollonius's *Plane Loci*, a work known only through comments by Pappus many centuries later. In fact, Fermat believed that Apollonius must have used some form of algebra in developing his theorems, which, as we saw in chapter 3, certainly had algebraic interpretations. Fermat thus viewed his task as one of uncovering the original analysis that underlay Apollonius's complicated geometric presentation. The germ of the subject we call analytic geometry thus came out of Fermat's application of Viète's notation and methods to these ancient problems in geometry.

Unfortunately, like Harriot, Fermat never published his work and so never felt compelled to develop it fully. Unlike the Englishman, however, Fermat maintained an extensive written correspondence with many mathematicians in France and elsewhere, in which he frequently stated results, outlined methods, and promised to fill in gaps "whenever I shall have the leisure to do so."[11] Although Fermat evidently never found the "leisure," a study of his correspondence as well as his manuscripts— the latter published by his son fourteen years after his father's death in 1665[12]—yields a reasonably complete picture of Fermat's methods.

Consider, for example, Fermat's treatment, carried out around 1635, of the following special case of a result of Apollonius: "If, from two points, straight lines are drawn to a point, and if the sum of the squares of the lines is equal to a given area, the point lies on a circle given in

---

[10] For Fermat's biography, see Mahoney, 1973.

[11] Mahoney, 1973, p. 70 and others.

[12] Fermat, 1679.

Figure 10.1. Pierre de Fermat (1601(?)–1665).

position."[13] He took the two given points A, B and bisected the line AB at E. With IE as radius (I yet to be determined) and E as center, he described a circle (figure 10.2). He then showed that, if I is chosen so that $2(\overline{AE}^2 + \overline{IE}^2) = M$ (the given area), then any point P on this circle satisfies the conditions of the theorem, namely, $\overline{AP}^2 + \overline{BP}^2$ equals M. (In modern notation, if the coordinates of A are $(-a, 0)$ and if r represents the length of the radius IE, then Fermat's condition is simply $r^2 + a^2 = \frac{m}{2}$, and therefore the equation of the circle is $x^2 + y^2 = \frac{m}{2} - a^2$.) The locus—the circle—was thus determined by the sum of the squares of two variable quantities, $\overline{AP}$ and $\overline{BP}$, and the point I was determined by its "coordinate" measured from the "origin" E.

Fermat's treatment of this problem reflected the two major ideas of analytic geometry: first, the correspondence between geometric loci and indeterminate algebraic equations in two or more variables and, second, the geometric framework for this correspondence, namely, a system of axes along which lengths are measured. It is important to note, however,

---

[13] Mahoney, 1973, pp. 102–103.

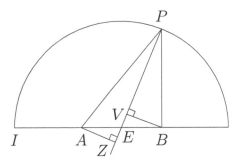

Figure 10.2.

that Fermat did not actually express the circle by means of an equation; he was likely trying to write his text in Apollonian style.

Apollonius's writings no longer confined Fermat two years later when he elaborated his new ideas on analytic geometry in his own *Introduction to Plane and Solid Loci*. He made that work's central theme clear when he wrote that "whenever two unknown magnitudes appear in a final equation, we have a locus, the extremity of one of the unknown magnitudes describing a straight line or a curve."[14] In other words, Fermat asserted that, if an equation in two unknowns results from solving a geometric problem algebraically, then the solution is a locus, either a straight line or a curve, the points of which are determined by the motion of one endpoint of a variable line segment, the other endpoint of which moves along a fixed straight line. He further asserted that if the moving line segment makes a fixed angle with the fixed line, and if neither of the unknown quantities occurs to a power greater than the square, then the resulting locus is a straight line, a circle, or one of the other conic sections.

To prove this result, Fermat considered various cases, starting with the straight line:

> Let NZM be a straight line given in position, with point N fixed. Let NZ be equal to the unknown quantity A, and ZI, the line drawn to form the angle NZI, the other unknown quantity E. If D in A equals B in E, the point I will describe a straight line given in position.[15]

[14] Smith, 1959, p. 389.
[15] Smith, 1959, p. 390. The next quotation is also on this page.

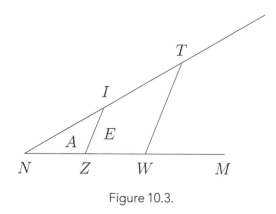

Figure 10.3.

He thus began with a single axis NZM and a linear equation (figure 10.3). (Note that he adhered closely to Viète's notation: vowels for unknowns; consonants for knowns; "in" to represent "times"; and no symbol for "equals.") To show that his equation, written in modern notation as $dx = by$, represents a straight line, he used basic geometry:

> Indeed, we have B to D as A to E. Therefore, the ratio A to E is given, as is also the angle at Z. So the triangle NIZ is given in species and the angle NIZ is also given. But the point N and the line NZ are given in position. Therefore the line NI is given in position. The synthesis is easy.

This was "easy" in the sense that it was straightforward to complete the argument by showing that any point T on NI determines a triangle TWN with $NW : TW = B : D$.

Although the basic notions of modern analytic geometry are apparent in Fermat's description, his ideas differ somewhat from those now current. First, Fermat used only one axis. He thought of a curve not as made up of points plotted with respect to two axes but as generated by the motion of the endpoint I of the variable line segment ZI as Z moves along the given axis. Fermat often took the angle between ZI and ZN as a right angle, although nothing compelled that choice. Second, for Fermat as for Viète and even Harriot (at least most of the time), the only proper solutions of algebraic equations were positive. Thus, Fermat's "coordinates" ZN and ZI—solutions to his equation D in A equals B in E—represented positive numbers, and he drew only the ray emanating from the origin into the first quadrant.

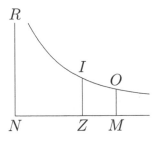

Figure 10.4.

Fermat's restriction to the first quadrant is also quite apparent in his treatment of the hyperbola. There, he assumed his readers' familiarity with Apollonius's *Conics* and its construction of the hyperbola in terms of its asymptotes:

> The second species of equations is of the form A in E equal to Z plane, in which case point I lies on a hyperbola [figure 10.4]. Draw NR parallel to ZI; through any point, such as M on the line NZ, draw MO parallel to ZI. Construct the rectangle NMO equal in area to Z plane. Through the point O describe a hyperbola between the asymptotes NR, NM. Its position is determined, and it will pass through the point I since the rectangle A in E, or NZI, is equal to the rectangle NMO.[16]

As before, Fermat did not do the synthesis, which would have amounted to showing that any point on the constructed hyperbola satisfied the original equation. And, his diagram again only showed the first quadrant. (Note that, in modern notation, Fermat was merely asserting that the equation $xy - b$ determined a hyperbola, the asymptotes of which were the coordinate axes.)

Fermat proceeded analogously to determine the curves represented by five other quadratic equations in two variables: in modern notation, $b^2 + x^2 = ay^2$ represents a hyperbola; $b^2 - x^2 = y^2$ is a circle; $b^2 - x^2 = ay^2$ an ellipse; $dy = x^2$ a parabola; and $x^2 \pm xy = ay^2$ a straight line. In each case, Fermat's argument hinged on the construction of a particular conic section according to Apollonius's procedures and showed that this conic represented the original equation. Finally, Fermat sketched a method—by

---

[16] Smith, 1959, p. 391.

showing how to change variables—for reducing any quadratic equation in two variables to one of his canonical forms. He could thus state that any quadratic equation in two variables represented a straight line, a circle, or a conic section.

In an appendix to the *Introduction*, Fermat began to explore the application of his ideas to the solutions of cubic and quartic equations. Like Archimedes and al-Khayyāmī, he showed how to find the solutions by intersecting two conic sections. Unlike his predecessors, however, since he could write out the conics using algebra, he was able to simplify the process considerably. For example, to solve the equation which we would write[17] as $x^3 + bx^2 = bc$, he set each side equal to bxy, "in order that by division of this solid, on the one hand by" x "and on the other by" b, "the matter is reduced to quadratic loci."[18] In other words, the two new equations were $x^2 + bx = by$ and $c = xy$, the first a parabola and the second a hyperbola. In this way, the x-coordinate of the intersection of the two curves (and Fermat only interested himself in a single intersection) determined a solution of the original equation.

ALBERT GIRARD AND THE FUNDAMENTAL THEOREM OF ALGEBRA

Harriot and Fermat were not the only mathematicians influenced by Viète. Albert Girard (1595–1632), who hailed originally from Lorraine but who had studied in Leiden in the Netherlands and had served as a military engineer in the army of Frederick Henry of Nassau, had also read Viète's work. Girard did not use Viète's notation, however. Probably influenced by his reading of Bombelli, he created some notation of his own, including fractional exponents ("the numerator is the power and the denominator the root"[19]), while, at the same time, continuing to employ the then-current notation for higher roots (for example, $\sqrt[3]{}$ for the cube root) as an alternative to the exponent $\frac{1}{3}$. His usage of fractional exponents, however, was not quite the same as ours. He wrote $(\frac{3}{2})49$ to

---

[17] Note that, like Viète, Fermat used vowels to represent variables and consonants to represent indeterminate magnitudes, unlike our modern usage.
[18] Mahoney, 1973, p. 126.
[19] Black, 1986, p. 107.

denote the cube of the square root of 49, that is, 343, and $49(\frac{3}{2})$ to mean what we now write as $49x^{3/2}$.

Girard also adopted this notation for polynomial equations. For example, in his 1629 work, *Invention nouvelle en l'algèbre* (or *A New Discovery in Algebra*), he wrote "let 1(3) be equal to 6(1)+40"[20] to represent what we would write as $x^3 = 6x + 40$. In this context, having previously written out the steps of Cardano's solution of this cubic, he presented a new technique. Noting that he could divide both sides by 1(1), that is, by x, to get, in modern notation, $x^2 = 6 + \frac{40}{x}$, he remarked that this equation meant that "6 with one of the aliquot parts of 40 will be the square of the other part." A simple check showed that $6 + 2$ is not the square of 20 and $6 + 4$ is not the square of 10, but since $6 + 10$ is the square of 4, 4 is a solution of the equation.

It is important to realize here that Cardano's method apparently gives a single solution, and Girard's technique of aliquot parts may or may not yield further information. Girard, however, developed an actual procedure for finding additional solutions. He began by systematizing some of the work of Viète (and, coincidentally, of Harriot). Specifically, he considered what he termed "factions" or what we would call elementary symmetric functions in n variables:

> When several numbers are proposed, the entire sum may be called the first faction; the sum of all the products taken two by two may be called the second faction; the sum of all the products taken three by three may be called the third faction; and always thus to the end, but the product of all the numbers is the last faction. Now, there are as many factions as proposed numbers.[21]

By way of example, Girard pointed out that for the numbers 2, 4, 5, the first faction is 11, their sum; the second is 38, the sum of the products of all possible pairs; and the third is 40, the product of all three numbers. He also noted that the Pascal triangle of binomial coefficients, which Girard called the "triangle of extraction," tells how many terms each of the factions contains. In the case of four numbers, the first faction contains 4 terms, the second 6, the third 4, and the fourth and last 1.

---

[20] Black, 1986, p. 132. The following quotation is also on this page.
[21] Black, 1986, p. 138.

Girard's basic result in the theory of equations was the following theorem—asserted confidently after giving all the appropriate definitions but ultimately stated without proof:

> Every algebraic equation... admits of as many solutions as the denomination of the highest quantity indicates. And the first faction of the solutions is equal to the [coefficient of the second highest] quantity, the second faction of them is equal to the [coefficient of the third highest] quantity, the third to the [fourth], and so on, so that the last faction is equal to the [constant term]—all this according to the signs that can be noted in the alternating order.[22]

By this last statement about signs, Girard meant that things first needed to be arranged so that the degrees alternate on each side of the equation. Thus, $x^4 = 4x^3 + 7x^2 - 34x - 24$ should be rewritten as $x^4 - 7x^2 - 24 = 4x^3 - 34x$. Since the roots of this equation are 1, 2, -3, and 4, the first faction equals 4, the coefficient of x^3; the second -7, the coefficient of x^2; the third -34, the coefficient of x; and the fourth -24, the constant term. Similarly, the equation $x^3 = 167x - 26$ can be rewritten as $x^3 - 167x = 0x^2 - 26$. Because he could find the solution -13 by checking divisors of 26 as above, his result implies that the product of the two remaining roots is 2, while their sum is 13. To find these simply required solving the quadratic equation $x^2 + 2 = 13x$, the roots of which are $6\frac{1}{2} + \sqrt{40\frac{1}{4}}$ and $6\frac{1}{2} - \sqrt{40\frac{1}{4}}$. In a different problem, Girard provided a geometric meaning of a negative solution of an equation: "The minus solution is explicated in geometry by retrograding; the minus goes backward where the plus advances."[23]

Even more importantly, the first part of Girard's theorem asserted what is now called the fundamental theorem of algebra, namely, that the number of solutions of every polynomial equation equals its degree ("the denomination of the highest quantity"). As his examples illustrated, Girard acknowledged that he had to deal with negative solutions and that a given solution could occur with multiplicity greater than one. He also fully realized that imaginary solutions (which he called impossible) had to be included in his count. In considering $x^4 + 3 = 4x$, for instance,

---

[22] Black, 1986, p. 139.
[23] Black, 1986, p. 145.

he noted that the four factions are 0, 0, 4, and 3. Since 1 is a solution of multiplicity 2, the two remaining solutions have the property that their product is 3 and their sum is -2. It follows that they are, in fact, $-1 \pm \sqrt{-2}$. Girard was categorical and unabashed in his acceptance of these roots. As he put it, "someone could ask what good these impossible solutions are. I would answer that they are good for three things: for the certainty of the general rule, for being sure that there are no other solutions, and for its utility."[24] Like Harriot, then, Girard had developed keen insight into how equations are constructed from their roots, even though there is no evidence that he knew of Harriot's work.

RENÉ DESCARTES AND *THE GEOMETRY*

Another key figure in the seventeenth-century development of algebra, René Descartes, *published* his ideas on analytic geometry in his 1637 text, *La Géométrie*, the very same year in which Fermat wrote his *unpublished* work on *Plane and Solid Loci*. Descartes was born in the west-central French city of La Haye (now La Haye–Descartes) near Tours into a family of the old French nobility.[25] Sickly throughout his youth and allowed to rise late during his school years, Descartes accustomed himself to morning meditation at the prestigious Jesuit *collège* of La Flèche in Anjou. Boarding there from 1606 until 1614, he imbued a classical curriculum that included languages, history, oratory, poetry, mathematics, ethics, theology, philosophy, law, medicine, and other sciences.[26] He soon proceeded to the University of Poitiers, where he graduated in civil and canon law in 1616.

Following tours of duty during the Thirty Years War—first in the Dutch Republic under Maurice of Nassau and later in central Europe in the service of Duke Maximilian of Bavaria—and after a sojourn in Paris from 1625 to 1628, Descartes settled in Holland to pursue what had become his lifelong goal of creating a new philosophy suited to discovering truth about the world. In particular, he resolved to accept as true only ideas so clear and distinct that they were beyond doubt and to follow a mathematical model of reasoning that discerned new truths through

[24] Black, 1986, p. 141.
[25] On Descartes's biography, see Gaukroger, 1995.
[26] Descartes, 1637/1968, p. 30.

Figure 10.5. René Descartes (1596–1650).

simple, logical steps. By 1630, he had enrolled as a student of Jacobus Golius (1596–1667), the professor of mathematics at the University of Leiden, and had embarked on a serious pursuit of properly mathematical ideas. It is thus surprising that later in his life Descartes claimed never to have read the works of Viète, which were surely available in the Netherlands in the 1630s, nor to know of Harriot's work, although there is evidence that some of Harriot's manuscripts circulated on the continent at that time.

Descartes's first major work combined both his philosophical and his mathematical lines of thought. Published in 1637 and entitled *Discours de la méthode pour bien conduire sa raison et chercher la vérité dans les sciences* (or *Discourse on the Method for Rightly Directing One's Reason and Searching for Truth in the Sciences*), it was accompanied by three essays on optics, meteorology, and geometry, respectively, designed to show the method's efficacy. In the third and most influential of these essays, *La Géométrie*, Descartes introduced both the algebraic symbolism we use today and the basic technique of relating this algebra to geometry, a technique the further development of which culminated in the modern subject of analytic geometry. In the latter regard, his version of analytic geometry was similar to that of his contemporary Fermat, but whereas Fermat tended to begin with an algebraic equation and interpret it geometrically in terms of a particular curve, Descartes started with a geometrical

DISCOURS
DE LA METHODE
Pour bien conduire fa raifon,& chercher
la verité dans les fciences.

Plus

LA DIOPTRIQVE.

LES METEORES.

et

LA GEOMETRIE.

Qui font des effais de cete Methode.

a Leyde
De l'Imprimerie de Ian Maire.

cIↄ Iↄ c xxxvii.

Auec Priuilege.

Figure 10.6. Title page of *Discours de la méthode* by René Descartes, from the Plimpton Collection of the Rare Book and Manuscript Library, Columbia University.

problem and cast it algebraically. As he put it, "any problem in geometry can easily be reduced to such terms that a knowledge of the lengths of certain straight lines is sufficient for its construction."[27]

---

[27] Descartes, 1637/1954, p. 2.

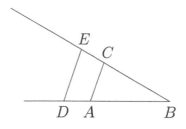

Figure 10.7.

Descartes opened his *Geometry* by showing how to perform the elementary arithmetic operations on geometrical line segments. He then allowed that "often it is not necessary thus to draw the lines on paper, but it is sufficient to designate each by a single letter"[28] and proceeded to denote line segments by letters—say, a and b—and to designate the sum, difference, product, and quotient by our standard expressions[29] $a + b$, $a - b$, ab, and $\frac{a}{b}$. He also used exponential notation—a^2, a^3, and so on (except that sometimes he wrote aa instead of a^2)—to represent powers.

In carrying out all of these algebraic operations, Descartes took another major step. He always thought of their results as line segments. That is, ab and a^2 were no longer surfaces, but lines; and a^3 was not a solid, but just a line. Higher powers no longer presented a problem, either. After all, a^4 did not go against three-dimensional nature; it was simply the product of the lines a and a^3. (See figure 10.7; here, we take AB as unity, BD as a^3, and BC as a. Then, with DE drawn parallel to AC, we have $BE : BC = BD : AB$, or $BE : a = a^3 : 1$, and BE is therefore the line representing a^4.) The only brief bow Descartes made to the homogeneity requirements carefully retained by both Viète and Harriot was his acknowledgment that any algebraic expression could be interpreted as including as many powers of unity as necessary for this purpose. In fact, he freely added algebraic expressions regardless of the power of the terms, and he did all of this manipulation in a notational convention still in use today, namely, one in terms of letters near the end of the alphabet for unknowns and those near the beginning for knowns.

Descartes's usage is exemplified in the constructions near the beginning of the first of his *Geometry*'s three books where he determines the

---

[28] Descartes, 1637/1954, p. 4.

[29] This made Descartes one of the first mathematicians, besides Harriot, to use juxtaposition rather than a word to represent multiplication.

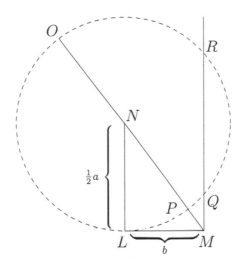

Figure 10.8.

(real) solutions of quadratic equations. For example, to find the solution of the quadratic equation $z^2 = az + b^2$, Descartes constructed a right triangle NLM with $LM = b$ and $LN = \frac{1}{2}a$ (figure 10.8).[30] Extending the hypotenuse to O, where $NO = NL$ and constructing the circle centered on N with radius NO, he concluded that OM is the required value z, because the value of z is given by the standard formula

$$ z = \frac{1}{2}a + \sqrt{\frac{1}{4}a^2 + b^2}. $$

Under the same conditions, MP is the solution of $z^2 = -az + b^2$, while if MQR is drawn parallel to LN, then MQ and MR are the two solutions of $z^2 = az - b^2$.

It was only at the end of the first book that Descartes introduced the ideas of analytic geometry, and he did so not like Fermat, who had considered just simple problems, but rather with a discussion of one of the classic problems of antiquity. In his problem of four lines, Apollonius had sought points from which lines drawn to four given lines at given angles satisfy the condition that the product of two of the line lengths bears a given ratio to the product of the other two. Descartes had

[30] Descartes, 1637/1954, pp. 12–15.

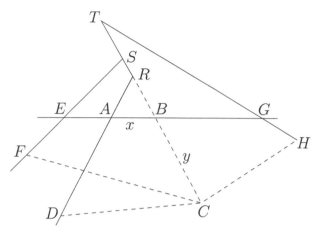

Figure 10.9.

encountered and discussed this problem and its generalization by Pappus to arbitrarily many lines with his professor, Golius, as early as 1631 and had arrived at the solution he presented in his *Geometry* sometime over the course of the winter of 1631–1632.[31] As he laid it out in print in 1637, "since there are always an infinite number of different points satisfying these requirements, it is also required to discover and trace the curve containing all such points."[32]

To solve the original four-line problem, Descartes introduced a coordinate axis to which all the lines as well as the locus of the solution are referred. That is, he set x as the length of segment AB along the given line EG and y as the length of segment BC to be drawn at the given angle to EG, where C is one of the points satisfying the requirements of the problem (figure 10.9). The lengths of the remaining line segments—CH, CF, and CD (drawn to the given lines TH, FS, and DR, respectively)— can each be expressed as a linear function of x and y.

For example, since all angles of the triangle ARB are known, the ratio $BR:AB = b$ is also known, and it follows that $BR = bx$ and $CR = y + bx$. Similarly, since the three angles of triangle DRC are known, so is the ratio $CD:CR = c$, and therefore $CD = cy + bcx$. Because the problem involves comparing the products of certain pairs of line lengths, it follows that the equation expressing the desired locus is quadratic in x and y. Descartes,

[31] Gaukroger, 1995, pp. 210–217.
[32] Descartes, 1637/1954, p. 22.

however, had shown how to construct solutions of quadratic equations. Thus, as many points of the locus as desired can be constructed, and the required curve can be drawn. It is Descartes's introduction of curves expressed algebraically as solutions to problems that marks the new, *dynamic* stage of algebra. Although Descartes himself did not deal with "motion," those who studied his work soon realized the importance of analytic geometry for expressing the paths of objects in motion.

Descartes returned to the Pappus problem in his *Geometry*'s second book and showed that the curve given by the quadratic equation in two variables is either a circle or one of the conic sections, depending on the values of the various constants involved.[33] This was not a new result; ancient Greek mathematicians already knew it. Descartes's novelty lay in showing that his new methods enabled him to solve the Pappus problem for *arbitrarily* many lines. He considered, for example, four equally spaced parallel lines and a fifth line perpendicular to each of the others (figure 10.10). He supposed that the lines drawn from the required point meet the given lines at right angles, and he sought the locus of all points such that the product of the lengths of the lines drawn to three of the parallel lines is equal to the constant spacing times the product of the lengths of the lines drawn to the remaining two lines.

To solve this problem, he let the four parallel lines be L_1, L_2, L_3 and L_4, the perpendicular line be L_5, and the constant spacing be a. For P a point satisfying the problem, he took d_1, d_2, d_3, d_4, and d_5 as the perpendicular distances from P to the five line segments. As before, he related all the line segments to two of them, namely, $d_5 = x$ and $d_3 = y$, to get $d_1 = 2a - y$, $d_2 = a - y$, and $d_4 = y + a$. The conditions of the problem translated to $d_1 d_2 d_4 = a d_3 d_5$, or $(2a - y)(a - y)(y + a) = ayx$, or, finally, to the cubic equation $y^3 - 2ay^2 - a^2 y + 2a^3 = axy$.

In developing techniques to trace the curve determined by this equation, as well as those determined by higher-degree equations, Descartes generalized the basic Euclidean tools of straightedge and compass, but he also wanted to construct actual points on these curves. Thus, for equations of degree greater than two, he needed to solve polynomial equations. Near the beginning of the third and final of his *Geometry*'s books, he stated almost verbatim Girard's result that "every equation

---

[33] See Descartes, 1637/1954, p. 48, as well as pp. 67–79. Bos, 2001, pp. 313–334 provides a detailed discussion of Descartes's solution to this problem.

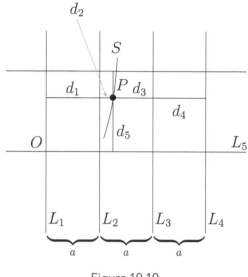

Figure 10.10.

can have as many distinct roots as the number of dimensions of the unknown quantity in the equation."[34] Descartes used "can have" rather than Girard's "admits of" because he only considered distinct roots and because, at least initially, he did not take imaginary roots into account. Later, however, he noted that roots are sometimes imaginary and that "while we can always conceive of as many roots for each equation as I have already assigned [that is, as many as the degree], yet there is not always a definite quantity corresponding to each root so conceived of."[35]

Just as Harriot had done, Descartes showed explicitly how equations are constructible from their solutions. Thus, if $x = 2$ or $x - 2 = 0$ and if $x = 3$ or $x - 3 = 0$, then the product of the two equations is $x^2 - 5x + 6 = 0$, an equation of degree 2 with the two roots 2 and 3. Again, if this latter equation is multiplied by $x - 4 = 0$, an equation of degree 3 results, namely, $x^3 - 9x^2 + 26x - 24 = 0$, with the three roots 2, 3, and 4. Multiplying further by $x + 5 = 0$, an equation with a "false" root 5, produces the fourth-degree equation $x^4 - 4x^3 - 19x^2 + 106x - 120 = 0$ with four roots, three "true" and one "false." Descartes concluded that

[34] Descartes, 1637/1954, p. 159.
[35] Descartes, 1637/1954, p. 175.

It is evident from the above that the sum of an equation having several roots [that is, the polynomial itself] is always divisible by a binomial consisting of the unknown quantity diminished by the value of one of the true roots, or plus the value of one of the false roots. In this way, the degree of an equation can be lowered. Conversely, if the sum of the terms of an equation is not divisible by a binomial consisting of the unknown quantity plus or minus some other quantity, then this latter quantity is not a root of the equation.[36]

Although this is the earliest statement of the modern factor theorem, Descartes did not give a complete proof; he just asserted that the result is "evident."

Similarly, he also stated without proof the result known today as Descartes's rule of signs:

An equation can have as many true [positive] roots as it contains changes of sign, from + to − or from − to +; and as many false [negative] roots as the number of times two + signs or two − signs are found in succession.[37]

By way of illustration, he considered the same fourth-degree equation as above. It has three changes of sign and one pair of consecutive negative signs. Thus, it can have up to three positive roots and one negative one. Descartes asserted, without explanation, that "we know" that these are exactly the number of positive and negative roots. Of course, he already knew the roots explicitly. Curiously, he made no further mention of this result in his book.

Although Descartes presented Cardano's solutions for cubic equations, he also suggested an alternative method of solving them when all the coefficients were integers. Namely, one must examine all the factors a of the constant term and try to divide the polynomial by $x \pm a$. If any of these divisions is successful, then one of the roots has been found, and the equation has been reduced to a quadratic solvable by the quadratic formula. He proposed a similar strategy for quartic equations. If, however, no binomial of the form $x \pm a$ divides the original

---

[36] Descartes, 1637/1954, pp. 159–160.
[37] Descartes, 1637/1954, p. 160.

polynomial, the coefficient of the third-degree term should be eliminated by an appropriate substitution to give $x^4 \pm px^2 \pm qx \pm r = 0$. Without explanation, he next suggested writing down the cubic equation in y^2,

$$y^6 \pm 2py^4 + (p^2 \pm 4r)y^2 - q^2 = 0,$$

and trying to solve it.[38] (He did give specific rules for when to use the plus sign and when to use the minus sign.) If this cubic can be solved for y^2, then the original equation can be written as a product of the two quadratics $x^2 - yx + \frac{1}{2}y^2 \pm \frac{1}{2}p \pm \frac{q}{2y} = 0$ and $x^2 + yx + \frac{1}{2}y^2 \pm \frac{1}{2}p \pm \frac{q}{2y} = 0$, again with appropriate decision rules for the signs. "It is then easy to determine the roots of the proposed equation, and consequently to construct the problem of which it contains the solution, by the exclusive use of circles and straight lines."[39]

For higher-degree polynomial equations, Descartes also recommended that one "try to put the given equation into the form of an equation of the same degree obtained by multiplying together two others, each of a lower degree. If, after all possible ways of doing this have been tried, none has been successful, then it is certain that the given equation cannot be reduced to a simpler one."[40] In this case, a geometrical construction was required, and, near the end of the third book, Descartes demonstrated explicitly some construction methods for equations of higher degree. In particular, for equations of degrees three and four, he used the intersection of a parabola and a circle. His procedures were similar to those of al-Khayyāmī, but, unlike his Islamic predecessor, Descartes realized that certain intersection points represented negative (false) roots of the equation and also that "if the circle neither cuts nor touches the parabola at any point, it is an indication that the equation has neither a true nor a false root, but that all the roots are imaginary."[41]

Descartes further showed how to solve equations of degree greater than four by intersecting a circle with a curve of degree greater than

---

[38] If we assume that the original quartic equation can be factored as $(x^2 - yx + f)(x^2 + yx + g)$, then multiplying out and equating coefficients yields three equations for r, q, and p in terms of f, g, and y. Eliminating f and g from these equations produces the sixth-degree equation for y.

[39] Descartes, 1637/1954, p. 184.

[40] Descartes, 1637/1954, p. 192.

[41] Descartes, 1637/1954, p. 200.

two, constructed in a particular way. Although he only briefly sketched his methods and applied them to just a few examples, Descartes believed that it was "only necessary to follow the same general method to construct all problems, more and more complex, ad infinitum; for in the case of a mathematical progression, whenever the first two or three terms are given, it is easy to find the rest."[42] His suggestion, however, of finding new geometrical means for solving equations, fell on deaf ears. After all, Descartes had contributed to making late seventeenth- and eighteenth-century mathematicians much more interested in solving equations—finding unknowns—algebraically. Moreover, one of the basic messages of his *Geometry* was that algebra should be used to solve geometric problems whenever possible and not the other way around. So even though Descartes's introduction to analytic geometry—not Fermat's—was published, it soon became clear that Fermat's approach, wherein an equation in two variables determines a curve, was the more fruitful for applying algebra in the solution of many of the problems then arising from the physics of motion. That is, if one knew that the solution to a physical problem was given by an equation, one just had to determine exactly what kind of curve the equation represented. It was no longer necessary actually to be able to construct the curve with straightedge, compass, or any other tool.

JOHANN HUDDE AND JAN DE WITT, TWO COMMENTATORS ON *THE GEOMETRY*

Descartes recognized that his *Geometry* was unconventional. It was written in French rather than Latin; it treated difficult examples; and it frequently had large gaps in its arguments. As he wrote, "I hope that posterity will judge me kindly, not only as to the things which I have explained, but also as to those which I have intentionally omitted so as to leave to others the pleasure of discovery."[43] A few years after the publication of his book, however, he encouraged others to translate it into Latin, to publish commentaries on it, and to carry some of its arguments

---

[42] Descartes, 1637/1954, p. 240. See Bos, 2001, pp. 363–382 for a detailed discussion of all of these constructions.
[43] Descartes, 1637/1954, p. 240.

further. Thus, Frans van Schooten (1615–1660) translated the *Geometry* into Latin and published it with commentary first in 1649 and then again in 1659–1661 with more extensive commentary and additions. Two of his students at the University of Leiden, Johann Hudde (1628–1704) and Jan de Witt (1623–1672), continued working in this spirit. Hudde dealt extensively with solving equations, while de Witt combined and systematized the approaches to analytic geometry of both Descartes and Fermat.

In his "De reductione aequationum" (or "On the Reduction of Equations") of 1657, Hudde expanded on Descartes's suggestion that the best way to handle equations of degree higher than three was to write them, if possible, as products of equations of lower degree. To this end, he explored various possibilities for factoring, and, although in limited cases he was able to determine when a fourth-degree polynomial could be divided by a quadratic, his attempts in higher-degree cases were largely fruitless. For example, when he tried to factor a sixth-degree polynomial as the product of a quartic and quadratic, he was left with the problem of determining the appropriate coefficients of an equation of degree fifteen. Similarly, when he attempted to factor a fifth-degree polynomial as the product of a quadratic and a cubic, he needed to solve a tenth-degree equation.

Hudde is best known for his rule for determining whether an equation has a repeated root:[44] if a polynomial $f(x) = a_0 + a_1x + a_2x^2 + \cdots + a_nx^n$ has a double root $x = \alpha$, and if p, $p + b$, $p + 2b$, ..., $p + nb$ is an arithmetic progression, then the polynomial $pa_0 + (p + b)a_1x + (p + 2b)a_2x^2 + \cdots + (p + nb)a_nx^n$ also has the root $x = \alpha$. If the arithmetic progression is $0, 1, 2, \ldots, n$, then this rule is easily verified using differential calculus, but Hudde did not justify it that way. In fact, he initially gave no explanation for it at all, sketching an argument only later in his "Epistola secunda, de maximis et minimis" (or "Second Letter on Maxima and Minima") of 1658. He also showed—without infinitesimal arguments—how to use his method to determine maxima and minima of polynomials.

Van Schooten had also instilled an interest in the mathematical ideas generated by Descartes and his contemporaries in Jan de Witt. In 1649 at

---

[44] Hudde's rule is explained more fully in Stedall, 2011, pp. 52–55.

the age of twenty-three, de Witt wrote the *Elementa curvarum linearum* (or *Elements of Curves*) in which he treated the subject of conic sections from both a synthetic and an analytic point of view.[45] He devoted the first of this work's two books to developing the properties of the various conic sections using the traditional methods of synthetic geometry. The second book extended Fermat's ideas into a complete algebraic treatment of the conics beginning with equations in two variables. Although the latter book's methodology resembled Fermat's, its notation was Cartesian. For example, de Witt proceeded, like Fermat, to show that $y^2 = ax$ represents a parabola. He also sketched the graphs of parabolas determined by such equations as $y^2 = ax + b^2$, $y^2 = ax - b^2$, and $y^2 = b^2 - ax$ as well as of the equations formed from these by interchanging x and y. De Witt pushed even further, however, by considering in detail more complicated equations such as

$$y^2 - \frac{bxy}{a} = -\frac{b^2x^2}{4a^2} + bx + d^2.$$

Setting $z = y - \frac{bx}{2a}$ or $y = z + \frac{bx}{2a}$ reduced this equation to $z^2 = bx + d^2$, clearly a parabola. He then showed how to use this transformation to draw the original locus.

De Witt also gave detailed treatments of both the ellipse and the hyperbola, presenting standard forms first and then showing how other equations can be reduced to one of these by appropriate substitutions. Although he did not state the conditions on the original equation that determine whether the locus is a parabola, ellipse, or hyperbola, it is easy enough to discover these by analyzing his examples. He concluded his work by remarking that any quadratic equation in two variables can be transformed into one of the standard forms and therefore represents a straight line, circle, or conic section. Although both Fermat and Descartes had stated this same result, de Witt was the first to provide all the details needed to solve the locus problem for quadratic equations.

---

[45] The two volumes of this work are now available in English translation as Grootendorst and Bakker, 2000 and Grootendorst et al., 2010, with introductions giving their context.

L I B. II. C A P. II.   263

explicari, determinari, ac demonſtrari queant; obſervatâ ſolum-
modo diversâ linearum poſitione, quæ ex ſignorum + & — differentia oriri debet, cumque omnes ſimilium locorum caſus mox
per generalem Regulam ſim exhibiturus.

Exempla reductionis æquationum ad formulam Theo-
 rematis V I I I.

Si æquatio ſit $yy - \frac{bxy}{a} \infty - \frac{bbxx}{4aa} + bx + dd$, aſſumpto juxta

Regulam $z \infty y - \frac{bx}{2a}$, erit $y \infty z + \frac{bx}{2a}$. quo ſubſtituto in locum
ipſius y, & ejuſdem quadrato loco yy, omiſſisque iis, quæ ſe in-
vicem tollunt, atque omnibus ritè ordinatis, æquatio ſuperior ſe-
quenti formâ erit induta: $zz \infty bx + dd$.

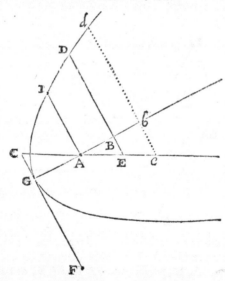

Vnde apparet, eandem eſſe reductam ad formulam Theore-
matis V I I I, ac proinde Locum quæſitum eſſe Parabolam. Ad
cujus particularem deſcriptionem eſto in adjuncta figura ipſius x
initium immutabile punctum A, atque eadem x à dicto puncto A
per

Figure 10.11. Problem on parabolas from *Elementa curvarum linearum* by Jan de Witt, in the 1659 edition of Descartes's *Geometria*, from the David Eugen Smith Collection of the Rare Book and Manuscript Library, Columbia University.

ISAAC NEWTON AND THE *ARITHMETICA UNIVERSALIS*

A member of the generation succeeding that of Hudde and de Witt, the Englishman Isaac Newton (1642–1727) had lectured on algebra as Cambridge's Lucasian Professor for over a decade until, in 1683, he finally complied with the rules of his professorship and deposited carefully dated copies of his lectures in the Trinity College library. Years later, his successor as Lucasian Professor, William Whiston (1667– 1752), prepared the lectures for publication over Newton's protestations, and they finally appeared in 1707 as the *Arithmetica universalis* (or *Universal Arithmetic*). Despite Newton's misgivings about the book, it proved very popular, going through numerous editions in Latin, English, and French, into the early nineteenth century.[46] Although the text begins at a very elementary level—with the basic rules for the arithmetic of signed numbers—it quickly develops into a detailed discussion of the solution of algebraic equations and, especially, on the use of such equations in solving interesting problems taken from geometry and physics.

Unfortunately, Newton had made no attempt in his lectures to justify the various theorems he asserted about equations. For example, he laid out a careful procedure for finding possible linear divisors of polynomials that involved first substituting consecutive terms of the arithmetic progression $3, 2, 1, 0, -1, -2, \ldots$ into the polynomial and then looking for arithmetic progressions among the divisors of the values resulting from those substitutions. What Newton found, however, were only "possible" divisors, and he noted that "if no divisor occur[s] by this method, ... we are to conclude that that quantity does not admit a divisor of one dimension."[47] Similarly, he discussed ways of finding quadratic divisors, again concluding that it might not be possible to uncover them.

Newton also explored the relationships between the coefficients of a polynomial and its roots and gave formulas to determine the sums of various integral powers of the roots of a polynomial equation. Stating

---

[46] Westfall, 1980, pp. 648–649.

[47] Newton, 1728, p. 40 = Newton, 1707, p. 45. The second Latin edition (1722) differs very little from the first edition, except for the ordering of the problems. In references to Newton's work, we give page numbers in both the 1728 English translation of the second Latin edition and the 1707 original Latin edition, which is available online.

Descartes's rule of signs, he produced his own rule for determining how many "impossible," that is, complex, roots an equation has, but he gave only examples illustrating it as opposed to justifications proving it. Likewise, he stated without real proof the fundamental theorem of algebra, in his words, "an equation may have as many roots as it has dimensions, and not more."[48] By way of justification, he repeated Descartes's (and Harriot's) technique for constructing equations by multiplying together polynomials of the form $x \pm a$. Evidently, techniques—not proofs—were what his audience desired, and Newton produced those in abundance, including the quadratic formula as well as Cardano's cubic formula.

One of Newton's new techniques dealt with "the transformation of two or more equations into one, in order to exterminate [that is, eliminate] the unknown quantities."[49] For example, "when the quantity to be exterminated is only of one dimension in both equations, both its values are to be sought. . . and the one made equal to the other."[50] Again, "when at least in one of the equations the quantity to be exterminated is only of one dimension, its value is to be sought in that equation, and then to be substituted in its room in the other equation."[51] Things were more difficult if the quantity to be eliminated was of a higher power in both equations. In that case, Newton suggested solving for the greatest power of the unknown in each equation and, if these were not the same, multiplying the equation involving the smaller power by an appropriate power of the unknown to make that power the same in both. At that point, the highest power of the unknown could be eliminated from the two equations, with the process repeated to eliminate the next lower power. Newton admitted that "when the quantity to be exterminated is of several dimensions, sometimes there is required a very laborious calculus to exterminate it out of the equations."[52] He thus offered several suggestions for reducing the labor.

Newton also expended considerable energy in the *Arithmetica universalis* in showing how to translate problems into algebraic equations. Many

[48] Newton, 1728, p. 191 = Newton, 1707, p. 237.

[49] Newton, 1728, p. 60 = Newton, 1707, p. 69.

[50] Newton, 1728, p. 61 = Newton, 1707, p. 70.

[51] Newton, 1728, p. 62 = Newton, 1707, p. 71.

[52] Newton, 1728, p. 65 = Newton, 1707, p. 74. Recall from chapter 5 that Zhu Shijie also developed a "very laborious calculus" to eliminate unknowns from systems of equations.

of his "word problems" are very familiar, since versions, such as the following two examples, still appear in algebra texts today:

> If two post-boys, A and B, 59 miles distance from one another, set out in the morning in order to meet. And A rides 7 miles in 2 hours, and B 8 miles in 3 hours, and B begins his journey one hour later than A; to find what number of miles A will ride before he meets B.[53]

> If a scribe can in 8 days write 15 sheets, how many such scribes must there be to write 405 sheets in 9 days?[54]

Unlike most current algebra textbook writers, however, Newton almost always generalized his problems. Thus, the problems above took the following forms:

> Having given the velocities of two moveable bodies, A and B, tending to the same place, together with the interval or distance of the places and times from, and in which they begin to move; to determine the place they shall meet in.[55]

> Giving the power of any agent, to find how many such agents will perform a given effect a in a given time b.[56]

In these cases, Newton produced a formula for the answer and then showed how to resolve specific problems using it.

Newton treated problems in geometry and physics differently. They were always cast quite generally—stated using letters rather than specific numbers for the coefficients—and they frequently required the solution of quadratic equations. Two examples will suffice:

> Suppose the intersection C of a given ellipse ACE with the right line CD, given in position, be sought [figure 10.12].[57]

> A stone falling down into a well, from the sound of the stone striking the bottom, to determine the depth of the well.[58]

[53] Newton, 1728, p. 72 = Newton, 1707, pp. 81–82.
[54] Newton, 1728, p. 75 = Newton, 1707, p. 85.
[55] Newton, 1728, p. 72 = Newton, 1707, p. 82.
[56] Newton, 1728, p. 75 = Newton, 1707, p. 85.
[57] Newton, 1728, p. 100 = Newton, 1707, p. 115.
[58] Newton, 1728, p. 161 = Newton, 1707, p. 195.

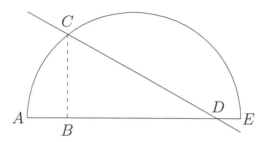

Figure 10.12.

In the first problem, as Newton noted, "if we do not use geometrical descriptions but equations to denote the curved lines by, the computations will thereby become as much shorter and easier as the gaining of those equations can make them."[59] Thus, Newton began by taking a standard equation of an ellipse, $rx - \frac{r}{q}x^2 = y^2$ and denoting the length AD by a. Then, if x is the coordinate of C, we know that $BD = a - x$. Also, since the line is "given in position," the angle that it makes with AD is fixed. Thus, if the ratio $BC : BD = e : 1$, then $y = BC = ea - ex$. Substituting the square of y in the equation for the ellipse and simplifying produces the quadratic equation

$$x^2 = \frac{2ae^2x + rx - a^2e^2}{e^2 + \frac{r}{q}},$$

the solution of which is

$$x = \frac{ae^2 + \frac{1}{2}r \pm e\sqrt{ar + \frac{r^2}{4e^2} - \frac{a^2r}{q}}}{e^2 + \frac{r}{q}}.$$

The second problem was perhaps the first "real-world" problem published in an algebra text that genuinely required the solution of a quadratic equation. Its resolution hinged on the important idea that the distance an object falls under constant acceleration is a quadratic function of the time. Newton assumed that if the stone descended under uniform

[59] Newton, 1728, p. 100 = Newton, 1707, p. 115.

acceleration through any given space a in a given time b, and if sound passed with a uniform motion through the same space a in the given time d, then the stone would descend through the unknown space x in time $b\sqrt{\frac{x}{a}}$, but the sound caused by the stone hitting the bottom would ascend through the same space x in time $\frac{dx}{a}$. Therefore, the time t, given observationally, from dropping the stone until the sound is heard can be expressed as $t = b\sqrt{\frac{x}{a}} + \frac{dx}{a}$. Rearranging terms and squaring yields

$$x^2 = \frac{2adt + ab^2}{d^2}x - \frac{a^2t^2}{d^2},$$

and the quadratic formula produces the answer:

$$x = \frac{adt + \frac{1}{2}ab^2}{d^2} - \frac{ab}{2d^2}\sqrt{b^2 + 4dt}.$$

These two problems, in addition to showing Newton's versatility, exemplify the changes that algebra had made in tackling physical problems since the beginning of the seventeenth century. Kepler, recall, used not algebraic but only geometrical descriptions of an ellipse. In his *Astronomia nova* (or *New Astronomy*) of 1609, he had to base his calculations demonstrating that the orbit of Mars was an ellipse on that curve's original Apollonian definition. They were thus considerably more involved than they would have been had he used algebra. As Newton remarked after solving this problem of an ellipse, with the use of equations "all the difficulties of problems proposed about it may be reduced hither."[60] Similarly, in relation to the second problem, Galileo was well aware, in his *Two New Sciences* of 1638, that "if a moveable [object] descends from rest in uniformly accelerated motion, the spaces run through in any times whatever are to each other as the duplicate ratio of their times; that is, are as the squares of those times."[61] His proof of this, however, was based on his definition of uniformly accelerated motion and so used Euclidean ratio theory. He was never able to express his result in the form $d = kt^2$, from which calculations such as Newton made would have been straightforward. Furthermore, even though Galileo was also able to demonstrate geometrically that the path of a projectile was

[60] Newton, 1728, p. 101 = Newton, 1707, p. 115.
[61] Galileo, 1638/1974, p. 166.

parabolic, he, like Kepler, used the basic Apollonian definitions—in terms of ratios—of a conic section. It was thus only with difficulty that he could make calculations that today, using equations, are so simple.

Descartes captured well the seventeenth-century tension between the traditional geometric style and the then-"modern" algebraic style when he wrote:

> Those things that do not require the present attention of the mind, but which are necessary to the conclusion, it is better to designate by the briefest symbols than by whole figures: in this way the memory cannot fail, nor will thought in the meantime be distracted by these things which are to be retained while it is concerned with other things to be deduced. . . .By this effort, not only will we make a saving of many words, but, what is most important, we will exhibit the pure and bare terms of the problem such that while nothing useful is omitted, nothing will be found in them which is superfluous and which vainly occupies the capacity of the mind, while the mind will be able to comprehend many things together.[62]

In the seventeenth century, and thanks to the work of Fermat and Descartes, algebra was brought into the service of geometry.

COLIN MACLAURIN'S *TREATISE OF ALGEBRA*

Although Newton was President of the Royal Society of London when the *Arithmetica universalis* appeared in 1707, his active days as a mathematician were long behind him, and he was devoting himself to his presidential duties as well as to his work in London as Master of the Mint. He never put the *Arithmetica universalis* into a truly polished form and bitterly complained about Whiston's edition of it.[63] As if in tacit agreement, the Scotsman Colin Maclaurin (1698–1746) designed his own *Treatise of Algebra* (1748), written before 1730 but not published until after his death, as a commentary on Newton's work. Maclaurin hoped that in his text "all those difficult passages in Sir Isaac's book, which have

---

[62] Quoted in Mahoney, 1971; his translation from Descartes's *Regulae ad directionem ingenii* (or *Rules for the Direction of the Mind*), Rule XVI.
[63] Westfall, 1980, pp. 648–649.

so long perplexed the students of algebra, [would be] clearly explained and demonstrated."[64] While he did manage to demonstrate and even to generalize some of Newton's assertions, he, too, never succeeded in proving the fundamental theorem of algebra.

Among his refinements of Newton's work, Maclaurin showed that a system of two linear equations in two unknowns could not only be resolved by Newton's procedure of solving each equation for the same unknown and equating the results, but also by means of determinants in the form that later became known as Cramer's rule. Given, for example, the system

$$ax + by = c,$$

$$dx + ey = f,$$

solving the first equation for x and substituting gives

$$y = \frac{af - dc}{ae - db},$$

and a similar answer for x. Analogously, the system of three linear equations

$$ax + by + cz = m,$$

$$dx + ey + fz = n,$$

$$gx + hy + kz = p$$

is dealt with by first solving each equation for x, thus reducing the problem to a system in two unknowns, and then using the earlier rule to find

$$z = \frac{aep - ahn + dhm - dbp + gbn - gem}{aek - ahf + dhc - dbk + gbf - gec},$$

with similar expressions for x and y. Maclaurin even extended the rule to systems of four equations in four unknowns, making him perhaps the first

---

to publish more general results on solving systems of linear equations, even though he did not generalize further.[65]

Newton had also merely *asserted* that the sum B of the squares of the roots of the polynomial $x^n - px^{n-1} + qx^{n-2} - \cdots$ was $p^2 - 2q$, but Maclaurin *proved* it by noting that "the square of the sum of any quantities is always equal to the sum of their squares added to double the products that can be made by multiplying any two of them."[66] Thus, since p is the sum of the roots and q is the sum of products of pairs of the roots, it follows that $p^2 = B + 2q$ or $B = p^2 - 2q$. Using similar arguments, Maclaurin gave formulas for the sums of the cubes and fourth powers of the roots.

Relative to the fundamental theorem of algebra, though, Maclaurin essentially just repeated Newton's argument for forming higher-degree polynomials by multiplying together polynomials of the form $x \pm a$. Thus, he noted that the product $(x - a)(x - b)(x - c)(x - d)$ vanishes precisely when $x = a$, $x = b$, $x = c$, or $x = d$. Therefore, although it is certainly true that "any other equation admits of as many solutions as there are simple equations multiplied by one another that produce it," Maclaurin evidently did not understand that it does not necessarily follow that there are as many solutions to any equation "as there are units in the highest dimension of the unknown quantity in the proposed equation."[67] Nor did he understand that, just because "roots become impossible in pairs," he could not conclude that "an equation of an odd number of dimensions has always one real root."[68] Maclaurin did remove one seeming anomaly. The fundamental theorem predicts that a cubic equation will have three roots, while Cardano's formula apparently only gives one. Maclaurin clarified the issue by explicitly writing down the three cube roots of unity and using them to generate formulas that produce all three roots of a cubic. (John Colson (1680–1760) had done the same thing in a paper of 1707.[69])

---

[65] See Boyer, 1966 and Hedman, 1999.
[66] Maclaurin, 1748, p. 142.
[67] Maclaurin, 1748, p. 134.
[68] Maclaurin, 1748, p. 138.
[69] Colson, 1707, p. 2356.

LEONHARD EULER AND THE FUNDAMENTAL
THEOREM OF ALGEBRA

By the eighteenth century, two equivalent versions of the fundamental theorem of algebra had entered into the mathematical literature. The first—essentially given by Girard, Descartes, and others—stated that *every polynomial of degree n with real coefficients has n roots, either real or complex*. Given the factor theorem, this is equivalent to the result that every polynomial has at least one real or complex root. The second, however, stemmed from Maclaurin's remark that complex roots of polynomial equations with real coefficients always occur in pairs. Thus, the fundamental theorem of algebra was also rendered as *every real polynomial can be factored into real polynomials of degree one or two*. In the latter form, the theorem implied—importantly and in light of the seventeenth-century development of the calculus at the hands of both Newton and Gottfried Leibniz (1646–1716)—that every rational function could be integrated using elementary functions via the method of partial fractions. It also enabled the Swiss mathematician, Leonhard Euler (1707–1783), to give a general method for solving linear ordinary differential equations. In fact, in a letter written on the topic in 1739 to his friend and fellow countryman Johann Bernoulli (1667–1748), Euler claimed that any real polynomial $1 - ap + bp^2 - cp^3 + \cdots$ could "always be put in the form of a product of factors either simple, $1 - \alpha p$, or of two dimensions $1 - \alpha p + \beta pp$, all real."[70]

At the beginning of the eighteenth century, some mathematicians continued to be skeptical of the fundamental theorem. In 1702, Leibniz, for one, believed that he had found a counterexample,[71] namely, the polynomial $x^4 + a^4$, which he factored as

$$x^4 + a^4 = (x^2 + a^2\sqrt{-1})(x^2 - a^2\sqrt{-1}).$$

The roots of the second factor, namely, $x = \pm a\sqrt{\sqrt{-1}}$, were not obviously complex numbers, thus violating the first version of the theorem. In fact, even though mathematicians of the day understood that solutions of cubic and quartic equations could be found by formulas involving radicals,

---

[70] Fauvel and Gray, 1987, p. 447.
[71] See Tignol, 2001, p. 74 for a discussion of Leibniz's original paper in the *Acta eruditorum*.

Figure 10.13. Leonhard Euler (1707–1783).

there was no consensus until the middle of the eighteenth century that roots of complex numbers always produced complex numbers. Relative to the second version of the fundamental theorem, Nicolaus Bernoulli (1687–1759) claimed that the polynomial $x^4 - 4x^3 + 2x^2 + 4x + 4$ could not be factored into two real quadratics.[72]

It turned out, of course, that both Leibniz and Bernoulli were wrong. As several mathematicians pointed out, $\sqrt{\sqrt{-1}} = \frac{\sqrt{2}}{2} + \frac{\sqrt{2}}{2}i$, a complex number, while Euler discovered that the two factors of Bernoulli's polynomial were

$$x^2 - \left(2 + \sqrt{4 + 2\sqrt{7}}\right)x + \left(1 + \sqrt{4 + 2\sqrt{7}} + \sqrt{7}\right)$$

and

$$x^2 - \left(2 - \sqrt{4 + 2\sqrt{7}}\right)x + \left(1 - \sqrt{4 + 2\sqrt{7}} + \sqrt{7}\right).$$

Although Euler successfully factored Bernoulli's polynomial, it seemed clear that an actual proof of the fundamental theorem would by no means

[72] See Dunham, 1999, p. 109 for a discussion of this problem, which comes from Euler's correspondence.

be trivial. Euler, however, understood a key point that Maclaurin did not: building higher polynomials from linear ones would not ensure the factorability of an arbitrary, given polynomial. That would require advance knowledge of the roots, but the problem of determining roots for polynomials of degree higher than four had thus far proven intractable. Could the proof of the theorem be approached in a different way? Could the *existence* of roots or factors of a polynomial be shown, without necessarily being able to *write down explicitly* those roots or those factors?

The first person actually to publish a proof of the fundamental theorem of algebra was the French mathematician Jean le Rond d'Alembert (1717–1783) in a 1746 article dealing with the integration of rational functions.[73] D'Alembert's argument turned on considering the real polynomial equation $F(y,z) = y^m + Ay^{m-1} + By^{m-2} + \cdots + Ky + z = 0$, which obviously has a solution $z = 0$, $y = 0$. He solved $F(y,z) = 0$ for y as a series $y(z)$ in z with increasing fractional exponents and then demonstrated that this series can be continued into a real or complex solution $y_1(z)$ as z ranges along the entire real axis. His argument involved the still sketchy geometric interpretation of complex numbers.

On reading d'Alembert's proof later in 1746, Euler wrote to the Frenchman to tell him that his proof "fully satisfies me, but as it proceeds from the resolution of the value of y in an infinite series, I do not know if everyone will be convinced."[74] Shortly thereafter, Euler himself apparently became one of those less "convinced." In an article entitled "Recherches sur les racines imaginaires des équations" (or "Investigations on the Imaginary Roots of Equations") published in 1751 although written two years earlier, he claimed that no one had yet demonstrated the theorem "with sufficient rigor"[75] and so presented his own proof. Intended to be devoid of any mention of ideas involving the calculus, that proof nevertheless made use of what we now term the intermediate value theorem (IVT), namely, the analytic fact that if a continuous function goes from being positive to negative (or vice versa) on a given interval, then it has a root in that interval. Euler had already used the IVT in his book, *Introductio in analysin infinitorum* (or *Introduction to Analysis of the Infinite*), of 1748 to show that any polynomial of odd degree has

---

[73] See Baltus, 2004 for a complete discussion of d'Alembert's proof.
[74] Baltus, 2004, p. 421.
[75] Euler, 1751, p. 225.

at least one real linear factor. At the same time, he had generalized Leibniz's example by showing that the irreducible quadratic factors of $a^n \pm z^n$ were $a^2 - 2az \cos \frac{2k\pi}{n} + z^2$ (for $1 \le k \le \frac{n}{2}$) when the sign is negative and $a^2 - 2az \cos \frac{(2k+1)\pi}{n} + z^2$ (for $0 \le k \le \frac{n-1}{2}$) when the sign is positive.[76]

Euler set the stage for his 1749 proof of the fundamental theorem with a discussion of a few particular examples, including one very similar to Nicolaus Bernoulli's example above. He then launched into his proof by using the IVT to show that any polynomial of even degree with negative constant term has at least two real roots. Next, he focused on fourth-degree polynomials, aiming to show that they always have two real quadratic factors. Since the cubic term can always be removed by a linear substitution, it sufficed to look at equations of the form $x^4 + Bx^2 + Cx + D = 0$. The two real factors of the polynomial must then be $x^2 + ux + \alpha$ and $x^2 - ux + \beta$, with u, α, and β to be determined. Multiplying and comparing coefficients shows that the equations for these three unknowns are

$$\alpha + \beta - u^2 = B, \qquad (\beta - \alpha)u = C, \qquad \alpha\beta = D.$$

It then follows that $\alpha + \beta = B + u^2$ and $\beta - \alpha = \frac{C}{u}$, so that

$$2\beta = u^2 + B + \frac{C}{u} \qquad \text{and} \qquad 2\alpha = u^2 + B - \frac{C}{u}.$$

Because $4\alpha\beta = 4D$, multiplying the last two equations gives

$$u^4 + 2Bu^2 + B^2 - \frac{C^2}{u^2} = 4D \qquad \text{or} \qquad u^6 + 2Bu^4 + (B^2 - 4D)u^2 - C^2 = 0.$$

By the result on polynomials of even degree with negative constant terms, Euler knew that the equation for u had at least two real nonzero roots. Whichever root he chose would then give real values for α and β and would thus determine the real factors of the fourth-degree polynomial. (Note that we are assuming here that $C \ne 0$; a much simpler argument will work if $C = 0$.) An immediate corollary of this result is that every fifth-degree polynomial also factors into one real linear factor and two real quadratic factors.

[76] Euler, 1748/1988, pp. 119–122.

In the remainder of the paper, Euler attempted to generalize the proof for fourth-degree polynomials to analogous results for polynomials of degree 2^n, from which he could then have derived the result for any degree. Unfortunately, the technical details of generalizing the argument above to polynomials of degree 8, 16, 32, and so on, were beyond even Euler's formidable skills. When he tried an alternate argument in the fourth-degree case and then the higher-degree cases, he unwittingly assumed that it was already known that the roots of the polynomial were complex numbers, and this, of course, was what he sought to prove.

Joseph-Louis Lagrange (1736–1813) made the same error in his 1772 attempt to complete Euler's proof. It was only Carl Friedrich Gauss (1777–1855) who, in his dissertation of 1799, gave a proof demonstrating the existence of a complex root of any real polynomial, using the geometric interpretation of complex numbers and certain analytic results on curves in the plane. Gauss later gave three additional proofs. In his fourth and final proof published in 1848, he even assumed that the coefficients of the polynomial were complex, since by that time mathematicians understood complex numbers and functions more deeply, and he could freely use various notions of complex function theory.[77] Even today, there is no proof of the fundamental theorem of algebra that does not involve some results from analysis. In fact, the most common proofs of the theorem are found as consequences of some important results in complex function theory, results available to neither d'Alembert nor Euler.

****

The seventeenth century witnessed two major changes in algebra. First, the rapid development of symbolism reached a climax in Descartes's *Geometry*, the algebraic symbolism of which is virtually the same as what we use today. Second, the invention of analytic geometry brought to the fore a new purpose for algebra, namely, the determination of entire curves— as opposed to single numbers—as solutions to problems. This second change—reflective of the shift to a dynamic stage in the evolution of algebra—tremendously increased the value of mathematics as the major problem-solving tool for astronomy and physics. After the publication of Newton's *Philosophiæ naturalis principia mathematica* (or *Mathematical Principles of Natural Philosophy*) in 1687, it became clear that the study of forces

---

[77] See van der Waerden, 1985, pp. 94–102 for a discussion of Gauss's proofs.

would occupy physicists over the next century. Naturally, force was related to acceleration and velocity, and these, in turn, were described by central ideas of the differential calculus. Some of the major problems studied in the eighteenth century thus concerned the paths of bodies under the influence of various types of forces and under various constraints. Such problems typically led to differential equations, the solutions of which turned out to be curves generally described algebraically through the use of Cartesian symbols. Interestingly, although Newton himself did not make great use of algebra in his *Principia* (he chose instead to write it as much as possible in the classic language of geometry), his interpreters in the next century translated his ideas into the Leibnizian language of calculus and used algebra in their analysis of the results. Viète may have been premature in stating that his ideas could "solve every problem." Nevertheless, his direct intellectual descendants in the century after his death transformed algebra into the incredibly efficacious problem-solving tool that it remains today.

11

Finding the Roots of Algebraic Equations

As we saw in chapter 9, Cardano and others had published methods for solving cubic and quartic polynomial equations by radicals in the middle of the sixteenth century. Over the next two hundred years, other mathematicians struggled to find analogous algebraic methods to solve higher-degree equations, that is, methods, beginning with the coefficients of the original polynomial, that consist of a finite number of steps and involve just the four arithmetic operations together with taking roots. Of course, certain particular equations could be solved, including those arising in circle-division problems, but, for more general equations, new methods were sought. Some mathematicians tinkered with Cardano's methods and found slightly different ways of solving cubics or quartics in the hope that these would generalize. Since it was well known that a simple substitution removed the term of degree $n-1$ from an nth-degree polynomial, others tried, using further substitutions, to remove more terms. If all the terms between the nth-degree term and the constant could somehow be removed, the equation would be solvable. Unfortunately, although some of the new ideas seemed promising, in the end, no one was able to find an algebraic solution to a general fifth- or higher-degree polynomial equation.

By the turn of the nineteenth century, then, many mathematicians began to believe that such a solution was actually impossible. It turned out that the key to proving this impossibility and, more generally, the key to determining which polynomial equations were, in fact, solvable algebraically, was a totally new construct, the group. This chapter highlights some of the struggles to solve fifth-degree polynomials as well as the development of the idea of a group and its application to answering the question of solvability of polynomial equations.

THE EIGHTEENTH-CENTURY QUEST TO SOLVE HIGHER-ORDER EQUATIONS ALGEBRAICALLY

We begin this story by surveying some of the methods by which mathematicians attempted to solve—and sometimes did solve—various algebraic equations of degree higher than four. It is a story filled with numerous ideas that initially seemed promising, but also with numerous dead ends. The latter frequently occurred only when mathematicians attempted to fill in all the details of their ideas in specific cases.

Certain special cases of higher-degree equations for which solutions were determined related to the binomial expansion. For example, the French Protestant emigré to England, Abraham de Moivre (1667–1754), in developing ideas surrounding the multiple-angle formulas for the sine and cosine, asserted in a paper in the *Philosophical Transactions of the Royal Society* in 1707[1] that an equation of the form

$$nx + \frac{n^2-1}{2 \times 3}nx^3 + \frac{n^2-1}{2 \times 3} \times \frac{n^2-9}{4 \times 5}nx^5 +$$
$$\frac{n^2-1}{2 \times 3} \times \frac{n^2-9}{4 \times 5} \times \frac{n^2-25}{6 \times 7}nx^7 + \cdots = a, \qquad (11.1)$$

where n is odd, had the solution

$$x = \frac{1}{2}\sqrt[n]{\sqrt{1+a^2}+a} - \frac{1}{2}\sqrt[n]{\sqrt{1+a^2}-a}.$$

To see that this is not an unreasonable assertion, take $n = 5$ and consider the expansion of $(m-n)^5$. With a little manipulation,[2] this can be written as

$$(m-n)^5 = -5mn(m-n)^3 - 5m^2n^2(m-n) + m^5 - n^5,$$

or, letting $x = m - n$, as

$$x^5 + 5mnx^3 + 5m^2n^2x - (m^5 - n^5) = 0. \qquad (11.2)$$

---

[1] De Moivre, 1707, p. 2368.

[2] De Moivre did not provide this justification in his short paper. We provide it here to illustrate how he may well have came up with his result.

Now, for $n = 5$, de Moivre's equation (11.1) is $5x + 20x^3 + 16x^5 = a$ or, equivalently, $x^5 + \frac{5}{4}x^3 + \frac{5}{16}x - \frac{a}{16} = 0$. Comparing this[3] with equation (11.2) then yields three equations for m and n:

$$mn = \frac{1}{4}, \quad m^2 n^2 = \frac{1}{16}, \quad \text{and} \quad m^5 - n^5 = \frac{a}{16}.$$

Given that the second equation automatically follows from the first and given that we can rewrite the first equation as $m^5 n^5 = \frac{1}{1024}$, we simply need to solve the quadratic equation

$$y^2 + \frac{a}{16}y - \frac{1}{1024} = 0$$

for m^5 and n^5 and then take fifth roots to determine m and n. The solutions are $m^5 = \frac{1}{32}(\sqrt{a^2 + 1} + a)$ and $n^5 = \frac{1}{32}(\sqrt{a^2 + 1} - a)$, so that a solution of the original equation is

$$x = \frac{1}{2}\sqrt[5]{\sqrt{a^2 + 1} + a} - \frac{1}{2}\sqrt[5]{\sqrt{a^2 + 1} - a},$$

exactly as de Moivre claimed. Although de Moivre did not say so explicitly, the second example in his paper, the analogue of equation (11.1) with alternating signs, is the multiple-angle formula relating $a = \sin n\theta$ to $x = \sin \theta$. In taking up this example, in fact, he noted that arithmetical solutions could be obtained through the use of sine tables, thus relating his equation-solving procedure to the circle-division problem that even Viète had been able to solve.[4]

These kinds of questions engaged other mathematicians working in the context of various Royal Academies of Science. At the Academy in St. Petersburg founded by Catherine I of Russia (1684–1727), Leonhard Euler made some conjectures, beginning in 1733, about the solution of more general polynomial equations, based on his knowledge of the procedures for degrees three and four.[5] In particular, he noted that a

---

[3] De Moivre actually considered precisely this example, with $a = 4$, and worked out the specific numerical solution for x according to his formula.

[4] See De Moivre, 1707, pp. 2370–2371.

[5] Euler, 1738. (Euler presented this paper to the St. Petersburg Academy in 1733, but it was not published until five years later.) See Stedall, 2011.

solution of the cubic equation $x^3 = ax + b$ could be expressed in the form $x = \sqrt[3]{A} + \sqrt[3]{B}$, where A and B are the roots of a certain quadratic equation, while, with some manipulation, a solution of the quartic equation $x^4 = ax^2 + bx + c$ could be expressed in the form $\pm\sqrt[4]{A} \pm \sqrt[4]{B} \pm \sqrt[4]{C}$, where A, B, and C are the roots of a particular cubic. Thus, he conjectured that a similar result would hold for equations of higher degree. Unfortunately, even in the fifth-degree case, he was unable to construct the auxiliary equation of degree four (called the *resolvent*), from the roots of which he could determine the roots of the original equation. Twenty years later, realizing that all the roots of a polynomial should be expressible in a particular form and understanding that the roots of unity would be involved, he made a more general conjecture about the form of the roots of an arbitrary polynomial as sums of radical expressions.[6] However, although he was able to find solutions in some special cases, his attempts to discover general solutions became bogged down in calculations that bested even the indefatigable Euler.

The *académicien* of the Paris Academy of Sciences, Étienne Bezout (1730–1783), read Euler's 1753 paper on the subject and pursued similar ideas in a memoir presented to the French Academy in 1763 and published two years later.[7] Bezout's basic idea was to solve an equation of degree m in x that lacked a term of degree $m - 1$ by considering it as the result of the elimination of y from the simultaneous equations

$$y^m - 1 = 0 \quad \text{and} \quad ay^{m-1} + by^{m-2} + \cdots + hy + x = 0.$$

To understand how this would work, consider the case of the cubic equation $x^3 + px + q = 0$. Solving this by elimination would mean eliminating y from the equations $y^3 - 1 = 0$ and $ay^2 + by + x = 0$. Multiplying the latter equation by 1, y, and y^2, respectively, gives three equations from which y and y^2 must be eliminated, namely,

$$ay^2 + by + x = 0,$$

$$by^2 + xy + a = 0,$$

$$xy^2 + ay + b = 0.$$

---

[6] Euler, 1764. The paper was presented to the St. Petersburg Academy in 1753, but because of the long delay in publication, was not printed until 1764.

[7] Bezout, 1765. For further discussion of Bezout's ideas, see Stedall, 2011, pp. 146–152.

Eliminating y^2 from any two pairs of equations leaves two equations in y and x, from which y can be eliminated by substitution, say. Some simplification then yields the equation $x^3 - 3abx + a^3 + b^3 = 0$. Comparing this with the original cubic results in the two equations $-3ab = p$ and $a^3 + b^3 = q$, from which we derive the sixth-degree resolvent in a: $a^6 - qa^3 - \frac{1}{27}p^3 = 0$. Since this quadratic equation in a^3 is easily solved, b may be found from $-3ab = p$, and we get the three values for x by setting $y = 1$, $y = \omega (= \frac{-1+\sqrt{-3}}{2})$, and $y = \omega^2$, respectively, in the equation $x = -ay^2 - by$. The important thing to note is that the resolvent, although of degree six, reduces to a quadratic.

Bezout used the same technique with the quartic equation $x^4 + px^2 + qx + r = 0$, using the two equations $y^4 - 1 = 0$ and $ay^3 + by^2 + cy + x = 0$ from which y must be eliminated.[8] Again, he multiplied the last equation by 1, y, y^2, and y^3, respectively, to get four equations from which it was necessary to eliminate y, y^2, and y^3. The three equations for a, b, and c that result from comparing the original equation with the fourth-degree equation coming from the elimination process are somewhat more complicated than in the degree-three case. Nevertheless, a and c can be eliminated from this system, leaving a resolvent for b of degree six, namely,

$$b^6 + \frac{1}{2}pb^4 + \left(\frac{1}{16}p^2 - \frac{1}{4}r\right)b^2 - \frac{1}{64}q^2 = 0,$$

but since this only contains even powers, it is essentially an equation of degree three.[9] Each of the values of b then determines values for a and c, and the four possible values of y give the four solutions for x. Trying to solve the three equations for a or c, however, results in an equation of degree $24 = 4!$. Since the degree of each term is a multiple of four, the equation reduces to one of degree six. A warning shot across the bow? If the resolvent is actually of degree *higher than* that of the original equation, would it likely be of use?

Bezout applied the same method to solving a fifth-degree polynomial equation and derived an exceedingly unwieldy fifth-degree equation from

[8] Both the third- and fourth-degree equations are discussed in Bezout, 1765, pp. 537–540.

[9] Bezout did not comment, either here or in the degree-three discussion, about the ambiguities in the selection of a set of values for a, b, and c, given that, for example, there are six solutions for b in this case and six solutions for a in the earlier case.

the elimination procedure.[10] He was unable to solve the resulting five equations to get a resolvent for any of the unknowns. While this might not be beyond the capabilities of a modern computer algebra program, the resulting equation would be of degree $120 = 5!$, reducing, perhaps, to degree $4! = 24$. Bezout believed strongly that the difficulty lay not in his method but in the problem itself. The belief that it was perhaps impossible to solve a fifth-degree equation algebraically thus began to percolate through the mathematical community.

Still, as late as 1770, both the Englishman Edward Waring (1736–1798) and the Frenchman Alexandre-Théophile Vandermonde (1735–1796) were attempting to find solutions. Neither one knew of the work of Euler or Bezout, at least until they had published their own discoveries, and so were basing their work mostly on seventeenth-century results. In his *Miscellanea analytica* (or *Analytic Miscellany*) of 1762 and *Meditationes algebraicæ* (or *Algebraic Meditations*) of 1770, Waring thus considered many of the same problems as his predecessors. For example, in the earlier volume, he showed that, in general, the degree of the resolvent would be higher than that of the original equation; he suggested that solutions of nth-degree polynomial equations would take the form of sums of nth powers of certain expressions; and he discussed methods for reducing sets of equations by elimination. In the latter volume, he expanded somewhat on these results, even discovering results similar to those of Euler and Bezout. Waring, however, never actually claimed to find any general solution of a higher-order polynomial equation, nor did his work influence any of the later researchers.

Vandermonde was somewhat more brash. In a paper written in 1770 but not published until 1774 in the *Mémoires* of the Paris Academy, he laid out a program that he believed would lead to such solutions. His first step was "to find a function of the roots, of which one may say, in a certain sense, that it is equal to each of the roots that one wishes."[11] For example, in the degree-two case $x^2 - (a + b)x + ab = 0$, the desired function is $\frac{1}{2}[(a + b) + \sqrt{(a + b)^2 - 4ab}]$, where the ambiguity of the square root enables both roots to be found. The next step is to "put this function into a form that is mostly unchanged when the roots are interchanged among themselves." In this case, it is easy to see that the function is,

---

[10] Bezout, 1765, pp. 543–544.
[11] Vandermonde, 1771, p. 370 (our translation). The next two quotations are also found here.

Figure 11.1. Joseph-Louis Lagrange (1736–1813).

in fact, unchanged under any permutation of the two roots. Finally, one must "write the values in terms of the sum of the roots, the sum of their products taken two by two, and so on." Here, of course, the values are already written in that form. Vandermonde argued that he could always accomplish the first and third steps, and while he could take the second step in the third- and fourth-degree cases, the calculations became too difficult for him to complete for higher degrees. Still, Vandermonde was the first to insist on the importance of functions that only took limited numbers of values under all possible permutations of the roots of the equation.

While Vandermonde was at work on his ideas, Joseph-Louis Lagrange (1736–1813) was drawing similar conclusions as the result of his detailed review of the existing solution methods for third- and fourth-degree equations. Published in 1770, just four years after he had left his native Turin to assume the directorship of mathematics at the Berlin Academy of Sciences, Lagrange's "Réflexions sur la théorie algébrique des équations" (or "Reflections on the Algebraic Theory of Equations") explored why the methods for cubics and quartics worked and pursued analogous methods for higher-degree equations. Although Lagrange, too, failed in his quest for general methods in higher degrees, he succeeded in sketching a new set of principles for dealing with such equations.

He began with a systematic study of the cubic equation $x^3 + nx + p = 0$.[12] Setting $x = y - \frac{n}{3y}$ transforms this equation into the sixth-degree equation $y^6 + py^3 - \frac{n^3}{27} = 0$ which, with $r = y^3$, reduces to the quadratic equation $r^2 + pr - \frac{n^3}{27} = 0$. (Of course, this is essentially the same equation that Cardano and Bezout had found.) It has two roots, r_1 and $r_2 = -(\frac{n}{3})^2 \frac{1}{r_1}$, but whereas Cardano took the sum of the real cube roots of r_1 and r_2 as his solution, Lagrange, like Bezout, knew that each equation $y^3 = r_1$ and $y^3 = r_2$ actually had three roots. There were thus six possible values for y, namely, $\sqrt[3]{r_1}$, $\omega\sqrt[3]{r_1}$, $\omega^2\sqrt[3]{r_1}$, $\sqrt[3]{r_2}$, $\omega\sqrt[3]{r_2}$, and $\omega^2\sqrt[3]{r_1}$, where ω is a complex root of $x^3 - 1 = 0$. Lagrange then showed that the three distinct roots of the original equation were given by

$$x_1 = \sqrt[3]{r_1} + \sqrt[3]{r_2},$$

$$x_2 = \omega\sqrt[3]{r_1} + \omega^2\sqrt[3]{r_2},$$

$$x_3 = \omega^2\sqrt[3]{r_1} + \omega\sqrt[3]{r_2}.$$

He next decided that rather than consider x as a function of y, he would reverse the procedure, since the solutions of y, the resolvent, would determine the solution of the original equation.[13] The idea, then, was to express those solutions in terms of the original ones. Thus, Lagrange noted that any of the six values for y could be expressed in the form $y = \frac{1}{3}(x' + \omega x'' + \omega^2 x''')$, where (x', x'', x''') was some permutation of (x_1, x_2, x_3). These expressions had also appeared in Vandermonde's work on the cubic, but Lagrange, unlike Vandermonde, was able to turn the idea of a permutation of the roots of an equation into the cornerstone not only of his own method but also of the methods others would use in the next century.

The case of the cubic highlights several important ideas. First, the six permutations of the x_i's lead to the six possible values for y and thus

[12] Lagrange, 1770, pp. 135–139.
[13] Lagrange, 1770, pp. 140–143.

show that y satisfies an equation of degree six. Second, the permutations of the expression for y can be divided into two sets, one consisting of the identity permutation and the two permutations that interchange all three of the x_i's and another consisting of the three permutations that interchange just two of the x_i's. (In modern terminology, the group of permutations of a set of three elements has been divided into two *cosets*.) For example, if $y_1 = \frac{1}{3}(x_1 + \omega x_2 + \omega^2 x_3)$, then the two nonidentity permutations in the first set change y_1 to $y_2 = \frac{1}{3}(x_2 + \omega x_3 + \omega^2 x_1)$ and $y_3 = \frac{1}{3}(x_3 + \omega x_1 + \omega^2 x_2)$, respectively. But then $\omega y_2 = \omega^2 y_3 = y_1$ and $y_1^3 = y_2^3 = y_3^3$. Similarly, if the results of the permutations of the second set are y_4, y_5, and y_6, it follows that $y_4^3 = y_5^3 = y_6^3$. Thus, because there are only two possible values for $y^3 = \frac{1}{27}(x' + \omega x'' + \omega^2 x''')^3$, the equation for y^3 is of the second degree. Finally, the sixth-degree equation satisfied by y has coefficients that are rational in the coefficients of the original equation.

Although other methods for solving cubics had been used earlier, Lagrange found that the same idea underlay them all. Each led to a rational expression in the three roots that took on only two values under the six possible permutations. The expression thus satisfied a quadratic equation.

Lagrange next considered the solutions of the quartic equation.[14] Ferrari's method for solving $x^4 + nx^2 + px + q = 0$, as noted in chapter 9, involved adding $2yx^2 + y^2$ to each side, rearranging, and then determining a value for y such that the right-hand side of the new equation,

$$x^4 + 2yx^2 + y^2 = (2y - n)x^2 - px + y^2 - q,$$

was a perfect square. He could then take square roots of each side and solve the resulting quadratic equations. For the right-hand side to be a perfect square, its discriminant must be zero, namely,

$$(2y - n)(y^2 - q) = \left(\frac{p}{2}\right)^2 \quad \text{or} \quad y^3 - \frac{n}{2}y^2 - qy + \frac{4nq - p^2}{8} = 0.$$

Therefore, the resolvent is a cubic, which can, of course, be solved. Given the three solutions for y, Lagrange then showed, as in the cubic case, that each is a permutation of a rational function of the four roots x_1, x_2, x_3, x_4

[14] Lagrange, 1770, pp. 173–182.

of the original equation. In fact, it turned out that $y_1 = \frac{1}{2}(x_1x_2 + x_3x_4)$ and that the twenty-four possible permutations of the x_i's lead to only three different values for that expression, namely, y_1, $y_2 = \frac{1}{2}(x_1x_3 + x_2x_4)$, and $y_3 = \frac{1}{2}(x_1x_4 + x_2x_3)$. Therefore, the expression must satisfy a third-degree equation, again, one with coefficients rational in the coefficients of the original equation.

Having studied the methods for solving both cubics and quartics, Lagrange felt prepared to generalize. First, as was clear from the discussion of cubic equations, the study of the roots of equations of the form $x^n - 1 = 0$ was important. Thus, he showed that if n is prime and $\alpha \neq 1$ is one of the roots, then α^m, for any $m < n$, can serve as a generator of all of the roots.[15] Second, however, Lagrange realized that to attack the problem of equations of degree n, he, like his predecessors, needed a way to determine a resolvent of degree $k < n$ that would be satisfied by certain functions of the roots of the original equation, functions that take on only k values when the roots are permuted by all $n!$ possible permutations. Because relatively simple functions of the roots did not work, Lagrange attempted to find general rules for determining both them and the degree of the equation that they would satisfy. While he made some progress in this regard—including showing that the resolvent would be of degree no greater than $(n - 1)!$—he was unable to show that this equation could be further reduced to one of degree smaller than n, even in the degree-five case. His further attempts to find a sequence of resolvents, where the solution of each one would lead to the solution of the next, also failed. Thus, like his predecessors, Lagrange was forced to give up his quest for algebraic solutions of polynomials of degree five or higher, and he clearly held out little hope that they could be found. Still, his introduction of the notion of permutations proved crucial to the ultimate proof that there was no general algebraic solution of a fifth-degree polynomial equation as well as to the methods for determining those equations for which such a solution *was* possible.

Among the first mathematicians to study Lagrange's work seriously was the Italian mathematician and physician, Paolo Ruffini (1765–1822). In his 1799 book, *Teoria generale delle equazioni, in cui si dimostra impossibile la soluzione algebrica delle equazioni generali di grado superiore al quarto* (or *The General Theory of Equations In Which It Is Shown that the Algebraic*

---

[15] Lagrange, 1770, pp. 169–171.

Solution of General Equations of Degree Greater Than Four Is Impossible), and then in several papers in the first decade of the nineteenth century, Ruffini carried Lagrange's ideas further, stating and giving a purported proof of the theorem that it is impossible to find an algebraic solution of general polynomial equations of degree greater than four.[16] Noting that, for degrees three and four, Lagrange had found rational functions of the roots that took on only two and three values, respectively, under all possible permutations of the roots, Ruffini showed that a similar result was impossible for the fifth degree. In particular, by carefully studying the "permutazioni" of a set of five elements (in modern terms, he considered the *symmetric group* of permutations S_5 and calculated all of its *subgroups*), he showed that there are no rational functions of five quantities that take on 8, 4, or 3 different values under all elements of the group. That meant that there could be no resolvent of degree three, four, or eight for a fifth-degree polynomial, and therefore the method that worked for degrees three and four could not work for degree five.[17]

Ruffini followed up on this result by claiming the unsolvability of the general quintic equation by radicals. The very involved arguments in his book and his subsequent articles failed, however, to convince his contemporaries. In fact, Lagrange himself not only had received a complimentary copy of Ruffini's book from the author but also had been one of the anonymous referees of a paper Ruffini had tried unsuccessfully to get published in the proceedings of the Paris Academy of Sciences in 1810. Although the Academy never formally rejected the paper, because Ruffini eventually withdrew it, the referees evidently believed that the effort to understand the paper was not worth the time it would have taken.[18] Modern readers of Ruffini's papers have concluded that his argument actually contained several gaps. These were eventually filled by the young Norwegian mathematician, Niels Henrik Abel (1802–1829). In his 1826 paper entitled "Beweis der Unmöglichkeit algebraische Gleichungen von höheren Graden als dem vierten allgemein auflösen"

[16] See Ruffini, 1799 as well as Ruffini, 1802a,b, and 1803, among other papers. On Ruffini's work in historical context, see, for example, Ayoub, 1980.

[17] Ruffini, 1799, pp. 243–287. There is a rational function that takes on only 2 different values, but the quadratic equation coming from those two values does not allow for the solution of the original equation.

[18] See Ruffini, 1803 for a briefer argument than in Ruffini, 1799, and see Bottazzini, 1994, p. 10 for Ruffini's failed attempts at reaching the French mathematical audience with his result. See Tignol, 2001, pp. 209–210 for more on the Academy's reaction to Ruffini's work.

(or "Proof of the Impossibility of Solving Algebraically General Equations of Degree Higher Than the Fourth"), Abel made essential use of the idea of permutations of the roots of an equation.[19] His proof, unlike Ruffini's, was soon generally accepted, although mathematicians over the years have found minor gaps in its argument.[20] In light of Abel's work, the new goal then became to determine which equations of degree higher than four were solvable algebraically.

THE THEORY OF PERMUTATIONS

Although Lagrange, Ruffini, and later Abel and others had used the idea of permutations, it was Augustin-Louis Cauchy (1789–1857) who began to formalize the theory in work published in 1815 and then again nearly thirty years later. Born at the outbreak of the French Revolution, Cauchy left Paris with his family as a young boy to avoid the immediate results of the political turmoil.[21] After his parents brought him back to Paris, Cauchy received his early education from his father, who was on friendly terms with a number of important mathematicians in the capital city. One of those, none other than Lagrange, helped Cauchy's father guide the boy's education. After earning an engineering degree, which enabled him to get his first job in Cherbourg, Cauchy began to study mathematics seriously. This ultimately led to his appointment to the faculty of the École Polytechnique in 1815, where he wrote his most important texts on calculus as accompaniments to his lectures.[22]

Cauchy published his initial work on permutations in 1815 just as he was assuming his new academic post. Continuing in the footsteps of Lagrange and Ruffini, but inspired by his recent reading of Gauss's *Disquisitiones arithmeticae* (or *Arithmetic Disquisitions*), he focused on the question of how many possible values a given rational (nonsymmetric) function of n variables could take under all possible permutations of the variables. As he noted, "one of the most remarkable consequences of the work of these various mathematicians is that, for a given number [n] of

---

[19] Abel, 1826, pp. 73–81.

[20] Sørensen, 1999, pp. 69–88 contains details of many of the criticisms of Abel's proof.

[21] On Cauchy's fascinating life, see, for example, Belhoste, 1991.

[22] For a brief overview of the place of the École Polytechnique in nineteenth-century French mathematics, see Grattan-Guinness, 2012.

Figure 11.2. Augustin-Louis Cauchy (1789–1857).

letters, it is not always possible to form a function that has a specified number of values."[23] Cauchy surpassed this result by proving that the number of different values such a function could take on "cannot be less than the largest odd prime p which divides n, without becoming equal to 2."

Cauchy wrote no more on these ideas until 1844. A strong supporter of the Bourbon restoration after Napoleon's defeat, Cauchy had refused to take a loyalty oath to the government of King Louis Phillipe in 1830 and had spent eight years in self-imposed exile in Switzerland, Italy, and elsewhere. Returning to Paris in 1838, he was able to resume his position as a member of the Académie des Sciences, but his political and religious convictions continued to present problems for him relative to securing a teaching position. He thus wrote his comprehensive 1844 "Mémoire sur les arrangements que l'on peut former avec des lettres données" (or "Memoir on the Arrangements That Can Be Formed with Given Letters") while working to reacquire his former stature within the French mathematical and scientific communities.

Prior to this work, the term "permutation" had generally referred to an arrangement of a certain number of objects, say letters. In his 1844

---

[23] Cauchy, 1815a, p. 9 (our translation). The following quotation (in our translation) is also found here.

treatise, however, Cauchy used the word *"arrangement"* to mean a certain ordering of n letters and focused on the action of changing from one arrangement to another. As he wrote, "we call *permutation* or *substitution* the operation that consists of displacing these variables, by substituting them for each other... in the corresponding arrangement."[24] He then introduced notation for representing a permutation that, say, changes xyz to xzy:

$$\begin{pmatrix} xzy \\ xyz \end{pmatrix} \quad \text{or} \quad \begin{pmatrix} zy \\ yz \end{pmatrix}.^{25}$$

Although this notation is the reverse of what is now standard, Cauchy was able to use it easily to show how to multiply permutations and to determine whether two permutations commute.

In addition to focusing on the functional aspect of a permutation, Cauchy often used a single letter, say S, to denote a given permutation and defined the product of two such permutations S, T, written ST, as the permutation determined by first applying S to a given arrangement and then applying T to the resulting one.[26] He dubbed the permutation that leaves a given arrangement fixed the "identity" permutation and noted that the powers S, S^2, S^3, ... of a given permutation must ultimately result in the identity. (Compare Jordan's definition of a (permutation) group below.) The "order" of a permutation S was then the smallest power n such that S^n is equal to the identity. He also defined the inverse of a permutation S in the obvious way, using the notation S^{-1}, and introduced the notation 1 for the identity. Cauchy even identified what he called a "circular" (that is, cyclic) permutation on the letters $x, y, z, \ldots, w$ as one that takes x to y, y to z, ..., and w to x, and then introduced the notation $(xyz \cdots w)$ for such a permutation.[27]

Given any set of substitutions on n letters, he next defined the "system of conjugate substitutions"—today called the *subgroup* generated by the given set—to be the collection of all permutations formed from the original ones by taking all possible products. Finally, he proved that the order k of this system H, that is, the number of elements in it,

[24] Cauchy, 1844, p. 152 (our translation).
[25] Cauchy, 1844, p. 152.
[26] This definition and the next several are found in Cauchy, 1844, pp. 163–165.
[27] Cauchy, 1844, p. 157.

always divides the order $n!$ of the complete system (today called a *group*) G of permutations on n letters.[28] His proof employed the by-now-usual method of multiplying each of the elements in H by an element of G not in H to get a new set of permutations of order k (what we would call a coset). Continuing to multiply the original system H by elements not already used, Cauchy was able to show that eventually G was divided into disjoint sets, all with k elements, thus proving the result.

DETERMINING SOLVABLE EQUATIONS

Cauchy did not directly apply his work to the study of solving equations, but, as we have seen, the notion of a permutation and, in particular, of a group of permutations, was already proving critical in trying to determine methods for solving equations of degree greater than four. Although Lagrange surmised that there could be no general solution algorithm—and Ruffini believed he had proved this—it was also true that there were many equations of degree higher than four that *could* be solved algebraically. The solution of one such class—the cyclotomic equations of the form $x^n - 1 = 0$—divided a circle in the complex plane into n equal parts.

Gauss had discussed cyclotomic equations in detail in the final chapter of his *Disquisitiones arithmeticae* of 1801 and had applied his results to the construction of regular polygons.[29] Familiar with the complex plane as a result of the researches on the fundamental theorem of algebra that he had presented two years earlier in his doctoral dissertation at the University of Helmstedt (recall the previous chapter), Gauss already knew that the solutions of this equation could be written as $\cos \frac{2\pi k}{n} + i \sin \frac{2\pi k}{n}$ for $k = 0, 1, 2, \ldots, n - 1$, but in the *Disquisitiones*, he aimed to determine them algebraically. Because the solution of the equation for composite integers follows immediately from that for primes, Gauss restricted his attention to the case where n is prime. Moreover, because $x^n - 1$ factors as $(x - 1)(x^{n-1} + x^{n-2} + \cdots + x + 1)$, the equation

$$x^{n-1} + x^{n-2} + \cdots + x + 1 = 0 \qquad (11.3)$$

---

[28] Cauchy, 1844, pp. 183–184. This result is a special case of what is now called Lagrange's theorem.

[29] Gauss, 1801/1966, pp. 407–460. For a fuller sense of Gauss's biographical details, see Bühler, 1981.

Figure 11.3. Carl Friedrich Gauss (1777–1855).

provided the focus for his work. To solve this $(n-1)$st-degree equation, Gauss sought to factor $n-1$ into prime factors and then to solve, in turn, each of a series of auxiliary equations of degree corresponding to these factors, where the coefficients of each equation are rational in the set of roots of the previous equation. Thus, for $n=17$, where $n-1=2\cdot 2\cdot 2\cdot 2$, he needed to determine four quadratic equations, while for $n=73$, three quadratics and two cubics were required.

Gauss began with what became a standard group-theoretic argument showing that the set Ω of the $n-1$ complex roots of $x^{n-1}+x^{n-2}+\cdots+x+1$ is a cyclic group:[30]

> If therefore r is any root in Ω, we will have $1=r^n=r^{2n}$ etc., and, in general, $r^{en}=1$ for any positive or negative integral value of e. Thus if λ, μ are integers that are congruent relative to n, we will have $r^\lambda=r^\mu$. But if λ, μ are noncongruent relative to n, then r^λ and r^μ will be unequal; for in this case we can find an integer v such that $(\lambda-\mu)v\equiv 1\pmod{n}$ so $r^{(\lambda-\mu)v}=r$ and certainly $r^{\lambda-\mu}$ does not equal 1. It is also clear that any power of r is also a root of the equation

[30] Gauss, 1801/1966, pp. 410–411.

$x^n - 1 = 0$. Therefore, since the quantities $1(=r^0), r, r^2, \ldots, r^{n-1}$ are all different, they will give us all the roots of the equation $x^n - 1 = 0$ and so the numbers $r, r^2, r^3, \ldots, r^{n-1}$ will coincide with Ω.

After proving that equation (11.3) was irreducible, he showed that if g is any primitive root modulo n (that is, a number g such that the powers $1, g, g^2, \ldots, g^{n-2}$ include all the nonzero residues modulo n), then the $n - 1$ roots of the equation can be expressed as $r, r^g, r^{g^2}, \ldots, r^{g^{n-2}}$ or even as $r^\lambda, r^{\lambda g}, r^{\lambda g^2}, \ldots, r^{\lambda g^{n-2}}$, for any λ not divisible by n. Then $r^{\lambda g^\mu}$ and $r^{\lambda g^\nu}$ are equal or unequal according to whether μ and ν are congruent or noncongruent modulo $n - 1$. (In modern terminology, Gauss had shown that the Galois group of the cyclotomic equation $x^n - 1 = 0$ is a *cyclic group* of order $n - 1$ *isomorphic* to the multiplicative group of *residue classes modulo n* and is generated by the *automorphism* taking r to r^g, for any *primitive* root g.)

To determine the auxiliary equations the solution of which would lead to the solution of the original equation, Gauss then constructed what he termed "periods," that is, certain sums of the roots r^j, which, in turn, were the roots of the auxiliary equations. In other words, if $n - 1 = ef$, he constructed a sum of f terms that took on exactly e values under the action of the group; each of these values would be a root of an auxiliary equation of degree e. (In fact, in modern terminology, each sum would be *invariant* under a subgroup H of G of order f, generated by the mapping $r \rightarrow r^{g^e}$, while the e values would be the images of this sum under the action of the cosets of H in G.)

An analysis of the case $n = 19$ will give the flavor of Gauss's work.[31] Since $n - 1 = 18$ can be factored as $18 = 3 \cdot 6$, Gauss first determined a period of six terms by choosing a primitive root g modulo 19 (here 2), setting $h = 2^3$, and computing

$$\alpha_1 = \sum_{k=0}^{5} r^{h^k}.$$

Since the permutations of the 18 roots of $x^{18} + x^{17} + \cdots + x + 1 = 0$ form a cyclic group G determined by the mapping $r \rightarrow r^2$, where r is any fixed root, this period is invariant under the subgroup H of G generated by the

---

[31] Gauss, 1801/1966, pp. 418–426. We have changed the notation somewhat to make the material easier to read.

mapping $r \to r^h$. It contains six elements because $h^6 = 2^{18} \equiv 1 \pmod{19}$, that is, H is a subgroup of G of order 6. In particular, since $H = \{r \to r, r^8, r^{64}, r^{512}, r^{4096}, r^{32768}\}$, it follows that

$$\alpha_1 = r + r^8 + r^7 + r^{18} + r^{11} + r^{12},$$

where the powers are reduced modulo 19. Furthermore, for $i = 2$ and $i = 4$, those mappings of the form $r \to r^{ih^k}$, for $k = 0, 1, \ldots, 5$, are precisely the three cosets of H in the group G. Thus, the two images of this initial period are

$$\alpha_2 = r^2 + r^{16} + r^{14} + r^{17} + r^3 + r^5 \text{ and } \alpha_4 = r^4 + r^{13} + r^9 + r^{15} + r^6 + r^{10},$$

and α_1, α_2, α_4 are the three values taken on by α_1 under all the permutations of G. These three values are thus the roots of a cubic equation, which Gauss easily found to be $x^3 + x^2 - 6x - 7 = 0$. It is also not hard to see that α_2 and α_4 can be expressed as a polynomial in α_1, so that the group of this equation is a cyclic group of order 3. In fact, $\alpha_2 = 4 - \alpha_1^2$, and $\alpha_4 = -5 - \alpha_1 + \alpha_1^2$. (In modern terms, the field $Q(\alpha_1)$ is a *splitting field* of the equation for these periods.)

The next step is to divide each of the three periods into three more periods of two terms each, where, again, the new periods satisfy an equation of degree 3 over the field determined by the original periods.[32] These periods,

$$\beta_i = \sum_{k=0}^{1} r^{im^k}, \quad \text{for} \quad i = 1, 2, 4, 8, 16, 13, 7, 14, 9,$$

where $m = 2^9$, are invariant under the subgroup M generated by the mapping $r \to r^m$. Because $m^2 \equiv 1 \pmod{19}$, M has order 2 and has nine cosets corresponding to the values of i given. For example,

$$\alpha_1 = \beta_1 + \beta_8 + \beta_7 = (r + r^{18}) + (r^8 + r^{11}) + (r^7 + r^{12}).$$

Given these new periods of length 2, it is not hard to show that β_1, β_8, and β_7 are all roots of the cubic equation $x^3 - \alpha_1 x^2 + (\alpha_1 + \alpha_4)x - 2 - \alpha_2 = 0$. As before, each of the other β_i can be expressed as a polynomial in β_1,

---

[32] Gauss, 1801/1966, pp. 429–432.

so that the group of this equation is also cyclic of order 3. Finally, Gauss broke up each of the periods with two terms into the individual terms of which it is formed and showed that, for example, r and r^{18} are the two roots of $x^2 - \beta_1 x + 1 = 0$. The sixteen remaining roots of the original equation are then simply powers of r. Because the equations of degree 2 and 3 involved in the above example are *solvable by radicals*, that is, because their roots may be expressed using the four basic arithmetic operations and the operation of root extraction applied to their original coefficients, Gauss had demonstrated that the roots of $x^{19} - 1 = 0$ are also all expressible in terms of radicals.

His more general result, applicable to any equation $x^n - 1 = 0$, only showed that a series of equations can be discovered (each of prime degree dividing $n - 1$), the solutions of which would determine the solution of the original equation. But, Gauss continued,

> everyone knows that the most eminent geometers have been inef-
> fectual in the search for a general solution of equations higher than
> the fourth degree, or (to define the search more accurately) for the
> reduction of mixed equations to pure equations. [A *pure equation* is
> one of the form $x^m - A = 0$, which can be solved by taking an mth
> root once the solutions of $x^m - 1 = 0$ are known.] And there is little
> doubt that this problem does not so much defy modern methods
> of analysis as that it proposes the impossible....Nevertheless, it
> is certain that there are innumerable mixed equations of every
> degree which admit a reduction to pure equations, and we trust that
> geometers will find it gratifying if we show that our equations are
> always of this kind.[33]

Gauss then sketched a proof, although with one minor gap, that the auxiliary equations involved in his solution of $x^n - 1 = 0$, for n prime, can always be reduced to pure equations.[34] Beginning with the assumption that all roots of unity of degree smaller than n can be expressed using radicals, he demonstrated by induction on n that the auxiliary equations for degree n are always solvable by radicals.

To understand why a rigorous proof of this result was not simple, consider the smallest n for which at least one of the necessary equations

---

[33] Gauss, 1801/1966, p. 445.
[34] See Tignol, 2001, pp. 192–200 for a discussion of Gauss's proof and the gap.

has degree greater than four, namely, the case $n = 11$, where $n - 1 = 5 \cdot 2$. Solving the corresponding cyclotomic equation

$$x^{10} + x^9 + x^8 + x^7 + x^6 + x^5 + x^4 + x^3 + x^2 + x + 1 = 0 \qquad (11.4)$$

(with roots denoted r, $r^2, \ldots, r^{10}$) requires finding a period of two elements that takes on exactly five values under the permutations of the roots of (11.4) (or, equivalently, is invariant under the subgroup $H = \{r \to r, r^{10}\}$ of the cyclic group G of order 10). It is not difficult to see that one period is $r + r^{10}$, while its four images under G are $r^2 + r^9$, $r^3 + r^8$, $r^4 + r^7$, and $r^5 + r^6$ (where in each case the two elements are inverses of each other). To find the fifth-degree equation these periods satisfy, substitute $y = -(x + x^{-1})$ into equation (11.4) (after dividing each term of the equation by x^5) to get

$$y^5 - y^4 - 4y^3 + 3y^2 + 3y - 1 = 0. \qquad (11.5)$$

Once we find a solution α of this equation by radicals, it only remains to solve the quadratic equation $x^2 + \alpha x + 1 = 0$ to find r, an eleventh root of unity and to express it by radicals.[35]

As Gauss correctly surmised, there is no general solution for quintic equations; the solution of equation (11.5) in radicals must thus be found using his proof technique, which requires using the already known expressions for fifth roots of unity. Although we will not reproduce Gauss's proof for this case, it turns out that one solution is $y = \frac{1}{5}(1 + A + B + C + D)$, where

$$A = \sqrt[5]{\frac{11}{4}\left(89 + 25\sqrt{5} - 5\sqrt{-5 - 2\sqrt{5}} + 45\sqrt{-5 + 2\sqrt{5}}\right)},$$

$$B = \sqrt[5]{\frac{11}{4}\left(89 + 25\sqrt{5} + 5\sqrt{-5 - 2\sqrt{5}} - 45\sqrt{-5 + 2\sqrt{5}}\right)},$$

---

[35] See Tignol, 2001, pp. 158–164 for further discussion of this equation and its solution. In fact, this solution was given by Vandermonde in Vandermonde, 1771, p. 416 without much explanation. There is a minor misprint in Vandermonde's solution, however, that has been corrected here. See also O. Neumann, 2006 for more details on Vandermonde's solution and on Gauss's proof.

$$C = \sqrt[5]{\frac{11}{4}\left(89 - 25\sqrt{5} - 5\sqrt{-5 + 2\sqrt{5}} - 45\sqrt{-5 - 2\sqrt{5}}\right)},$$

$$D = \sqrt[5]{\frac{11}{4}\left(89 - 25\sqrt{5} + 5\sqrt{-5 + 2\sqrt{5}} + 45\sqrt{-5 - 2\sqrt{5}}\right)}.$$

It is far simpler, of course, to solve the degree n cyclotomic equation if $n - 1$ is a power of 2, because in that case, all of the auxiliary equations are quadratic. In this situation, however, since the roots of $x^n - 1 = 0$ can be interpreted as the vertices of a regular n-gon (in the complex plane), Gauss wrote that "the division of the circle into n parts or the inscription of a regular polygon of n sides can be accomplished by geometric [that is, Euclidean] constructions."[36] The only such primes n known to Gauss— and to us today—for which $n - 1$ is a power of 2 are 3, 5, 17, 257, and 65,537. (It has been alleged that Gauss's discovery of the construction of the regular 17-gon in 1796 was instrumental in his decision to pursue a career in mathematics.[37]) Gauss concluded with a caution to his readers: "Whenever $n - 1$ involves prime factors other than 2, we are always led to equations of higher degree....We can show with all rigor that these higher-degree equations cannot be avoided in any way nor can they be reduced to lower degree equations. The limits of the present work exclude this demonstration here, but we issue this warning lest anyone attempt to achieve geometric constructions for sections [of the circle] other than the ones suggested by our theory (e.g. sections into 7, 11, 13, 19, etc. parts) and so spend his time uselessly." Interestingly, although he asserted that he had a proof that he was "excluding," the young French engineer, Pierre Wantzel (1814–1848), actually provided the proof in 1837 that these polygons could not be constructed by straightedge and compass.[38]

As noted earlier, Abel had established the impossibility of solving algebraically the general fifth-degree equation in 1826, twenty-five years after Gauss had published his *Disquisitiones*, eleven years before Wantzel's results, and three years before his own untimely and tragic death.[39] In

[36] Gauss, 1801/1966, p. 458. For the next quote, see p. 459.
[37] Bell, 1937, p. 67.
[38] See Wantzel, 1845.
[39] For more on Abel's biography, see Ore, 1974 and Stubhaug, 2000.

light of this negative result, Abel changed his research goals to attempt to solve the following problems instead: "To find all equations of any given degree that are algebraically solvable" and "To decide whether a given equation is algebraically solvable or not."[40] Although he was unable, in what remained of his brief life, to solve either of these questions in its entirety, he did make progress on a particular type of equation.

In a paper published in 1829 in Crelle's *Journal für die reine und angewandte Mathematik* (or *Journal for Pure and Applied Mathematics*), Abel generalized Gauss's solution method for the cyclotomic equations $x^n - 1 = 0$. Recall that for these equations, every root is expressible as a power of one root. Abel showed that "if the roots of an equation of any degree are related so that all of them are rationally expressible in terms of one, which we designate as x, and if, furthermore, for any two of the roots θx and $\theta_1 x$ [where θ and θ_1 are rational functions], we have $\theta\theta_1 x = \theta_1\theta x$, then the equation is algebraically solvable."[41] (In modern terms, these conditions imply that the Galois group of the equation is a *commutative group* of order equal to its degree.) He then argued that, in this situation, as in the cyclotomic case, the solution could always be reduced to that of auxiliary equations of prime degree, each of which is then algebraically solvable. It is because of this result that commutative groups are often referred to as "Abelian" today.

THE WORK OF GALOIS AND ITS RECEPTION

Although Abel never completed his research program, it was largely realized by another genius who died young, Évariste Galois (1811–1832). Born in Bourg La Reine near Paris, Galois discovered his interest in mathematics while attending the prestigious Lycée Louis-le-Grand.[42] Failing twice to gain admission to the École Polytechnique, he finally took a baccalaureate degree from the École Normale Supérieure and

---

[40] Abel, 1828, p. 219, as quoted in Wussing, 1984, p. 98.

[41] Abel, 1829, p. 131–132, as quoted in Wussing, 1984, p. 100.

[42] The story of Galois's life has been told many times. See, for example, Dupuy, 1896; Dalmas, 1956; the more recent (and more controversial) comparative biographical study, Alexander, 2010; and Ehrhardt, 2011. The two-hundredth anniversary of Galois's birth in 2011 saw a flurry of new research on Galois and his mathematical legacy.

Figure 11.4. Évariste Galois (1811–1832).

repeatedly tried to publish the fruits of his mathematical researches.[43] Politics and not mathematics quickly came to dominate Galois's life, however. An ardent republican, he was caught up in the events of the revolution of July 1830 and imprisoned at Sainte-Pélagie. He died in a duel only a few months after his release in the spring of 1832. Despite this turbulent and short life, Galois outlined his thoughts on the solvability of algebraic equations not only in a manuscript he submitted to the Paris Academy of Sciences in 1831 but also in the letter he wrote to his friend, Auguste Chevalier, just before the duel that ended his life.

He opened the manuscript, entitled "Mémoire sur les conditions de résolubilité des équations par radicaux" (or "Memoir on the Conditions for Solvability of Equations by Radicals"), by clarifying the idea of rationality. Since an equation has coefficients in a certain "domain of

---

[43] Although he did publish some of his work, he sought to put his ideas directly before the academicians of the Paris Academy of Sciences. His first submission there in 1829 was refereed by Cauchy, who could not understand it and recommended that Galois explain his ideas more fully in a revised version for reconsideration. Galois's resubmission appropriately—but, as it turned out, unfortunately—landed on the desk of the corresponding secretary of the Academy, Joseph Fourier (1768–1830), who died before handling the paper. The submission was subsequently lost among Fourier's *Nachlaß*. On the saga, specifically, of the rejection of his third and final submission to the Academy in 1831, see Ehrhardt, 2010. We discuss the contents of Galois's ill-fated paper below.

rationality" (today we would call this a *field*)—for example, the ordinary rational numbers—to say that it is solvable by radicals means, as noted above, that any root of it can be expressed using the four basic arithmetic operations and the operation of root extraction, all applied to elements of this original domain. It turns out, however, that it is usually convenient to solve an equation in steps, as Gauss did in the cyclotomic case. For example, having solved $x^n = \alpha$, solutions expressible as $\sqrt[n]{\alpha}$, $r\sqrt[n]{\alpha}$, $r^2\sqrt[n]{\alpha}, \ldots$, where r is an nth root of unity, are available as coefficients in the next step. Galois noted that such quantities are adjoined to the original domain of rationality and that any quantity expressible by the four basic operations in terms of these new quantities and the original ones can then also be considered as *rational*. (In modern terminology, one begins with a particular field, then constructs an *extension field* by adjoining certain quantities not in the original field.) Of course, "the properties and the difficulties of an equation can be altogether different, depending on what quantities are adjoined."[44]

In addition to concepts associated with rationality, that is, in addition to building his new theory in a critical way on the new field concept, Galois also fundamentally built it on the idea of a group. In his introduction, he discussed the notion of a permutation, in terms of Cauchy's somewhat ambiguous language, and he used the word "group," although not always in a strictly technical sense. For Galois, a group was sometimes a set of permutations closed under composition and other times a set of arrangements of letters determined by applying certain permutations.

Galois laid out the manuscript's main result in its Proposition I:

> Let an equation be given of which a, b, $c, \ldots$ are the m roots.[45] There will always be a group of permutations of the letters $a, b, c, \ldots$ which has the following property: 1. that every function of the roots, invariant under the substitutions of the group, is rationally known; 2. conversely, that every function of the roots which is rationally known is invariant under the substitutions.[46]

---

[44] Galois, 1831, as translated in Edwards, 1984, p. 102. In this work, Edwards discussed Galois's work in detail and included a complete translation of Galois's "Mémoire sur les conditions de résolubilité des équations par radicaux." For new English translations of almost all of Galois's materials (with mathematical commentary), see Neumann, 2011.

[45] Galois tacitly assumed that this equation is irreducible and that all the roots are distinct.

[46] Galois, 1831, as translated in Edwards, 1984, p. 104.

Galois called this group of permutations the "group of the equation." (In modern usage, the group of the equation—that is, the *Galois group*—is a group of automorphisms acting on the entire field created by adjoining the roots of the equation to the original field of coefficients. Galois's result then stated that the group of the equation is that group of automorphisms of the extension field that leaves invariant precisely the elements of the original field, that is, the elements that are "rationally known.")

Besides giving a brief proof of his result, Galois presented Gauss's example of the cyclotomic polynomial $x^{p-1} + x^{p-2} + \cdots + x + 1$, for p prime.[47] In that case, as Gauss had already proved, the group of the equation is the cyclic group of $p - 1$ permutations generated by the mapping $r \rightarrow r^g$, where g is any primitive root modulo p. On the other hand, the group of the general equation of degree n, that is, of the equation with literal coefficients, is the group of all $n!$ permutations of n letters.

Having stated the main theorem, Galois explored its application to the solvability question. His second proposition showed what happens when one or all of the roots of some auxiliary equation (or of the original equation) are adjoined to the original field. In general, G will contain another group H, the group of the equation over the extended field. (In modern terminology, because any automorphism that leaves the new field invariant certainly leaves the original field invariant, the group H of the equation over the new field is a subgroup of the group G over the original field.) As Galois noted in his letter to Chevalier, G can then be decomposed either as $G = H + HS + HS' + \cdots$ or as $G = H + TH + T'H + \cdots$, where $S, S', T, T', \ldots$ are appropriately chosen permutations of G. He further noted that ordinarily these two decompositions do not coincide. When they do, however, and this always happens when all the roots of an auxiliary equation are adjoined, he called the decomposition "proper": "If the group of an equation has a proper decomposition, so that it is divided into m groups of n permutations, one may solve the given equation by means of two equations, one having a group of m permutations, the other one of n permutations."[48] (In modern terminology, a proper decomposition occurs when the subgroup H is *normal*, that is, when the right cosets $\{HS\}$ coincide with the left cosets

---

[47] Galois, 1831, as translated in Edwards, 1984, pp. 104–106.
[48] Galois, 1831, in Galois, 1962, p. 175 (our translation).

$\{TH\}$.) Under these circumstances, the question of solvability reduces to the solvability of two equations, each having groups of order less than the original one.

Gauss had already shown that the roots of the polynomial $x^p - 1$, for p prime, can be expressed in terms of radicals. It follows that if the pth roots of unity are assumed to be in the original field, then the adjunction of one root of $x^p - \alpha$ amounts to the adjunction of all of the roots. If G is the group of an equation $p(x)$, this adjunction of a pth root therefore leads to a normal subgroup H of the group G such that the index of H in G (that is, the order of G divided by the order of H) is p. Galois also proved the converse, namely, if the group G of an equation has a normal subgroup of index p, then there is an element α of the original field (assuming that the pth roots of unity are in that field) such that adjoining $\sqrt[p]{\alpha}$ reduces the group of the equation to H.[49] Galois concluded, both in his manuscript and in his letter, that an equation is solvable by radicals as long as one can continue the process of finding normal subgroups until all of the resulting indices are prime. For example, in the case of the general equation of degree four, the group of the equation of order 24 has a normal subgroup of order 12, which, in turn, contains one of order 4, which, in turn, contains one of order 2, which, in turn, contains the identity. It follows that the solution can be obtained by first adjoining a square root, then a cube root, and then two more square roots. Galois observed that the standard solution of the quartic equation uses precisely those steps. He concluded with two additional results applicable to solving irreducible equations of prime degree.

First, Galois showed that such an equation is solvable by radicals if and only if each of the permutations in the Galois group transforms a root x_k into a root $x_{k'}$, with $k' \equiv ak + b$ (mod p) (under a suitable ordering of the roots).[50] For example, the Galois group of an irreducible cubic with one real and two complex roots is the group of all six permutations on three letters. If we use 0, 1, and 2 to identify the three roots, then these six permutations can be expressed as $k \to ak + b$, with $a = 1, 2$ and $b = 0, 1, 2$. As another example, note that if $p = 5$, the group of permutations identified by Galois has 20 elements, while the group of the

[49] Galois, 1831, as translated in Edwards, 1984, p. 108.
[50] Galois, 1831, as translated in Edwards, 1984, pp. 111–112.

general fifth-degree polynomial has 120 elements. Therefore, Galois's result shows that the general quintic is not solvable by radicals.

Second, Galois proved that an irreducible equation of prime degree is solvable if and only if all of its roots can be expressed rationally in terms of any two of them.[51] In other words, the adjunction of two of the roots to the original field amounts to adjoining all of them. As a consequence, if an irreducible fifth-degree polynomial has three real roots and two complex ones, the condition is not met, and the equation is not solvable by radicals.

Following Galois's untimely death in 1832, his manuscripts lay unread—and so completely unappreciated—by a French mathematical community that had failed to understand and appreciate his work during his lifetime. By the early 1840s, however, Joseph Liouville (1809–1882) had begun to lecture on Galois's papers at the Collège de France in an effort to understand them prior to their publication in his *Journal de mathématiques pures et appliquées* (or *Journal of Pure and Applied Mathematics*) in 1846. Among his auditors were Charles Hermite (1822–1901), Joseph Serret (1819–1885), and Joseph Bertrand (1822–1900), all of whom published mathematical research related to ideas they encountered in Galois's work.[52]

In particular, Hermite engaged in the set of questions surrounding the solvability of equations. Abel had shown that equations of the fifth degree or higher *were not*, in general, solvable by radicals; Galois had characterized, at least for prime degrees, those higher-degree equations that *were* solvable under that restriction. Pursuing the general question of solvability raised by Galois's work, however, Hermite recognized "the need to introduce some new analytic element into the search for the solution" and showed how to solve the quintic in terms not of radicals but of elliptic functions.[53]

In that same vein, Serret gave the first textbook treatment of Galois's theory in the third edition of the *Cours d'algèbre supérieure* (or *Course on Higher Algebra*) that he had first given while in the chair of higher algebra at the Sorbonne in the late 1840s. There, Serret stated and gave

---

[51] Galois, 1831, as translated in Edwards, 1984, p. 113.

[52] Belhoste and Lützen, 1984.

[53] For a very nice exposition of this work in historical context, see Goldstein, 2011, especially pp. 226–250. The quote—translated by Goldstein from Hermite, 1858, p. 508—appears on p. 245 in Goldstein's paper.

reasonable proofs of all of Galois's results, but, like Hermite, he did not interpret the idea of a group—in the modern sense—as the basis of the work.[54] Instead, he emphasized Galois's final results on the solvability of equations of prime degree, finding his predecessor's main contribution to have been the answer to Abel's question about which equations could be solved algebraically.[55]

Four more years would pass before Galois's work would be cast brightly in a group-theoretic light by a member of the next mathematical generation. In 1870, Camille Jordan (1838–1922), a *polytechnicien* and engineer who only later would teach at both his *alma mater* and at the Collège de France, published his monumental *Traité des substitutions et des équations algébriques* (or *Treatise on Substitutions and Algebraic Equations*). There, he not only fully explained Galois's ideas of basing the theory of solvability on the notions of domains of rationality and of groups and their properties, but he also incorporated the fruits of a decade of his own group-theoretic researches in the context of groups of permutations (substitutions).[56] An instant classic, his *Traité* set a new research agenda of creating a *theory* of groups as opposed to the older agenda of devising ways to *calculate* the solutions of polynomial equations. In a group-theoretic guise, this is a manifestation of the new *abstract* stage of algebra that developed over the course of the last half of the nineteenth and into the twentieth century and that we trace in this and the remaining chapters.[57]

Jordan defined many of the concepts now viewed as central to group theory in Book II of the *Traité*. For him, a *group* was a system of permutations of a finite set with the condition that the product (that is, composition) of any two such permutations belongs to the system.[58] The unit permutation 1 leaves every element unchanged, while the inverse a^{-1} of any permutation a is the permutation such that $aa^{-1} = 1$. He then defined the *transform* of a permutation a by a permutation b to be the permutation $b^{-1}ab$ and the transform of the group $A = \{a_1, a_2, \ldots, a_n\}$ by

---

[54] See Serret, 1866.

[55] For a discussion of the reception of Galois's work in Europe, and especially in Italy, see Martini, 2002.

[56] For more on the connections between Galois's work—especially in his so-called "Second Mémoire" of 1830 (unpublished until 1846)—and Jordan's ideas, see P. Neumann, 2006.

[57] Compare Goldstein, 2011, pp. 257–262.

[58] Jordan, 1870, p. 22. The next several definitions are on pp. 22–24.

b to be the group $B = \{b^{-1}a_1b, b^{-1}a_2b, \ldots, b^{-1}a_nb\}$ consisting of all the transforms. If B coincides with A, then A is said to be *permutable* with b.

Although Jordan did not explicitly delineate the notion of a normal subgroup of a group, he did define a *simple* group as one that contains no subgroup (other than the identity) permutable with all elements of the group.[59] For a nonsimple group G, there must then exist a *composition series*, that is, a sequence of groups $G = H_0$, H_1, H_2, $\ldots$, $\{1\}$ such that each group is contained in the previous one and is permutable with all its elements (that is, is normal), and such that no other such group can be interposed in this sequence. Jordan further proved that if the order of G is n and the orders of the subgroups are successively $\frac{n}{\lambda}$, $\frac{n}{\lambda\mu}$, $\frac{n}{\lambda\mu\nu}$, $\ldots$, then the integers $\lambda, \mu, \nu, \ldots$ are unique up to order, that is, any other such sequence has the same *composition factors*.[60]

Jordan was then able to use some of these group-theoretic concepts to restate the results of Abel and Galois. Starting with Galois's group of an equation, Jordan defined a *solvable group* to be one that belongs to an equation solvable by radicals.[61] Thus, a solvable group is one that contains a composition series with all composition factors prime. Because a commutative group always has prime composition factors, Jordan showed that an *Abelian equation*, one "of which the group contains only substitutions which are interchangeable among themselves,"[62] is always solvable by radicals. On the other hand, because the alternating group on n letters, which has order $\frac{n!}{2}$, is simple for $n > 4$, it follows immediately that the general equation of degree n is not solvable by radicals. Jordan's work made it abundantly clear that—as Galois had tried to contend—the theory of permutation groups was intimately connected with the solvability of equations.

THE MANY ROOTS OF GROUP THEORY

Until well into the nineteenth century, algebra was defined in terms of solving equations. As we just saw, as late as 1866, Serret still conceived of

[59] Jordan, 1870, p. 41.
[60] Jordan, 1870, p. 42.
[61] Jordan, 1870, p. 385.
[62] Jordan, 1870, p. 287 (our translation).

it that way in his *Cours d'algèbre supérieure*. Yet, at the same time, the notion of abstracting some important ideas out of various concrete situations was beginning to take hold, and the idea of a mathematical structure was starting to appear. We have seen how important the concept of a permutation group had become by 1870, but group-theoretic ideas had also begun to appear in other contexts over the course of the nineteenth century.

One place such ideas arose was in Section V of Gauss's *Disquistiones*, where he discussed the theory of quadratic forms,[63] that is, functions of the form $ax^2 + 2bxy + cy^2$. There, he aimed primarily to determine whether a given integer can, in some sense, be represented by a particular form. Gauss defined the *determinant* of a form (today called the *discriminant*) to be the quantity $b^2 - ac$, and, as a tool in the solution of the representation problem, he defined the notion of the equivalence of two forms. First, he wrote that a quadratic form $F = ax^2 + 2bxy + cy^2$ *contains* a form $F' = a'x'^2 + 2b'x'y' + c'y'^2$, if there exists a linear substitution $x = \alpha x' + \beta y'$, $y = \gamma x' + \delta y'$ that transforms F into F'. As Gauss showed explicitly, the coefficients of F' under the given substitution are

$$a' = a\alpha^2 + 2b\alpha\gamma + c\gamma^2,$$

$$b' = a\alpha\beta + b(\alpha\delta + \beta\gamma) + c\gamma\delta,$$

$$c' = a\beta^2 + 2b\beta\delta + c\delta^2.$$

Simplifying this by "multiplying the second equation by itself, the first by the third, and subtracting we get

$$b'b' - a'c' = (b^2 - ac)(\alpha\delta - \beta\gamma)^2."[64] \tag{11.6}$$

In other words, the discriminants of F and F' are equal up to a power of the determinant of the linear substitution. In terminology that would only be developed later (see the next chapter), Gauss had shown that the discriminant is an *invariant* of these forms. Next, he defined F to

---

[63] See Gauss, 1801/1966, pp. 108–406.
[64] Gauss, 1801/1966, pp. 111–112.

be *equivalent* to F' if F' also contains F. In this case, $(\alpha\delta - \beta\gamma)^2 = 1$, so the discriminants of the two forms are equal. Gauss then noted that, for any given value D of the discriminant, there were finitely many classes of equivalent forms. In particular, there was a distinguished class, the so-called principal class, consisting of those forms equivalent to the form $x^2 - Dy^2$.

To investigate these classes, Gauss developed an operation on them, namely, composition. He first defined composition of forms, then showed that if F and F' are equivalent to G and G', respectively, the composition of F and F' is equivalent to the composition of G and G'. Moreover, he showed that the order of the composition was irrelevant and that the associative law holds as well. Finally, he proved that "if any class K is composed with the principal class, the result will be the class K itself,"[65] that for any class K there is a class L (opposite to K) such that the composite of the two is the principal class, and that "given any two classes K, L of the same [discriminant], . . .we can always find a class M with the same [discriminant] such that L is composed of M and K." Given that composition enjoys the basic properties of addition, Gauss noted that "it is convenient to denote composition of classes by the addition sign, $+$, and identity of classes by the equality sign."

With the addition sign as the sign of operation, Gauss designated the composite of a class C with itself by $2C$, the composite of C with $2C$ as $3C$, and so on. He then proved that for any class C, there is a smallest multiple mC that is equal to the principal class and that, if the total number of classes is n, then m is a factor of n. Recalling that he had achieved similar results earlier in his book in the context of powers of residue classes of numbers, he noted that "the demonstration of the preceding theorem is quite analogous to the demonstrations [on powers of residue classes] and, in fact, the theory of the [composition] of classes has a great affinity in every way with the subject treated [earlier]."[66] He could therefore assert, without proof, various other results coming from this analogy, in terms of what is now the theory of Abelian groups.

---

[65] See Gauss, 1801/1966, pp. 264–265 for this and the quotes that follow in this paragraph. See Gauss, 1801/1966, pp. 232–233 for the rather involved definition of composition of forms. Initially, Gauss did not symbolize this operation, but just wrote it in words, only using symbols once he showed that the operation was an operation on classes.

[66] Gauss, 1801/1966, p. 366.

Although he explicitly signaled the analogy between his two treatments, Gauss did not attempt to develop an abstract theory of groups.
Another who noticed such similarities, Leopold Kronecker (1823–1891),
pushed the analogy further in 1870. Kronecker, who had studied under
Ernst Kummer (1810–1893) first at the Liegnitz Gymnasium and later at
the University of Breslau, earned his doctoral degree at the University
of Berlin under the guidance of Peter Lejeune Dirichlet (1805–1859) in
1845. Independently wealthy, Kronecker moved permanently to Berlin
in 1855 in order to participate personally in the vibrant mathematical environment engendered by the university, the Berlin Academy of
Sciences, and Crelle's *Journal für die reine und angewandte Mathematik*.
That same year, his former teacher, Kummer, succeeded his adviser,
Dirichlet, on the Berlin faculty; Carl Borchardt (1817–1880), already
on the Berlin faculty, succeeded August Crelle (1780–1855) as editor of
the *Journal*; and one year later, Karl Weierstrass (1815–1897) moved to
the faculty of the Industry Institute in Berlin and accepted an associate
professorship at the university. Almost overnight, Berlin had become
unquestionably *the* place in Europe to pursue mathematical research at
its cutting edge. By 1870, when Kronecker recognized commonalities
between Gauss's work and Kummer's development of the theory of ideal
complex numbers, Weierstrass had left the Industry Institute for an
actual chair at Berlin, and he, Kummer, and Kronecker had established
an unrivaled mathematical triumvirate at Berlin recognized as such
internationally.[67]

Kummer had introduced ideal complex numbers in an effort to
address the problem that domains of algebraic integers did not have
the property of unique factorization (see chapter 13 for more on this).
In fact, he had defined an equivalence of ideal complex numbers that
partitioned them into classes, the properties of which were analogous to
those of Gauss's classes of forms.[68] In his 1870 paper, Kronecker not only
developed certain properties of these classes but also recalled Gauss's
work on quadratic forms:

The very simple principles on which Gauss's method rests are
applied not only in the given context but also frequently elsewhere,

---

[67] For more on the Berlin mathematical scene in the nineteenth century, see Biermann, 1988.
[68] See Kummer, 1847 and Wussing, 1984, p. 62.

in particular in the elementary parts of number theory. This circumstance shows, and it is easy to convince oneself, that these principles belong to a more general, abstract realm of ideas. It is therefore appropriate to free their development from all unimportant restrictions, so that one can spare oneself from the necessity of repeating the same argument in different cases. This advantage already appears in the development itself, and the presentation gains in simplicity, if it is given in the most general admissible manner, since the most important features stand out with clarity.[69]

Kronecker thus began to develop those "very simple principles": "Let θ', θ'', θ''', ... be finitely many elements such that to each pair there is associated a third by means of a definite procedure."[70] Kronecker required that this association, which he first wrote as $f(\theta', \theta'') = \theta'''$ and later as $\theta' \cdot \theta'' = \theta'''$, be commutative and associative and that, if $\theta'' \neq \theta'''$, then $\theta'\theta'' \neq \theta'\theta'''$. From the finiteness assumption, Kronecker then deduced the existence of a unit element 1 and, for any element θ, the existence of a smallest power n_θ such that $\theta^{n_\theta} = 1$. Finally, Kronecker stated and proved what is now called the fundamental theorem of Abelian groups, namely, there exists a finite set of elements $\theta_1, \theta_2, \ldots, \theta_m$ such that every element θ can be expressed uniquely as a product of the form $\theta_1^{h_1}\theta_2^{h_2}\cdots\theta_m^{h_m}$, where, for each i, $0 \leq h_i < n_{\theta_i}$. Furthermore, the θ_i can be arranged so that each n_θ is divisible by its successor, and the product of these numbers is precisely the number of elements in the system.[71]

Having demonstrated the abstract theorem, Kronecker interpreted the elements in various ways, noting that analogous results in each case had been proved previously by others. Kronecker did not, however, give a name to the system he defined, nor did he interpret it in terms of the permutation groups arising from Galois theory, perhaps because he was dealing solely with commutative groups. He was also probably unaware that, sixteen years earlier, the English mathematician, Arthur Cayley (1821–1895), had developed a similar abstract theory based on the notion of groups of substitutions.

Cayley had published "On the Theory of Groups" in 1854, while earning his livelihood as a conveyancing lawyer in London and some nine

---

[69] Kronecker, 1870, pp. 882–883, as quoted in Wussing, 1984, p. 64.

[70] Kronecker, 1870, p. 884, as quoted in Wussing, 1984, p. 65.

[71] Kronecker, 1870, pp. 885–886, and see Wussing, 1984, pp. 66–67.

Figure 11.5. Arthur Cayley (1821–1895).

years before he would assume his first academic position, the Sadleirian chair of pure mathematics at Cambridge University.[72] In this paper, Cayley credited Galois with the idea of a group of permutations and proceeded to generalize it to any set of operations, or functions, on a set of quantities. He used the symbol 1 to represent the function that leaves all quantities unchanged and noted that for functions, there is a well-defined notion of composition, which is associative but not, in general, commutative. He then abstracted the basic ideas from those of operations and defined a *group* to be a "set of symbols, 1, α, β, . . . , all of them different, and such that the product of any two of them (no matter in what order), or the product of any one of them into itself, belongs to the set." "It follows," he added, "that if the entire group is multiplied by any one of the symbols, either [on the right or the left], the effect is simply to reproduce the group."[73]

From a modern point of view, however, Cayley left out a significant portion of the definition, without which it is not clear why the last statement "follows." However, at the same time that he wrote his paper on groups, he wrote another on what he termed the theory of caustics, in

---

[72] For more on Cayley and his algebraic work, see the next chapter.
[73] Cayley, 1854a, p. 41.

which he considered a set of six functions on a particular set that formed, under composition, what he now called a group.[74] It thus appears that in writing his initial paper on groups, he was thinking—despite his "..." in the definition of his set of symbols—about *finite* collections of symbols, the composition of which was always assumed to be associative. He even stated explicitly in the introduction to the group theory paper that "if $\theta = \phi$, then, whatever the symbols α, β may be, $\alpha\theta\beta = \alpha\phi\beta$, and conversely."[75] This "converse" statement is usually referred to as the cancellation law and implies, assuming finiteness, that every symbol has an inverse, that is, there is another symbol whose product with the first is 1. With these caveats, then, it does "follow" that multiplication by any symbol produces a permutation of the group elements.

To study different abstract groups, Cayley introduced the group multiplication table

$$
\begin{array}{c|cccc}
 & 1 & \alpha & \beta & \cdots \\
\hline
1 & 1 & \alpha & \beta & \cdots \\
\alpha & \alpha & \alpha^2 & \beta\alpha & \cdots \\
\beta & \beta & \alpha\beta & \beta^2 & \cdots \\
\vdots & \vdots & \vdots & \vdots & \vdots
\end{array}
$$

in which, as he claimed in his definition, each row and each column contains all the symbols of the group.[76] Furthermore, he noted that each element θ satisfies the symbolic equation $\theta^n = 1$, if n is the number of elements in the group.

With these definitional aspects out of the way, he next proved, using a now-familiar argument, that if n is prime, then the group is necessarily of the form 1, α, $\alpha^2, \ldots, \alpha^{n-1}$, and if n is not prime, there are other possibilities. In particular, he displayed the group multiplication tables of the two possible groups of four elements and the two possible groups of six elements. In a further paper of 1859,[77] he described all five groups of order eight by listing their elements and defining the relations among them in addition to determining the smallest power (or *order*) of

[74] Cayley, 1853.
[75] Cayley, 1854a, p. 40.
[76] Cayley, 1854a, p. 41.
[77] Cayley, 1859.

each element that equals 1. For example, one of these groups contains the elements 1, α, β, $\beta\alpha$, γ, $\gamma\alpha$, $\gamma\beta$, $\gamma\beta\alpha$ with the relations $\alpha^2 = 1$, $\beta^2 = 1$, $\gamma^2 = 1$, $\alpha\beta = \beta\alpha$, $\alpha\gamma = \gamma\alpha$, and $\beta\gamma = \gamma\beta$. Each element in this group, except the identity, thus has order 2.

Although Cayley wrote an article for the *English Cyclopedia* in 1860 in which he again explained the term "group," no continental mathematician paid attention to this (nearly) abstract definition. At this point in time, mathematical communication across the English Channel was hit or miss, despite efforts, especially of the British, to make their work known abroad.[78] Coincidentally, Richard Dedekind (1831–1916), in his (unpublished) lectures on Galois theory at Göttingen University in 1856, abstracted the identical notion of a finite group out of the somewhat imprecise definition Galois had given solely in the context of permutations. As Dedekind wrote in his lecture notes, "the results [derived about permutations] are therefore valid for any finite domain of elements, things, concepts θ, θ', θ'', ... admitting an arbitrarily defined composition $\theta\theta'$, for any two given elements θ, θ', such that $\theta\theta'$ is itself a member of the domain, and such that this composition satisfies [associativity and right and left cancelability]. In many parts of mathematics, and especially in the theory of numbers and in algebra one often finds examples of this theory."[79] Following this definition, then, Dedekind continued his lectures by discussing some of the major ideas connected to the theory of groups, such as subgroups, cosets, normal subgroups, and so on. He made use of these in his elaboration of Galois's ideas on solving equations.[80]

Only two students attended Dedekind's lectures, so it is perhaps understandable that his group-theoretic ideas were not developed further in 1860s Germany.[81] In 1878, however, Cayley again tried to disseminate his ideas on this abstract concept, this time successfully, through the publication of four new papers on group theory in the inaugural volume of the *American Journal of Mathematics*. He began by repeating his basic definitions of 1854, writing that "a group is defined by means of the

---

[78] For more on this theme, see Parshall and Seneta, 1997.

[79] As quoted in Corry, 1996/2004, p. 77/78.

[80] While Dedekind's was the first university course on Galois theory, others soon followed. Sylow lectured on the topic at the University of Cristiania (Oslo) in 1862 (see Birkeland, 1996); Cesare Arzelà (1847–1912) gave the first such course in Italy only in 1886–1887 (see Martini, 2002).

[81] One of the students in the course, Paul Bachmann (1837–1920), did eventually publish on Galois theory in 1881. See Bachmann, 1881.

laws of combinations of its symbols."[82] Further, "although the theory [of groups] as above stated is a general one, including as a particular case the theory of substitutions, yet the general problem of finding all the groups of a given order n is really identical with the apparently less general problem of finding all the groups of the same order n which can be formed with the substitutions upon n letters."[83] Cayley took this result, today known as Cayley's theorem, as nearly obvious. He merely noted that any element of a group may be thought of as acting upon all the elements of the group by the group operation and that that operation induces a permutation of the group elements. However, Cayley noted, this "does not in any wise show that the best or the easiest mode of treating the general problem is thus to regard it as a problem of substitutions; and it seems clear that the better course is to consider the general problem in itself, and to deduce from it the theory of groups of substitutions." Thus, Cayley, like Kronecker, realized that problems in group theory could often best be attacked by considering groups in the abstract, rather than in their concrete realizations.

Although mathematicians in the 1870s and 1880s began to work on Cayley's problem of finding all groups of a given order, they did so not abstractly, but by following Jordan's lead and considering permutation groups. For example, the Norwegian mathematician, Ludwig Sylow (1832–1918), explicitly recorded his indebtedness to Jordan's work in his 1872 paper, "Théorèmes sur les groupes de substitutions" (or "Theorems on Substitution Groups"). There, he generalized the by-now-familiar result that if the order of a permutation group is divisible by a prime n, then the group always contains a permutation of order n: "If n^α designates the greatest power of the prime number n which divides the order of the group G, the group contains a subgroup H of order n^α. Furthermore, ... the order of G will be of the form $n^\alpha v(np + 1) \ldots$.[Under these conditions], the group G contains precisely $np + 1$ distinct groups of order n^α."[84] These results, for arbitrary finite groups, are today collectively known as the Sylow theorems, but Sylow himself proved them using ideas about permutation groups.[85] In fact, hearkening back to Galois as interpreted by Jordan, Sylow used his theorems to prove that

[82] Cayley, 1878a, p. 127.
[83] Cayley, 1878b, p. 52. The next quotation is also found here.
[84] Sylow, 1872, pp. 586–587 (our translation).
[85] For more on the history of the Sylow theorems, see Waterhouse, 1979–1980.

if the order of the group of an equation is a power of a prime, then the equation is solvable by radicals.

Also in a permutation-theoretic vein, Eugen Netto (1848–1919) wrote a major textbook on what the Germans termed "Substitutionentheorie" (or "theory of substitutions") in 1882. There, he noted the analogies between permutation groups and the Abelian groups of number theory, but he was unable to generalize these ideas to more general types of groups. He did, however, determine the possible permutation groups of orders p^2 and pq, where p and q are distinct primes.[86]

At the same time that the group concept was arising in permutation-theoretic settings, it was also cropping up in geometry. In 1872, the German mathematician Felix Klein (1849–1925) delivered his inaugural lecture as professor of mathematics at the University of Erlangen and introduced a research program that became known as the *Erlanger Programm*.[87] Klein explored the notion that the various geometrical studies of the nineteenth century could all be unified and classified by viewing geometry in general as the study of those properties of figures that remained invariant under the action of what he termed a group of transformations (see below). Thus, ordinary two-dimensional Euclidean geometry corresponded to the group of all rigid motions of the plane along with similarity transformations and reflections; projective geometry corresponded to the transformations that take lines into lines; and Lobachevskian geometry corresponded to projective transformations that left the cross ratio unaltered.[88] Evidently, Klein was heavily influenced by the study of permutation groups in defining a group of transformations. As he wrote, "the ideas, as well as the notation [for this theory], are taken from the theory of substitutions, with the difference merely that there, instead of the transformations of a continuous region, the permutations of a finite number of discrete quantities are considered."[89] He understood the notion of a transformation group as a set of transformations closed under composition. It was unnecessary to state explicitly the associative axiom or the existence of identities and

---

[86] Netto, 1882, pp. 133–137.

[87] For a self-contained overview of Klein's life and early career, see Parshall and Rowe, 1994, pp. 147–187.

[88] If segment AB contains two points C and D, then the *cross ratio* is defined to be the ratio of $\frac{AC}{CB}$ to $\frac{AD}{DB}$.

[89] Klein, 1893, p. 218.

inverses, since these were implicit in the sets of transformations he was considering.

That Klein referred to Jordan's work is not surprising given that he had spent time in Paris in the summer of 1870 and had met Jordan in person. He had been accompanied by the Norwegian mathematician, Sophus Lie (1842–1899), whom he had met earlier in Berlin.[90] Lie, too, was influenced by Jordan's work on group theory and began to develop his own ideas on groups of transformations. Unfortunately, when the Franco-Prussian War broke out in July of 1870, their studies were interrupted. As a Prussian subject, Klein had to return to the Rhineland immediately. Lie, a Norwegian, stayed for another month, but with the imminent defeat of the French army, he decided to hike to Italy. Arrested by the French as a spy because his mathematical manuscripts (in German) were thought to have been coded messages, Lie was released thanks to the intervention of the influential mathematician and *académicien*, Gaston Darboux (1842–1917), whom he had also met in Paris. Lie was allowed to continue on to Italy and eventually made his way back to Germany.

During the three years following his stay in Paris, Lie developed his ideas on continuous transformation groups, intending to apply them to the theory of differential equations. He outlined his goals in a paper published in 1874. As he explained, "the concept of a group of transformations, first developed in number theory and in the theory of substitutions, has in recent times been applied in geometric and general analytic investigations, respectively. One says that a set of transformations $x_i' = f_i(x_1, \ldots, x_n, \alpha_1, \ldots, \alpha_r)$ (where the x are the original variables, the x' are the new variables, and the α are parameters that are always supposed to vary continuously) is an r-parameter group, if when any two transformations of the set are composed, one gets a transformation belonging to the set."[91] Thus, like his German colleague, Klein, Lie did not feel compelled to state explicitly the associative axiom, given that transformations were automatically associative; he assumed the existence of an identity and inverse without comment.

In this initial paper, Lie noted that, in the case of a single variable, the only one-parameter group was the group of translations of the form $x' = x + \alpha$. Similarly, there was only one two-parameter group, namely,

---

[90] On Lie and Klein, see Yaglom, 1988.
[91] Lie, 1874, p. 1 (our translation).

the transformations combining a translation with a similarity expressed by $x' = \alpha x + \beta$. Finally, the three-parameter group is the group of linear fractional transformations $x' = \frac{\alpha x + \beta}{\gamma x + \delta}$, with $\alpha\delta - \beta\gamma \neq 0$. Interestingly, however, in a paper two years later, Lie explicitly proved the existence of identities and inverses for these groups and showed that all transformation groups of a one-dimensional manifold are (1) of these three types and (2) contain at most three parameters.[92]

THE ABSTRACT NOTION OF A GROUP

Following the successful reception of Cayley's ideas after their 1878 publication, many mathematicians began to realize that it was worthwhile to combine Kronecker's and Cayley's definitions into a single abstract group concept.[93] One of them, Walther von Dyck (1856–1934), had earned his doctorate under Felix Klein in Munich in 1879 during Klein's brief sojourn there as professor at the Technische Hochschule and before both of them moved to Leipzig University in 1880. As a student of Klein, Dyck was thoroughly familiar with contemporaneous continental, and especially German, work in the theory of algebraic equations, number theory, and geometry, as well as with Cayley's 1878 papers.

In 1882, Dyck combined this knowledge in his groundbreaking "Gruppentheoretische Studien" (or "Group-Theoretic Studies"). There, he formulated the basic problem: "To define a group of discrete operations, which are applied to a certain object, while one ignores any special form of representation of the individual operations, regarding these only as given by the properties essential for forming the group."[94] Although he alluded to the associative and inverse properties, Dyck did not give these as defining properties of a group. Instead, he showed how to construct a group using generators and relations, that is, beginning with a finite number of operations $A_1, A_2, \ldots, A_M$, he built the "most general"

---

[92] Lie, 1876.

[93] Group-theoretic concepts, especially symmetries as in Klein's work, also arise in diverse physical settings. As early as the late 1840s, for example, the French polymath Auguste Bravais (1811–1863) made seminal contributions in the pages of the publications of the Paris Academy of Sciences, in which he explored the structure of crystals in terms of the symmetry of molecular polyhedra. Although Cauchy applauded this work, the actual application of group theory to chemistry, as developed by the mathematicians, would be a twentieth-century development. For more on Bravais's work, see Partington, 1952, 3:52–58 and Birembaut, 1970.

[94] Dyck, 1882, p. 1, as quoted in Wussing, 1984, p. 240.

group G on these by considering all possible products of powers of them and their inverses. This group, today called the *free group* on $\{A_i\}$, automatically satisfies the modern group axioms. Dyck constructed new groups by assuming various relations of the form $F(A_1, A_2, \ldots, A_m) = 1$. In particular, he showed that if the group $\bar{G}$ is formed from operations $\bar{A}_1$, $\bar{A}_2, \ldots, \bar{A}_m$ satisfying the given relations, then "all these infinitely many operations of the group G, which are equal to the identity in $\bar{G}$, form a [sub]group H and this...commutes with all operations S, S', ... of the group G."[95] Dyck then proved that the mapping $A_i \to \bar{A}_i$ defined what he called an *isomorphism* from G onto $\bar{G}$. (In modern terminology, he had shown that the subgroup H is normal in G and that $\bar{G}$ is isomorphic to the *quotient group* G/H.)

In a continuation of this paper the following year, Dyck made explicit the connection between his construction and some of the already well-known groups. Thus, he showed how to construct permutation groups and transformation groups by carefully choosing particular relations among the generators. He wrote further that "the following investigations aim to continue the study of the properties of a group in its abstract formulation. In particular, this will pose the question of the extent to which these properties have an invariant character present in all the different realizations of the group, and the question of what leads to the exact determination of their essential group-theoretic content."[96] In other words, Dyck understood that while his concept of a group was fully abstract, that did not mean totally avoiding the specific nature of a given group. For any given algebraic, geometric, or number-theoretic problem dealing with groups, it was always necessary to discuss "the extent to which they are based on purely group-theoretic as against other properties of the posed problem."

Dyck's definition began with "operations" probably because most of the examples he knew were of this sort. His countryman Heinrich Weber (1842–1913), on the other hand, gave a complete axiomatic description of a finite group—without any reference to the nature of the elements composing it—in an 1882 paper on quadratic forms. He wrote,[97]

---

[95] Dyck, 1882, p. 12, as quoted in Wussing, 1984, p. 291 (note 226).

[96] Dyck, 1883, p. 70, as quoted in Wussing, 1984, p. 242. The next quotation is from the same source.

[97] Weber, 1882, p. 302 (our translation).

A system G of h elements of any sort $\theta_1, \theta_2, \ldots, \theta_h$ is called a group of order h if it satisfies the following conditions:

    I. Through some rule, which is called composition or multiplication, one derives from any two elements of the system a new element of the same system. In signs, $\theta_r \theta_s = \theta_t$.

    II. Always $(\theta_r \theta_s)\theta_t = \theta_r(\theta_s \theta_t) = \theta_r \theta_s \theta_t$.

    III. From $\theta\theta_r = \theta\theta_s$ and from $\theta_r\theta = \theta_s\theta$, there follows $\theta_r = \theta_s$.

Weber then derived the existence and uniqueness of units and inverses from the given axioms and from the finiteness of the group. Furthermore, he explicitly defined a group to be *Abelian* if the multiplication is commutative and proved the fundamental theorem of Abelian groups by essentially the same method Kronecker had used.

With several abstract definitions of a finite group in the literature by the 1880s, mathematicians began to develop abstract versions of results that had been previously proved just for permutation groups. For example, Georg Frobenius (1849–1917)—officially a student of Weierstrass but a product of the entire Berlin school of mathematics as animated by Kummer, Kronecker, and Weierstrass—re-proved abstractly the results that Sylow had demonstrated in 1872. In this new guise, they read as follows: if p^n is the largest power of a prime p that divides the order of a finite group, then the group has at least one subgroup of order p^n, and, in fact, the number of such subgroups divides the order of the group and is congruent to 1 modulo p. He admitted, moreover, that since every finite group can be considered as a group of permutations, Sylow's results must hold for abstract groups. "However," he wrote, "in view of the nature of the theorem to be proved, we shall not use this interpretation in what follows."[98] Frobenius thus began his proof by restating the definition of a finite group, essentially repeating Weber's definition.

Similarly, Otto Hölder (1859–1937), in a paper published in 1889, carried Jordan's ideas further by defining the notion of a *quotient group* (or *factor group*) and showing how these are Galois groups of the auxiliary equations that may come up in the process of solving a particular solvable equation. He also showed that in the resulting composition series, the actual quotient groups—and not just their orders—are unique, a result

---

[98] Frobenius, 1887, p. 179, as quoted in Wussing, 1984, p. 244.

now called the Jordan–Hölder theorem.[99] Four years later, Hölder re-proved abstractly Netto's results on the structure of groups of orders p^2 and pq, where p and q are primes and then generalized the results to determine the possible groups of order p^3, p^4, pq^2, and pqr, where r is an additional prime.[100] For example, by using the abstract Sylow theorem, he determined that in a group of order pq with $p > q$, there is one subgroup of order p and, if q does not divide $p - 1$, one subgroup of order q. In that case, the group is cyclic. On the other hand, if q does divide $p - 1$, then there is a second group that is generated by two elements S and T, with $S^q = 1$, $T^p = 1$, and $S^{-1}TS = T^r$, for $r \not\equiv 1 \pmod{p}$ and $r^q \equiv 1 \pmod{p}$.

Although all of the published definitions of a group through the 1880s required the group to be finite, mathematicians soon began to realize that the basic notions also applied to infinite sets. Thus, in 1893, Weber published a definition that included infinite groups. He repeated the three conditions he had given in 1882 and noted that if the group is finite, these suffice to ensure that if any two of the three group elements A, B, C are known, there is a unique solution of the equation $AB = C$.[101] He also pointed out that this conclusion is no longer valid for infinite groups. In that case, one must assume the existence of unique solutions of $AB = C$ as a fourth axiom. This fourth axiom, even without finiteness, implies a unique identity and unique inverses for every element of the group.

After defining the modern notion of isomorphism of groups, Weber illuminated the basis for his abstract approach:

> We can...combine all isomorphic groups into a single class of groups that is itself a group, whose elements are the generic characters obtained by making one general concept out of the corresponding elements of the individual isomorphic groups. The individual isomorphic groups are then to be regarded as different representatives of the generic concept, and it makes no difference which representative is used to study the properties of the group.[102]

[99] See Hölder, 1889 and Wussing, 1984, pp. 244–245. For more on Hölder's work on quotient groups, see Nicholson, 1993.
[100] Hölder, 1893.
[101] Weber, 1893, p. 524.
[102] Weber, 1893, p. 524, as quoted in Wussing, 1984, p. 248.

He proceeded to present many examples of groups, including the additive group of vectors in the plane, the group of permutations of a finite set, the additive group of residue classes modulo m, the multiplicative group of residue classes modulo m relatively prime to m, and the group of classes of binary quadratic forms of a given discriminant under Gauss's law of composition.

Interestingly, although Weber incorporated his abstract definition of groups in his two-volume algebra text, *Lehrbuch der Algebra* (or *Textbook on Algebra*), he did not present the notion of a group as a central concept of algebra. Because, for Weber, the central concept of the subject was still the solution of polynomial equations, he built his exposition of algebra from what, in his view, was the ground up. He carefully developed the properties of the integers, discussed the concept of a set and its various ramifications, and introduced the rational, real, and complex numbers.[103] The properties of these number systems fundamentally underlay all of the algebraic theory that he then went on to develop; groups were but a tool for effecting that process. As he wrote,

> One of the oldest questions, which the new algebra has chiefly addressed, is that of the so-called algebraic solution of equations, under which one seeks a representation of the roots of an equation through a series of radicals, or their calculation through a finite chain of root extractions. On this question, group theory sheds the brightest light.[104]

Thus, he discussed permutation groups in the first volume of the text in connection with his presentation of Galois theory, and it was only in the second volume that he presented the general definition of a group.

Although Weber was, of course, well aware of many instances of groups, he was evidently not entirely sure of the purpose of his general definition. He only knew that it would stimulate further research:

> The general definition of group leaves much in darkness regarding the nature of the concept... The definition of group contains more than appears at first sight, and the number of possible groups that can be defined given the number of their elements is quite limited.

---

[103] Weber, 1895, pp. 1–20.
[104] Weber, 1895, p. 644, as quoted in Corry, 1996/2004, p. 42/40.

The general laws concerning this question are barely known, and thus every new special group, in particular of a reduced number of elements, offers much interest and invites detailed research.[105]

Here, Weber would appear to be on the cusp of what was to become the structural revolution in algebra (see chapter 14). He had published a fully abstract definition of a group, the first major mathematical structure to be so defined, and he certainly understood a central consequence of the structural approach, namely, that two isomorphic groups were in essence the same mathematical object. Still, in putting the ideas of algebra into a form for students, he continued to contextualize them in terms of solving equations. As Leo Corry so astutely argued, "here one confronts one of the central images which dominated both nineteenth-century algebraic research, and Weber's own conception of algebra, in spite of the technical ability to formulate mathematical concepts in abstract terms, *algebraic knowledge in its totality derives from the fundamental properties of the basic systems of numbers.*"[106]

Subsequent nineteenth- and early twentieth-century textbook authors basically shared Weber's "image of algebra," although, as researchers, they were fully cognizant of the value of an abstract approach. For example, William Burnside (1852–1927) published the *Theory of Groups of Finite Order* in 1897, the next major textbook published after Weber's.[107] Even though Burnside referred in the preface to Weber's 1895 abstract definition of a group, he still began his own definition with, "Let A, B, C, ... represent a set of operations, which can be performed on the same object or set of objects."[108] Since group elements were always "operations," it was natural for Burnside to spend most of his book dealing with permutation groups. In France, however, Jean-Armand de Séguier (1862–1935) published his *Éléments de la théorie des groupes abstraits* (or *Elements of the Theory of Abstract Groups*)[109] in 1904 and attempted to base his discussion on a purely abstract definition of a group. That definition was, in some sense, too abstract; it hinged on notions from set

---

[105] Weber, 1896, p. 121, as quoted in Corry, 1996/2004, p. 42/41.

[106] Corry, 1996/2004, p. 39/38 (his emphasis).

[107] Four years later in 1901, the American mathematician, Leonard Eugene Dickson (1874–1954), published another major group-theoretic text, *Linear Groups: With an Exposition of the Galois Field Theory*. See Dickson, 1901.

[108] Burnside, 1897, p. 11.

[109] Séguier, 1904.

theory and on the idea of direct products of sets to produce composition of elements. As a result, Séguier's work had little influence. Nor, for other reasons, did *Abstract Group Theory*, published in 1916 in Russian by the Russian mathematician, Otto Schmidt (1891–1956). As we shall see below, a very different "image" of algebra would develop over the course of the first two decades of the twentieth century (see chapter 14).

****

In 1800, algebra still meant solving equations. Yet, although certain equations of degree higher than four were, in fact, solved *by radicals* during the nineteenth century, it soon became quite clear that there was no possibility of solving all equations in that way. The struggles of mathematicians to prove this result and then to determine what equations could, in fact, be solved by radicals eventually led, as we have seen, to the study of groups. By the end of the century, the very notion of "algebra" had begun to change, and algebra was moving into what we have called the abstract stage. Research mathematicians had begun to generalize the term "algebra" far beyond "the solution of equations." They had begun to apply it to the study of abstract groups, an example of what became known as a mathematical structure, that is, a set of elements with well-defined operations satisfying certain specified axioms. Other emergent structures were also under study during that time period, as we shall see in the next two chapters. What had become, by the nineteenth century, the mathematical subdiscipline of "algebra" was shifting.

12

Understanding Polynomial Equations in

n Unknowns

A s mathematicians in the sixteenth, seventeenth, eighteenth, and nineteenth centuries worked to understand the solvability of polynomial equations in one unknown of degree three and higher, they also naturally encountered systems of equations for which they sought simultaneous solutions. This should come as no surprise, since, as we have seen, such systems had arisen in mathematical problem solving since ancient times. Mesopotamian mathematicians simultaneously solved two equations in two unknowns using their method of false position. Diophantus developed a whole series of techniques for tackling more complicated systems involving squares and higher powers and treated examples that were both determinate and indeterminate, that is, examples with unique and with infinitely many solutions, respectively. Chinese, medieval Indian and Islamic, and Renaissance European mathematicians also all encountered and provided solutions to particular systems.

By the seventeenth century, however, and owing at least in part to the notational advances that followed as a result of the adoption of the Cartesian conventions introduced in *La Géométrie*, it became increasingly possible to see patterns and to search for more general techniques for solving systems of equations simultaneously. Over the course especially of the eighteenth and nineteenth centuries, this search, like that for the solution of polynomial equations in a single unknown, led to the creation of whole new algebraic constructs, in this case determinants, matrices, and invariants, among others. These, like groups, came to define new algebraic areas of inquiry in and of themselves.

Figure 12.1. Gottfried Wilhelm Leibniz (1646–1716).

SOLVING SYSTEMS OF LINEAR EQUATIONS IN n UNKNOWNS

By the seventeenth century, the simplest case of a system of simultaneous linear equations—two equations in two unknowns, say, x and y—was easy to handle. Solving one equation for y in terms of x and substituting that value for y into the other equation yielded a value for x that could then be substituted back into the other equation to get a value for y (provided the two original lines were neither identical nor parallel). But, what if more than two equations were involved? Was there some general way of determining the solution of such a system? This was precisely the question that Gottfried Leibniz addressed in the course of his correspondence with the Parisian mathematician, the Marquis de l'Hôpital (1661–1704), in 1693.

Writing from Hannover where he served as privy councillor to the Hannoverian Elector, Leibniz, who would soon become embroiled in an international controversy with Isaac Newton over priority for the discovery of the calculus, rued his isolation at the same time that he brought some of his mathematical thoughts to the attention of his French contemporary. In particular, he alluded to a method that he had developed in which he "used numbers instead of letters... treating these numbers as if they were just letters."[1] When he provided no details, l'Hôpital's curiosity was piqued. "I find it difficult to believe," he wrote in reply to Leibniz, "that [your method] is general enough and suitable enough to make use of numbers rather than letters in ordinary analysis." When Leibniz responded on 28 April 1693, he provided an example in which he demonstrated his new notation as well as its efficacy in dealing "especially with long and difficult calculations in which it is easy to make a mistake." That method and that example boiled down to calculating what we would call a 3×3 determinant.

Leibniz considered three equations, which he wrote in this way:

$$10 + 11x + 12y = 0,$$
$$20 + 21x + 22y = 0,$$
$$30 + 31x + 32y = 0.$$

As he explained, "the number supposes being of two characters, the first indicates for me which equation it is, the second indicates for me to which letter it belongs."[2] In other words, Leibniz used his numerical notation to stand in for the specific coefficients of the equations, which, depending on their values, could have resulted in "long and complicated calculations in which it [might have been] easy to make a mistake." Today, we might write a_{ij} for the coefficient in the jth place in the ith equation that Leibniz would have represented by two numbers corresponding to ij.

Given this notation, Leibniz wanted a general method by which to eliminate the two unknowns. As he explained to L'Hôpital, the procedure involved first eliminating y from the first and second equations to get a fourth equation in x alone and then doing the same thing relative to the first and third equations to get a fifth equation also in x alone.[3] Keeping Leibniz's numerical notation but modifying his presentation slightly, the

---

[1] Leibniz, 1849, p. 229 (our translation). For the next two quotes, see pp. 234 and 239, respectively.
[2] See Leibniz, 1849, p. 239 or Muir, 1906–1923, 1:7 (our translation). This system can be thought about this way (respecting Leibniz's notation): as three equations $a_{10}z + a_{11}x + a_{12}y = 0$, $a_{20}z + a_{21}x + a_{22}y = 0$, $a_{30}z + a_{31}x + a_{32}y = 0$ in three unknowns z, x, y, where $z = 1$.
[3] See Leibniz, 1849, pp. 239–240 or Muir, 1906–1923, 1:7–8 for the discussion and for the quotations below (our translation).

fourth and fifth equations are

$$(10.22 - 12.20) + (11.22 - 12.21)x = 0,$$

$$(10.32 - 12.30) + (11.32 - 12.31)x = 0,$$

respectively. Now, eliminating x from these two equations (and canceling out the common factor of 12—tacitly assumed to be nonzero—that appears in each of the resulting summands) yields

$$10.21.32 + 11.22.30 + 12.20.31 - 10.22.31 - 11.20.32 - 12.21.30 = 0,$$

that is, the 3×3 determinant, which Leibniz expressed as

$$
\begin{array}{ccc}
10.21.32 & & 10.22.31 \\
11.22.30 & = & 11.20.32 \\
12.20.31 & & 12.21.30,
\end{array}
$$

with the positive summands on the left-hand side of the equation and the negative summands on the right. This determinant, which would later be called the *resultant* of the system, vanishes precisely when there is an x and a y that satisfy all three equations simultaneously.[4] In Leibniz's view, this highlighted the many "harmonies" that "one would have had difficulty discovering by employing letters... especially when the number of letters and of equations is large." "One part of the secret of analysis," he continued, "consists in the characteristic, that is, in the art of employing well the marks that one uses, and you see, Sir, by this little example that Viète and Descartes did not yet know all of the mysteries." He then proceeded to give a general (if cumbersome and less than clear) rule for forming an $n \times n$ determinant.

Although Leibniz was proud of this technique and used it on other occasions in his correspondence, his statement of it remained unknown to the mathematical community until well after the publication of his collected works in the nineteenth century. It did not take long, however,

---

[4] In other words, when there are z, x, y (where Leibniz has taken $z = 1$) that satisfy the three equations simultaneously.

for others independently to discover it in the course of their algebraic explorations. As we saw in chapter 10, the Scottish mathematician Colin Maclaurin may have been the first to publish general results on simultaneously solving systems of linear equations in multiple unknowns in his 1748 book, *Treatise of Algebra*, but it was his Swiss contemporary, Gabriel Cramer, who independently introduced such techniques into the mathematical mainstream. In his *Introduction à l'analyse des lignes courbes algébriques* (or *Introduction to the Analysis of Algebraic Curved Lines*) of 1750, Cramer not only gave a remarkably readable textbook account of the mid-eighteenth-century state of the art of analyzing algebraic plane curves but also presented what has since come to be called Cramer's rule.

Cramer opened his text with an explicit statement of the importance to science that a well-articulated theory of algebraic curves had come to have by the mid-eighteenth century. "The theory of curved lines comprises a considerable part of mathematics," he stated. "It begins where the *Elements* end. . . .One would not know how to do without it in the sciences, the perfection of which depends on geometry, such as mechanics, astronomy, physics."[5] For Cramer, the key to unlocking this theory, and thus to unlocking the secrets of these natural sciences, was algebra. He deemed it "an ingenious means for reducing calculational problems to the simplest and easiest level to which the question posed may admit. This universal key of mathematics has opened the door to several minds [*Esprits*], to whom it was forever closed without this aid. One can say that this discovery has produced a veritable revolution in the sciences that depend on calculation." Cramer laid out one aspect of that "revolution" in the third chapter of his book. There, in the context of a system of five linear equations in five unknowns,[6] he illustrated the method for solving n linear equations in n unknowns that he treated in full generality in his book's first appendix.

For Cramer, the problem had arisen in the context of the number of points required to determine a plane algebraic curve of a given degree. As he explained, if v is the order (Cramer's term for what we call the *degree*) of a given plane algebraic curve, then the equation of the curve will, in

---

[5] Cramer, 1750, p. v. For the quote that follows, see p. vii (our translations).
[6] Cramer, 1750, p. 59.

general, have $\frac{1}{2}v^2 + \frac{3}{2}v$ independent coefficients, since "the number of coefficients a, b, c, d, e, &c. of each general equation is the same as the number of its terms. But this number of coefficients can be decreased by one, because the second side [*membre*] of these equations being zero, one can divide all of the first side by any one of the coefficients, such as the highest power of y or x, which, after this division, has one as its coefficient." "It follows from this," he noted, "that one can make a [curved] line of order v pass through the number $\frac{1}{2}v^2 + \frac{3}{2}v$ of given points, or that a [curved] line of order v is determined and its equation given when one has fixed $\frac{1}{2}v^2 + \frac{3}{2}v$ points through which it must pass." "Thus," he continued, "a [curved] line of the first order is determined by two given points; one of the second by 5; one of the third by 9; one of the fourth by 14; one of the fifth by 20, etc."[7] Taking the degree-two case, then, he had to solve a system of five linear equations in five unknowns in order to determine the curve.

In general, though, it was a matter of solving n equations in n unknowns, and to tackle this, he wrote the general system as

$$A^1 \;=\; Z^1z + Y^1y + X^1x + V^1v + \text{etc.},$$
$$A^2 \;=\; Z^2z + Y^2y + X^2x + V^2v + \text{etc.},$$
$$A^3 \;=\; Z^3z + Y^3y + X^3x + V^3v + \text{etc.},$$
$$A^4 \;=\; Z^4z + Y^4y + X^4x + V^4v + \text{etc.},$$

[etc.]

"where the letters A^1, A^2, A^3, A^4, etc. do not denote, as usual, the powers of A, but the first member, supposed known, of the first, second, third, fourth, etc. equation. Similarly, Z^1, Z^2, etc. are the coefficients of z; Y^1, Y^2, etc. those of y... in the first, second, etc. equation."[8]

He then proceeded to write out solutions explicitly for the unknowns in the cases of one equation in one unknown, two equations in two unknowns, and three equations in three unknowns. Using the notation

[7] Cramer, 1750, p. 58 (our translation).
[8] Cramer, 1750, p. 657 (our translation). The example that follows of the $n = 3$ case is also on this page.

established above, he expressed the solution of the $n = 3$ case this way:

$$Z = \frac{A^1Y^2X^3 - A^1Y^3X^2 - A^2Y^1X^3 + A^2Y^3X^1 + A^3Y^1X^2 - A^3Y^2X^1}{Z^1Y^2X^3 - Z^1Y^3X^2 - Z^2Y^1X^3 + Z^2Y^3X^1 + Z^3Y^1X^2 - Z^3Y^2X^1},$$

$$Y = \frac{Z^1A^2X^3 - Z^1A^3X^2 - Z^2A^1X^3 + Z^2A^3X^1 + Z^3A^1X^2 - Z^3A^2X^1}{Z^1Y^2X^3 - Z^1Y^3X^2 - Z^2Y^1X^3 + Z^2Y^3X^1 + Z^3Y^1X^2 - Z^3Y^2X^1},$$

$$X = \frac{Z^1Y^2A^3 - Z^1Y^3A^2 - Z^2Y^1A^3 + Z^2Y^3A^1 + Z^3Y^1A^2 - Z^3Y^2A^1}{Z^1Y^2X^3 - Z^1Y^3X^2 - Z^2Y^1X^3 + Z^2Y^3X^1 + Z^3Y^1X^2 - Z^3Y^2X^1},$$

that is, translating into our now-usual notational conventions, exactly as we would write out the solution today.

Cramer's discussion then concluded with a general, narrative statement of what we now call Cramer's rule:

> The number of equations and of unknowns being n, one finds the value of each unknown by forming the n fractions, the common denominator of which has as many terms as there are different arrangements of n different things. Each term is composed of the letters $ZYXV$, etc. always written in the same order, but to which one distributes as exponents, the first n numbers arranged in all possible ways....One assigns to these terms the signs $+$ or $-$, according to the following rule. When an exponent is followed in the same term...by an exponent less than it, I will call this a derangement [*dérangement*]. Count the number of derangements for each term: if it is even or zero, the term will have a $+$ sign; if it is odd, the term will have a $-$ sign....
>
> The common denominator so formed, one obtains the value of z by giving to the denominator the numerator formed by changing, in all of its terms, Z into A .... And similarly one finds the values of the other unknowns.[9]

Cramer even noted explicitly that this procedure can be carried out only when the common denominator is nonzero. To put into modern notation and terminology what Cramer said perfectly clearly in words: if D is a nonsingular $n \times n$ matrix (that is, if it has nonzero determinant) and if $\bar{A}$

---

[9] Cramer, 1750, p. 658 or Muir, 1906–1923, 1:12–13 (our translation).

is an $n \times 1$ column vector composed of the A^i's and X is an $n \times 1$ column vector of the unknowns relabeled as x_1, x_2, x_3, etc., then $x_i = \frac{\det \bar{A}_i}{\det D}$, where $\bar{A}_i$ is the $n \times n$ matrix obtained by replacing the ith column of D by $\bar{A}$. (On the concept of matrices, see below; on vectors, see the next chapter.)

While Cramer used what became known as Cramer's rule to excellent advantage in tackling the problem of determining curves of given degree in terms of $\frac{1}{2}v^2 + \frac{3}{2}v$ points, the problem itself raised what seemed to be a contradiction in mathematics. Writing to Euler as early as 30 September 1744, Cramer had noted that one theorem on curves stated that a curve of degree n can be determined by $\frac{1}{2}n^2 + \frac{3}{2}n$ points so that a curve of degree three can be determined by $\frac{1}{2} \cdot 3^2 + \frac{3}{2} \cdot 3 = 9$ points, yet another theorem on curves held that two curves of degrees m and n, say, have mn points in common so that two curves of degree three can intersect in $3 \cdot 3 = 9$ points. How can this be? Do nine points uniquely determine a curve of degree three, or not?

By 1750, Euler had had some key insights into this apparent conundrum—now called Cramer's paradox—that he published in a paper entitled "Sur une contradiction apparente dans la doctrine des lignes courbes" (or "On an Apparent Contradiction in the Doctrine of Curved Lines"). There, in the context of several specific examples, he showed that some systems of equations may have not one unique solution but rather infinitely many solutions. As he put it in considering the case of a plane algebraic curve of order four (like Cramer he used the term "order" instead of our term "degree"), "when two lines of the fourth order intersect in 16 points—since 14 points, *when they lead to equations all different from one another*, are sufficient to determine a line of this order— these 16 points will always be such that three or several of the equations that result from them, *are already comprised in the others*."[10] Although he was unable to express his insight more clearly or to prove it more generally, Euler had put his finger on a key issue—what boils down to what is now called the *rank* of a system of equations and its implications for the number of solutions of that system—that would only be fully understood with the development in the nineteenth and twentieth centuries of linear algebra (see the next chapter).

---

[10] Euler, 1750, p. 233 (our translation and our emphasis). For a nice discussion of Euler's paper, see Sandifer, 2007, chapter 7 and Dorier, 1995, pp. 228–230.

Cramer's work, published as it was in French, fairly quickly attracted the attention of mathematicians linked through the publications of the Academy of Sciences in Paris. Over the course of the last four decades of the eighteenth century, Étienne Bezout, Alexandre-Théophile Vandermonde, and Pierre Simon de Laplace (1749–1827), among others, began to engage in research in what they came to call the *theory* of elimination, that is, the theory concerned with determining when systems of equations are solvable and what their solutions are.[11] In the case particularly of Laplace, moreover, that research was done in the dynamic, applied context of celestial mechanics, once again showing the power that algebra came to have as a mathematical tool as it shifted into its dynamic stage into the eighteenth century. By 1812, another Parisian mathematician, the same Augustin-Louis Cauchy whom we encountered in the previous chapter in the context of the theory of permutations, had adapted (from Gauss's *Disquisitiones arithmeticae*) the terminology *determinant* for the construct det D that Cramer had articulated in words, had (like Vandermonde) considered it an entity worthy of study in and of itself, and had established some of its basic properties. In fact, this work was a natural outgrowth of his researches on permutations and was published in tandem with them in 1815.

As Cramer had laid it out in his statement of Cramer's rule, forming the $n \times n$ determinant involves taking all possible permutations of the integers $1, 2, \ldots, n$. Determinants, then, represented a natural application of the theory of permutations, but Cauchy saw even more. Whereas his work on permutation theory primarily focused on nonsymmetric rational functions in n variables and on the number of values they could assume when their variables were permuted in all possible ways, he had noted that symmetric functions have the property that they remain unchanged under any permutation of their variables (for example, $x^2 + y^2 + z^2 + xy + xz + yz$ is symmetric).[12] But what about functions that are, in some sense, almost symmetric? They have the property that they take on only

---

[11] For a sense of their work, see Muir, 1906–1923, 1:14–79. Recall, however, from chapter 5 that the late thirteenth-century Chinese mathematician, Zhu Shijie, had also displayed some techniques for elimination of systems of nonlinear equations. Isaac Newton, too, had discussed elimination theory. It was, in some sense, a natural topic.

[12] Albert Girard and, independently, Isaac Newton had pursued the connections between symmetric functions of roots of polynomial equations and the coefficients of those equations as early as the seventeenth century. See Funkhouser, 1930 for a brief overview of the early history of symmetric functions.

two values when their variables are permuted, and those two values are the same up to sign. Because such functions have the additional property that their terms alternate signs between plus and minus, Cauchy dubbed them "alternating symmetric functions" and sought to understand them more fully.

At the close of the first part of his paper, he noted that determinants were a special case of alternating symmetric functions and proceeded, in his paper's lengthy second part, to develop a general theory of them.[13] Among the many facts he established were (1) the now-standard realization of an $n \times n$ determinant in terms of $(n-1) \times (n-1)$ square arrays, namely, if D_{ij} denotes the square array of terms taken from the $n \times n$ square array D with entries a_{ij} by deleting the ith row and jth column, then $\det D = \sum_{i=1}^{n} (-1)^{i+j} a_{ij} \det D_{ij}$,[14] and (2) the determinant of a product (of two matrices) is the product of the determinants.[15]

With this work, Cauchy began the exploration of an actual *theory* of determinants. As Thomas Muir, the early twentieth-century chronicler of the history of that theory, put it, Cauchy "presented the entire subject from a perfectly new point of view, added many results previously unthought of, and opened up a whole avenue of fresh investigation."[16] Over the course of the nineteenth century, in fact, determinants came to define a major area of original, algebraic research pursued by such nineteenth-century mathematical luminaries as Carl Jacobi (1804–1851) and Otto Hesse (1811–1874) in Germany, James Joseph Sylvester (1814–1897) and Arthur Cayley in England, Augustin-Louis Cauchy and Charles Hermite in France, and Francesco Brioschi (1824–1897) and Francesco Fàa di Bruno (1825–1888) in Italy.[17]

---

[13] Cauchy, 1815b, p. 51 or Muir, 1906–1923, 1:98.

[14] Cauchy expressed this in terms of deleting the nth row and the nth column. See Cauchy, 1815b, p. 68 or Muir, 1906–1923, 1:104–105.

[15] In the absence of the terminology of matrices, Cauchy wrote, "When a system of quantities is determined symmetrically via two other systems, the determinant of the resulting system is always equal to the product of the determinants of the two composing systems." See Cauchy, 1815b, p. 82 or Muir, 1906–1923, 1:109 (our translation).

[16] Muir, 1906–1923, 1:131.

[17] The German mathematical abstracting journal, the *Jahrbuch über die Fortschritte der Mathematik* (or *Yearbook on the Development of Mathematics*), recognized "elimination and substitution, determinants, and symmetric functions" as defining one of the three principal areas of algebraic research in its first edition in 1870 as well as in its remaining nineteenth-century editions in 1880, 1890, and 1900. (The other two main areas were "equations" and the "theory of forms.")

LINEARLY TRANSFORMING HOMOGENEOUS POLYNOMIALS
IN n UNKNOWNS: THREE CONTEXTS

Solving systems of linear equations was just one of the analytical, that is, algebraic problems that occupied mathematicians in the eighteenth and into the nineteenth century. Another, that of transforming homogeneous polynomials in n unknowns, arose in such disparate contexts as mechanics both terrestrial and celestial, analytic geometry, and the exploration of the properties of numbers.[18] From there, it was explored on its own terms from a purely algebraic point of view. Cauchy was again one of the key figures in these developments.

In their belief that the natural world could be fully understood, eighteenth-century natural philosophers confronted thorny phenomena—such as the behavior of a swinging string hung with a finite number of masses, the rotational motion of a rigid body, and the long-term or secular perturbations of orbiting celestial objects—and sought rational, mathematical explanations for them. In this quest, Johann and Daniel (1700–1782) Bernoulli, Jean d'Alembert, Leonhard Euler, Joseph-Louis Lagrange, and Pierre Simon de Laplace all encountered particular systems both of first- and second-order differential equations in n unknowns for which they sought (real) solutions.[19] Although they made progress on this front, their arguments often came up short from a mathematical point of view, as they tended to invoke physical considerations at critical junctures rather than rely solely on the conclusions of algebraic analysis.

It was precisely that "impurity" of the analysis that Lagrange addressed in his *Méchanique analitique* of 1788. As he put it, his goal was to reduce mechanics "and the art of solving problems pertaining to it, to general formulas, the simple development of which gives all the equations necessary for the solution of each problem....No figures will be found in this work. The methods I expound in it require neither constructions nor geometrical nor mechanical reasoning, but only algebraic operations subjected to a systematic and uniform progression. Those who like

The work of the many nineteenth-century mathematicians who developed the new field is surveyed there as well as in the four volumes of Muir, 1906–1923.

[18] The transformation of homogeneous polynomials in n unknowns also arose in the theory of elliptic functions, but we will not consider that here.

[19] For more on this, see Hawkins, 1975a and compare Hawkins, 1975b.

analysis will be pleased to see mechanics become a branch of it and will be grateful to me for having thus extended its domain."[20]

One of the analyses that Lagrange reformulated in light of this philosophy was due to Euler and concerned the rotational motion of rigid bodies. Euler had interpreted rigid bodies in three-space in terms of what he called a set of "principal axes," that is, three, mutually perpendicular axes in terms of which the body's product moments of inertia vanish.[21] Euler started with the "right" axis system;[22] Lagrange wanted to demonstrate that such an axis system could always be obtained analytically, that is, that it is always possible to effect a suitable rotation of the axis system. In particular, Lagrange showed how to transform a quadratic form in three variables into a sum of squares.[23]

He considered the quadratic form

$$\Phi(p,q,r) = \frac{1}{2}(Ap^2 + Bq^2 + Cr^2) - Fqr - Gpr - Hpq \qquad (12.1)$$

and sought a linear transformation of its variables

$$\begin{aligned} p &= p'x + p''y + p'''z, \\ q &= q'x + q''y + q'''z, \\ r &= r'x + r''y + r'''z, \end{aligned} \qquad (12.2)$$

that transforms it into a *canonical* form

$$\Psi(x,y,z) = \frac{1}{2}(\alpha x^2 + \beta y^2 + \gamma z^2)$$

in a such a way that

$$p^2 + q^2 + r^2 = x^2 + y^2 + z^2$$

---

[20] Lagrange, 1788, p. i, as quoted in Hawkins, 1975a, pp. 15–16.

[21] Given a rigid body with mass m, density ρ, and volume V, the *product moment of inertia* relative to the x- and y-axes is $\int_V \rho xy \, dV$ (similarly, for the x- and z-axes and for the y- and z-axes).

[22] See Euler, 1765a and Euler, 1765b.

[23] In what follows, we adopt the notation used in Hawkins, 1975a, p. 17. Compare Lagrange, 1788, 2:Section 6, which is the same as 2:Section 9 in later editions, or see Lagrange, 1867–1892, 12:201ff.

(that is, he sought what would today be termed an *orthogonal* transformation). As Lagrange noted, the latter condition implies that the coefficients in (12.2) must also satisfy these six relations:

$$(p')^2 + (q')^2 + (r')^2 = 1, \qquad p'p'' + q'q'' + r'r'' = 0,$$

$$(p'')^2 + (q'')^2 + (r'')^2 = 1, \qquad p'p''' + q'q''' + r'r''' = 0, \qquad (12.3)$$

$$(p''')^2 + (q''')^2 + (r''')^2 = 1, \qquad p''p''' + q''q''' + r''r''' = 0.$$

He proceeded to determine the desired coefficients of (12.2) explicitly. In terminology that would only be developed much later (see below), his solution boiled down, first, to finding the three *eigenvalues* α, β, γ that are the (real) roots of the *characteristic equation* $\det(M - \lambda I) = 0$, where the matrix

$$M = \frac{1}{2} \begin{bmatrix} A & -H & -G \\ -H & B & -F \\ -G & -F & C \end{bmatrix}$$

and, then, to determining the corresponding *eigenvectors* (p', q', r'), (p'', q'', r''), (p''', q''', r'''), respectively, which form the coefficients of the desired linear transformation. Lagrange was able to prove that, in fact, the eigenvalues were all real by assuming that two of them were complex and deriving a contradiction. His argument—as well as the various symmetries reflected in the structure of the quadratic form and in the transformation—proved very suggestive to Cauchy.

Working within the analytic geometric tradition of the École Polytechnique, Cauchy well understood the connections between Lagrange's work on the rotational motion of rigid bodies in three-space and the more abstract study of quadric surfaces represented, for example, by the equation $\Phi(p, q, r) = k$ derived from (12.1).[24] In particular, he had studied quadric surfaces in his two-volume *Leçons sur les applications du calcul infinitésimal à la géométrie* (or *Lessons on the Applications of the Infinitesimal Calculus to Geometry*) of 1826–1828 and had considered there the same kind of eigenvalue problem that Lagrange had encountered in his work. Moreover, as a key figure in the creation of a *theory* of

---

[24] Hawkins, 1975a, pp. 18–24.

determinants, Cauchy was perfectly equipped to treat Lagrange's result in full generality and in terms of that key construct.

In a paper published in the *Exercices de mathématiques* (or *Mathematical Exercises*) in 1829, Cauchy considered a (real) quadratic form in n variables, which he wrote as

$$s = A_{xx}x^2 + A_{yy}y^2 + A_{zz}z^2 + \cdots + 2A_{xy}xy + 2A_{xz}xz + \cdots + 2A_{yz}yz + \cdots .^{25}$$

Assuming the variables satisfied the additional condition

$$x^2 + y^2 + z^2 + \cdots = 1, \tag{12.4}$$

he sought to solve the system

$$A_{xx}x + A_{xy}y + A_{xz}z + \cdots = sx,$$

$$A_{yx}x + A_{yy}y + A_{yz}z + \cdots = sy,$$

$$A_{zx}x + A_{zy}y + A_{zz}z + \cdots = sz,$$

$$\vdots$$

(where $A_{xy} = A_{yx}$, $A_{xz} = A_{zx}$, $A_{yz} = A_{zy}$, etc.) which he rewrote as

$$(A_{xx} - s)x + A_{xy}y + A_{xz}z + \cdots = 0,$$

$$A_{yx}x + (A_{yy} - s)y + A_{yz}z + \cdots = 0, \tag{12.5}$$

$$A_{zx}x + A_{zy}y + (A_{zz} - s)z + \cdots = 0,$$

$$\vdots$$

As he next explained, that solution is effected by forming the determinant—call it D—of the square array of coefficients in (12.5) (that is, the *characteristic polynomial* which is analogous to the construct $\det(M - \lambda I)$ that Lagrange had considered in the 3×3 case).[26]

---

[25] See Cauchy, 1829, pp. 140–146 for the discussion that follows.

[26] Note, however, that Cauchy did not call the construct the "determinant," rather he wrote in words how to form it.

To illustrate his point, Cauchy explicitly wrote out the determinant in the cases of two, three, and four variables. In the three-variable case, for example, he obtained the third-degree polynomial

$$(A_{xx} - s)(A_{yy} - s)(A_{zz} - s) \ - \ A_{yz}^2(A_{xx} - s) - A_{xz}^2(A_{yy} - s)$$
$$- \ A_{xy}^2(A_{zz} - s) + 2A_{xy}A_{xz}A_{yz}$$

in s.[27] In the general case, however, D is an nth-degree polynomial in s, and the problem was to solve the equation $D = 0$ for the eigenvalues s of the characteristic equation. By the fundamental theorem of algebra (recall chapter 10), $D = 0$ has n real and/or complex roots $s_1, s_2, \ldots, s_n$, each of which corresponds to an eigenvector $(x_i, y_i, z_i, \ldots, u_i)$, for $1 \leq i \leq n$. Without too much algebraic manipulation, Cauchy was then able to show that the n-dimensional analogues of the equations in (12.3) must obtain. Finally, employing a purely determinant-theoretic argument, he surpassed Lagrange by giving a perfectly general proof that the n roots $s_1, s_2, \ldots, s_n$ (assumed to be distinct) of (12.5) must actually all be *real*, and hence that it is always possible, in modern terminology, to diagonalize a quadratic form using an orthogonal transformation.

Cauchy's result sparked the interest of Jacobi and his student Carl Borchardt (1817–1880) in Germany as well as that of Sylvester in England. In particular, Sylvester demonstrated what he dubbed "the law of inertia for quadratic forms," namely, given a quadratic form s in n variables as above and a linear transformation T of the variables such that "only positive and negative squares of the new variables...appear" in $T(s)$, then "the number of such positive and negative squares respectively will be constant for a given function whatever of the linear transformations employed."[28]

While Cauchy's determinant-theoretic work on what would today be termed the spectral theory of linear transformations was also picked up on and developed further in Germany by Karl Weierstrass, Georg Frobenius, and others (see below), some researchers in the British Isles ultimately drew different inspiration from the problem of linearly

---

[27] Cauchy, 1829, p. 142.
[28] Sylvester, 1853, pp. 480–481. Sylvester had stated and named the theorem earlier in Sylvester, 1852a, p. 142.

Figure 12.2. George Boole (1815–1864).

transforming homogeneous polynomials in n unknowns.[29] One of them, the largely self-taught Englishman, George Boole (1815–1864), took as his point of departure the same transformation from Lagrange's *Méchanique analitique* that had attracted Cauchy's attention and about which Boole had read further in the work of Cauchy and Jacobi, among others. In a two-part paper published in 1841 and 1842 in the *Cambridge Mathematical Journal*, however, he sought not to treat just the linear transformation of a quadratic form in n variables, but rather, as his title attested, a "general theory of linear transformation."[30]

Boole opened with a statement (in terms almost like those above) of the general result for quadratic forms and noted that "the method commonly employed in this investigation has been, to substitute... a series of linear functions of the variables involved..., to equate coefficients, and to eliminate the unknown constants....It is in the effecting of this elimination," he continued, "that the principal difficulty of the problem consists; a difficulty... so great, that no one has yet shewn how it is to be overcome, when the degree of the function to be transformed rises above the second." This more general problem was what Boole proposed to tackle in his paper.

---

[29] Thomas Hawkins coined the phrase *spectral theory* to encompass, as he put it, "the concept of an eigenvalue, the classification of matrices into types (symmetric, orthogonal, Hermitian, unitary, etc.), the theorems on the nature of the eigenvalues of the various types and, above all, those on the canonical (or normal) forms for matrices." See Hawkins, 1975a, p. 1.

[30] Boole, 1841, p. 1. For the quotation that follows in the next paragraph, see p. 2. See also Wolfson, 2008 for a discussion of Boole's work in this vein.

Boole considered a form Q of degree n in the m unknowns $x_1, x_2, \ldots, x_m$ and asked,[31] when can Q be transformed into a form R of degree n in the m unknowns $x'_1, x'_2, \ldots, x'_m$, by the linear transformation T defined by $x_i \mapsto \sum_{j=1}^{m} t_{ij} x'_j$? If

$$Q(T(x_1, x_2, \ldots, x_m)) = R(x'_1, x'_2, \ldots, x'_m), \tag{12.6}$$

then, as Boole remarked, it must be the case that

$$\frac{\partial Q(T(x_1, x_2, \ldots, x_m))}{\partial x'_j} = \frac{\partial R(x'_1, x'_2, \ldots, x'_m)}{\partial x'_j},$$

or, expanding the left-hand side and expressing it in terms of the linear transformation T,

$$\sum_{i=1}^{m} \frac{\partial Q(x_1, x_2, \ldots, x_n)}{\partial x_i} \frac{\partial T(x_i)}{\partial x'_j} = \sum_{i=1}^{m} t_{ij} \frac{\partial Q(x_1, x_2, \ldots, x_n)}{\partial x_i}$$

$$= \frac{\partial R(x'_1, x'_2, \ldots, x'_m)}{\partial x'_j}. \tag{12.7}$$

Thus, provided the determinant of T is nonzero (a condition that Boole did not articulate explicitly but assumed tacitly), $\frac{\partial Q}{\partial x_i} = 0$ for all i, precisely when $\frac{\partial R}{\partial x'_j} = 0$ for all j, a necessary condition for (12.6) to hold. Moreover, Boole showed that taking the equations in (12.7) pairwise and successively eliminating the variables between them ultimately results in one homogeneous equation in one unknown. Assuming further that the variables x_i (respectively, x'_i) do not all vanish simultaneously, this elimination shows that the partial derivatives $\frac{\partial Q}{\partial x_i}$ (respectively, $\frac{\partial R}{\partial x'_i}$) are dependent if and only if a certain combination $\theta(Q)$ of the coefficients of Q (respectively, $\theta(R)$ in the coefficients of R) vanishes. Boole's fellow countryman, Sylvester, would somewhat later term this construct θ the *discriminant* of the form.[32]

Boole then proceeded to illustrate this in the cases of quadratic forms in two and three variables. Consider the simplest case of

---

[31] The discussion here follows Wolfson, 2008, pp. 39–40 (with some necessary changes being made) and thus somewhat modernizes and standardizes the notation in Boole, 1841, pp. 2–6.

[32] See Sylvester, 1852b, p. 52.

$Q = Ax_1^2 + 2Bx_1x_2 + Cx_2^2$ and calculate $\frac{\partial Q}{\partial x_1}$ and $\frac{\partial Q}{\partial x_2}$. Setting each equal to zero yields the two equations

$$\frac{\partial Q}{\partial x_1} = 2(Ax_1 + Bx_2) = 0,$$

$$\frac{\partial Q}{\partial x_2} = 2(Bx_1 + Cx_2) = 0.$$

Eliminating the variables between these two equations pairwise and assuming that x_1 and x_2 do not vanish simultaneously then gives

$$\theta(Q) = B^2 - AC = 0,$$

the discriminant of the quadratic binary form. Similarly, Boole noted, the same process could be applied to $T(Q(x_1, x_2)) = R(x_1', x_2') := A'x_1'^2 + 2B'x_1'x_2' + C'x_2'^2$, where the linear transformation T is defined as

$$x_1 \mapsto t_{11}x_1' + t_{12}x_2',$$

$$x_2 \mapsto t_{21}x_1' + t_{22}x_2'.$$

It is easy to see, then, that the analogous elimination yields

$$\theta(R) = B'^2 - A'C'$$

and, moreover, that

$$\theta(Q) = \frac{\theta(R)}{(t_{11}t_{22} - t_{12}t_{21})^2},$$

where $t_{11}t_{22} - t_{12}t_{21}$ is the (nonzero) determinant of T. In other words, $\theta(Q)$ and $\theta(R)$ are equal up to a power of the determinant of T. As Boole stated in full generality—but on the basis of his inductive analysis of a number of examples—"if Q_n be a homogeneous function of the nth degree, with m variables, which by the linear [transformation] is transformed into R_n, a similar homogeneous function; and if γ represent the degree of $\theta(Q_n)$ and $\theta(R_n)$, then

$$\theta(Q_n) = \frac{\theta(R_n)}{E^{\frac{m}{m}}},$$
(12.8)

E being the result obtained by the elimination of the variables from the second members of the linear [transformation], equated to 0."[33]

Boole also showed how his methods provided a less calculationally involved path to the result that Lagrange had obtained in considering the orthogonal transformation of the ternary quadratic form (12.1) into the sum of squares.[34] That had been Boole's main motivation all along, namely, devising a more general and less calculationally involved method for effecting the linear transformation of homogeneous polynomials in n unknowns. Others, like Sylvester and Cayley, would find something else in Boole's work, namely, especially in (12.8), the germ of a whole new theory that they would call *invariant theory* (see below).

The linear transformation of homogeneous polynomials arose in at least one other key and influential context in early nineteenth-century mathematics. As we saw in chapter 11, Gauss, in his systematic study of quadratic forms in the *Disquisitiones arithmeticae* of 1801, had developed the notion of the *equivalence* of two quadratic forms from that of a linear transformation. In so doing, he had discovered that, given a binary quadratic form F and a linear transformation of its variables that takes F to F', the discriminants of F and F' (Gauss called them "determinants," but, as we saw, Cauchy appropriated that term for a more general construct) are equal up to a power of the determinant (in Cauchy's new sense of the word) of the linear substitution (recall equation (11.6)). This is the same result that Boole independently framed, but in the context of his general study of linear transformations, forty years later in his paper in the *Cambridge Mathematical Journal*.

After developing an extensive theory of quadratic forms in two variables, Gauss next made a "brief digression" to extend some of those results to ternary quadratic forms, that is, to quadratic forms in

[33] Boole, 1841, p. 19.
[34] Boole, 1841, p. 12.

three variables.[35] Again, he considered a linear transformation, but this time one of the form

$$x = \alpha y + \beta y' + \gamma y'',$$
$$x' = \alpha' y + \beta' y' + \gamma' y'', \qquad (12.9)$$
$$x'' = \alpha'' y + \beta'' y' + \gamma'' y''$$

(where the various α's, β's, and γ's are all integers), that takes the ternary quadratic form $f = ax^2 + a'x'^2 + a''x''^2 + 2bxx' + 2b'xx'' + 2b''x'x''$ to $g = Ay^2 + A'y'^2 + A''y''^2 + 2Byy' + 2B'yy'' + 2B''y'y''$. As Gauss noted, "for brevity [in (12.9)] we will neglect the unknowns and say simply that f is transformed into g by the substitution (S)

$$\begin{matrix} \alpha, & \beta, & \gamma \\ \alpha', & \beta', & \gamma' \\ \alpha'', & \beta'', & \gamma'' \end{matrix}$$

... From this supposition will follow six equations for the six coefficients in g, but it is unnecessary to transcribe them here."

Instead, Gauss laid out a number of consequences that follow from this setup.[36] First, he noted that, as in the case of the binary quadratic form, if k denotes the determinant of the linear substitution, that is, if $k = \alpha\beta'\gamma'' + \beta\gamma'\alpha'' + \gamma\alpha'\beta'' - \gamma\beta'\alpha'' - \alpha\gamma'\beta'' - \beta\alpha'\gamma''$ and if D and E denote the discriminants of f and g, respectively, then $E = k^2 D$. As Gauss explicitly remarked, "it is clear that with regard to transformations of ternary forms the number k is similar to the number $\alpha\delta - \beta\gamma$ with respect to transformations of binary forms." Moreover, he noted that if S transforms f into g and the substitution

$$\begin{matrix} \delta, & \epsilon, & \zeta \\ \delta', & \epsilon', & \zeta' \\ \delta'', & \epsilon'', & \zeta'' \end{matrix}$$

(call it T) transforms g into h, then f will be transformed into h by

$$\alpha\delta + \beta\delta' + \gamma\delta'', \quad \alpha\epsilon + \beta\epsilon' + \gamma\epsilon'', \quad \alpha\zeta + \beta\zeta' + \gamma\zeta'',$$
$$\alpha'\delta + \beta'\delta' + \gamma'\delta'', \quad \alpha'\epsilon + \beta'\epsilon' + \gamma'\epsilon'', \quad \alpha'\zeta + \beta'\zeta' + \gamma'\zeta'',$$
$$\alpha''\delta + \beta''\delta' + \gamma''\delta'', \quad \alpha''\epsilon + \beta''\epsilon' + \gamma''\epsilon'', \quad \alpha''\zeta + \beta''\zeta' + \gamma''\zeta''.$$

This notation would soon prove quite suggestive. The composition of the linear transformations S and T (or $S \circ T$ as we would now write it) corresponded precisely to what would be termed the multiplication of the 3×3 matrices of S and T in that order (see below). In the course of his discussion of ternary forms, Gauss also defined and used other concepts—such as the symmetric matrix, the adjoint, and the transpose—that would become key parts of a theory of matrices.

Gauss's work on ternary quadratic forms—and the use he made in it of the notion of linearly transforming the three unknowns—piqued the interest of a young Gotthold Eisenstein (1823–1852) in Germany. In exploring the properties of ternary quadratic forms in 1844 while still a student at the University of Berlin, Eisenstein developed further the notation that Gauss had employed, denoting the composition of two linear transformations S and T as $S \times T$ and introducing the notation $\frac{1}{S}$ for the inverse of S. At the same time, he recognized that nothing was sacred about three variables in all of this as long as a modicum of care was taken. As he put it, "an algorithm for calculation can be based on this; it consists in applying the usual rules for the operations of multiplication, division and exponentiation to symbolical equations between linear systems; correct symbolical equations are always obtained, the sole consideration being that the order of the factors, i.e., the order of the composing systems, may not be altered."[37] Although he never really pursued this line of thought during a life cut short by tuberculosis, Eisenstein had recognized that linear transformations could be manipulated much like ordinary numbers, except that their "multiplication," that is, their composition, was not commutative.

---

[37] For the notation, see Eisenstein, 1844e, pp. 324 and 328; for the quote, see p. 354 (or Hawkins, 1977a, p. 86 (his translation)). For more on Eisenstein's work in this regard, see Hawkins, 1977a, pp. 85–86.

THE EVOLUTION OF A THEORY OF MATRICES AND
LINEAR TRANSFORMATIONS

By 1855, Arthur Cayley had offered a formalization of at least one aspect of the notational conventions that had been emerging in the study of linear transformation, namely, to use a term coined by his friend and mathematical colleague, Sylvester, that of a *matrix*.[38] In a short paper published in the *Journal für die reine und angewandte Mathematik* (also known as Crelle's *Journal*), he stated that he used "the notation

$$\begin{vmatrix} \alpha, & \beta, & \gamma, & \cdots \\ \alpha', & \beta', & \gamma', & \cdots \\ \alpha'', & \beta'', & \gamma'', & \cdots \\ \vdots & \vdots & \vdots & \end{vmatrix}$$

to denote what I call a *matrix*; that is, a *system* of quantities arranged in the form of a square but otherwise completely *independent*." "I will not speak here of *rectangular matrices*," he added, but he offered the opinion that "this notation seems very convenient for the theory of *linear* equations" and proceeded to use it to express the system of equations

$$\xi = \alpha x + \beta y + \gamma z + \cdots ,$$

$$\eta = \alpha' x + \beta' y + \gamma' z + \cdots ,$$

$$\zeta = \alpha'' x + \beta'' y + \gamma'' z + \cdots ,$$

$$\vdots$$

as

$$(\xi, \eta, \zeta, \ldots) = \left(\begin{vmatrix} \alpha, & \beta, & \gamma, & \cdots \\ \alpha', & \beta', & \gamma', & \cdots \\ \alpha'', & \beta'', & \gamma'', & \cdots \\ \vdots & \vdots & \vdots & \end{vmatrix} \right) (x, y, z, \ldots).^{39}$$

---

[38] See Sylvester, 1850, p. 369 and Sylvester, 1851a, p. 302.

[39] Cayley, 1855a, p. 282 (his emphases, our translation). For the quotes that follow, see pp. 283–284. Crilly, 2006 presents Cayley's life and work in its full Victorian context.

Alternately, he noted that to express "the system of equations which gives x, y, z, ... in terms of ξ, η, ζ, ...," he was "led to the notation

$$
\begin{vmatrix}
\alpha, & \beta, & \gamma, & \dots \\
\alpha', & \beta', & \gamma', & \dots \\
\alpha'', & \beta'', & \gamma'', & \dots \\
\vdots & \vdots & \vdots &
\end{vmatrix}^{-1}
$$

of the *inverse* matrix." "The terms of this matrix," he continued, "are fractions, having the *determinant* formed by the terms of the original matrix as a common *denominator*; the *numerators* are the minors formed by the terms of this same matrix by suppressing any one of the rows and any one of the columns." He also explicitly described the composition of two linear transformations in terms of what we would call the multiplication of their respective matrices and remarked that "there would be a lot of things to say about this theory of matrices, which, it seems to me, should precede the theory of *determinants*."

Although this paper served mainly to introduce Cayley's suggestion for a new notational convention, another that followed almost immediately on its heels in Crelle's *Journal* showed the efficacy of that new notation. In it, among other things, Cayley expressed mathematical results that he had considered as early as 1846 and that his friend and French contemporary, Charles Hermite, had extended in 1854.[40] The question at hand was similar to the "principal axis" problem considered by Euler and Lagrange and generalized in determinant-theoretic terms by Cauchy. In 1846, Cayley had analyzed a particular $n \times n$ determinant (we would call it *skew symmetric*) and had shown that, when $n = 3$, it corresponded to a linear transformation that effected a particular change of coordinates in three-space. In 1854, Hermite had come to the same problem—but in the context of number-theoretic work on ternary quadratic forms motivated by results of both Gauss and Eisenstein—and had formulated and treated a generalization of it (now called the Cayley–Hermite problem), namely, to find all linear transformations of the variables of a quadratic form in n unknowns of nonzero determinant that leaves the form invariant.

---

[40] See Cayley, 1846b and Hermite, 1854a. Hawkins, 1977a, pp. 87–91 puts this work in fuller historical context.

In 1855, Cayley showed how to translate Hermite's argument succinctly into his new matrix-theoretic notation and, specifically, in terms of matrix multiplication and inverse matrices.[41] By November of 1857, he had seen how to take matrices even further. In a jubilant letter to Sylvester, he laid out a *"very remarkable"* result, which he published in 1858 in the context of "A Memoir on the Theory of Matrices" in the *Philosophical Transactions of the Royal Society of London.*[42]

Cayley's paper represented a significant step in the evolution of algebra, for it showed how matrices could be understood as *constructs* subject to many of the usual operations of algebra. In other words, as Eisenstein had realized relative to linear transformations, matrices, too, behaved in a certain sense like rational numbers. In particular, it made sense to talk about additive and multiplicative identities for matrices, just like it made sense to consider 0 and 1 the additive and multiplicative identities, respectively, for the rational numbers, and Cayley wrote down those matrix identities. He also gave the rules for matrix addition, scalar multiplication, and matrix multiplication, and he proved that the latter satisfies the associative but not the commutative law. In language that would be developed only later, Cayley had shown that the collection of $n \times n$ matrices forms an *algebra* (see the next chapter). Moreover, he demonstrated that, unlike the rational numbers, the product of two nonzero $n \times n$ matrices could be the zero matrix (that is, the collection of matrices contains *zero divisors*); he considered the conditions under which a matrix has a multiplicative inverse; he explored when two $n \times n$ matrices do commute; he studied the properties of the transpose of a matrix; and he noted that since matrices can be multiplied repeatedly by themselves, it made sense to consider raising them to powers and, therefore, to think about polynomial *functions* of *matrices*.[43] It was the latter realization that formed the context for the *"very remarkable"* result that Cayley claimed in general but cast in terms of 2×2 matrices, a result known today as the Hamilton–Cayley theorem.[44]

---

[41] Cayley, 1855b.

[42] For the quote, see Parshall, 1998, p. 95 (Cayley's emphasis).

[43] See Cayley, 1858a, and compare the discussion in Parshall, 1985, pp. 237–239. In a paper that followed his 1858a in the *Philosophical Transactions*, moreover, Cayley freely employed a single-letter notation to denote matrices and to carry out matrix calculations such as, for two square matrices Ω and Υ, $\Omega(\Omega - \Upsilon)^{-1}(\Omega + \Upsilon)\Omega^{-1}\Omega\Omega^{-1}(\Omega - \Upsilon)(\Omega + \Upsilon)^{-1}\Omega = \Omega$. See Cayley, 1858b, p. 43.

[44] On Sir William Rowan Hamilton (1805–1865), see the next chapter.

"Imagine," Cayley asked his readers, "a matrix

$$M = \begin{pmatrix} a, b \\ c, d \end{pmatrix}$$

and form the determinant

$$\begin{vmatrix} a - M, & b \\ & c, & d - M \end{vmatrix},$$

the developed expression of this determinant is

$$M^2 - (a+d)M^1 + (ad - bc)M^0;$$

the values of M^2, M^1, M^0 are

$$\begin{pmatrix} a^2 + bc, & b(a+d) \\ c(a+d), & d^2 + bc \end{pmatrix}, \quad \begin{pmatrix} a, b \\ c, d \end{pmatrix}, \quad \begin{pmatrix} 1, 0 \\ 0, 1 \end{pmatrix},$$

and substituting these values the determinant becomes equal to the matrix zero, viz., we have

$$\begin{vmatrix} a - M, & b \\ & c, & d - M \end{vmatrix} = \begin{pmatrix} a^2 + bc, & b(a+d) \\ c(a+d), & d^2 + bc \end{pmatrix} - (a+d)\begin{pmatrix} a, b \\ c, d \end{pmatrix} + (ad - bc)\begin{pmatrix} 1, 0 \\ 0, 1 \end{pmatrix}$$

$$= \cdots = \begin{pmatrix} 0, 0 \\ 0, 0 \end{pmatrix}."^{45}$$

In other words, Cayley had noted that the 2×2 matrix M satisfies the (characteristic) equation $\det(M - \lambda I) = 0$, where I and 0 represent the identity and zero matrices, respectively.[46] In a different context, this was precisely the equation (recall (12.5) above) that Cauchy had encountered in 1829 in his proof that a quadratic form is diagonalizable via an orthogonal transformation or, in other words, that the characteristic polynomial of a real symmetric matrix has n real roots. The emergent

[45] Cayley, 1858a, p. 23. For the quote in the next paragraph, see p. 24.
[46] Today, we generally define the characteristic equation as $\det(\lambda I - M) = 0$.

Figure 12.3. Karl Weierstrass (1815–1897).

spectral theory in the context of linear transformations and the emergent theory of matrix algebra were, in some sense, just two sides of the same coin, although Cayley did not realize that.

Published as it was in the *Philosophical Transactions*, a journal that covered the broadest sweep of (mostly) British scientific research, Cayley's work of 1858 remained largely unknown to his contemporaries, especially those in Germany, who were developing the spectral theory from the point of view of linear transformations. In fact, Cayley's approach was largely antithetical to theirs. He followed up his justification of the Hamilton–Cayley theorem in the 2×2 case with the statement that he had "verified the theorem, in the next simplest case of a matrix of the order 3...but I have not thought it necessary to undertake the labour of a formal proof of the theorem in the general case of a matrix of any degree." This so-called "generic reasoning"—that is, argumentation in terms of general symbols and without concern for any possible pitfalls that special cases might present—was typical of nineteenth-century British mathematics as well as of much mathematics being developed elsewhere in Europe. It was, however, a style of mathematical reasoning that Cauchy had resisted and that the Berlin school of mathematics, as animated particularly after mid-century by Karl Weierstrass and Leopold Kronecker (recall the previous chapter), had, in some sense, striven to supplant.[47] In fact, Weierstrass worked on the problem of linear transformation from a Berliner perspective in two key papers, one published in 1858 and the other in 1868, in which he explored, first, quadratic forms and, then, the more general situation of bilinear forms, that is, expressions $\sum_{i,j=1}^{n} a_{ij}x_iy_j$, where $a_{ij} \in \mathbb{R}$. His aim in these papers was to develop techniques that proved theorems in full generality, that is, in recognition of exceptional cases.

Weierstrass's entrée into this seems to have been Cauchy's 1829 determinant-theoretic paper as well as Jacobi's reworking of some of Cauchy's results.[48] Weierstrass cast things a bit more generally in his 1858 paper (since he wanted ultimately to apply his results in the setting of a system of linear differential equations). He considered two quadratic forms $\phi = \sum_{i,j=1}^{n} a_{ij}x_ix_j$ (where $a_{ij} = a_{ji}$ and where ϕ is assumed to be positive definite, that is, $\phi(x_1, x_2, \ldots, x_n) > 0$ for all nonzero $(x_1, x_2, \ldots, x_n) \in \mathbb{R}^n$) and $\psi = \sum_{i,j=1}^{n} b_{ij}x_ix_j$ (where $b_{ij} = b_{ji}$), respectively, and he showed that there exists an orthogonal linear transformation of the variables x_i to x_i' that takes ϕ to $\sum_{i=1}^{n} (x_i')^2$ and ψ to $\sum_{i=1}^{n} s_i(x_i')^2$, where the s_i are the roots of $f(s) = \det(sA - B) = 0$ and A and B are the $n \times n$ matrices with coefficients a_{ij} and b_{ij}, respectively.[49] Taking $A = I$, that is, the $n \times n$ identity matrix, $\det(sA - B) = 0$ is the same equation that Cauchy had encountered in his 1829 paper and essentially the same one that had arisen in Cayley's 1858 paper. Weierstrass, however, considered a symmetric matrix A more arbitrary than the identity matrix and analyzed the equation $\det(sA - B) = 0$ per se. In particular, unlike Cauchy, he explicitly did *not* assume that its roots are all distinct. As he put it,

It does not appear that special attention has been given to peculiar circumstances that arise when the roots of the equation $f(s) = 0$ are not all different; and the difficulties which they present...were not properly cleared up. I also at first believed that this would not be possible without extensive discussions in view of the large number of different cases that can occur. I was all the more pleased to find

---

[47] On this notion of generic reasoning, see Hawkins, 1977a, pp. 94–95.

[48] Hawkins, 1977b deals in great detail with Weierstrass's work and its historical context. The discussion here follows his account.

[49] Weierstrass, 1858, p. 207. For the quote that follows, see p. 208, and Hawkins, 1977b, pp. 128–129 (his translation). Today, this result in various guises is called the principal axis theorem.

that the solution to the problem given by the above-mentioned mathematicians [Cauchy and Jacobi] could be modified in such a way that it does not at all matter whether some of the quantities $s_1, s_2, \ldots, s_n$ are equal.

What exactly did Weierstrass show?[50]

Consider $f_0(s) := f(s) = \det(s A - B)$, and let $f_1(s)$ denote the greatest common divisor of the $(n-1) \times (n-1)$ minors of $f(s)$ considered as polynomials in s. Now, define $f_i(s)$ to be the greatest common divisor of the $(n-i) \times (n-i)$ minors of $f(s)$. It is not hard to show that $f_i(s)$ divides $f_{i-1}(s)$. Weierstrass then considered the polynomials $e_i(s) := \frac{f_{i-1}(s)}{f_i(s)}$. The divisibility properties of the polynomials $f_i(s)$ imply that $e_i(s)$ divides $e_{i-1}(s)$ and, moreover, that $f(s)$ differs from the product of the $e_i(s)$'s by a nonzero constant. If, then, $s_1, s_2, \ldots, s_k$ are the distinct roots of $f(s)$, it is the case that

$$e_i(s) = \gamma_i \prod_{j=1}^{k} (s - s_j)^{m_{ij}},$$

where γ_i is a constant and $m_{ij} \in \mathbb{Z}^+$. Weierstrass would later call the factors $(s - s_j)^{m_{ij}}$ (for $m_{ij} > 0$) the elementary divisors of $f(s)$.[51]

Weierstrass showed[52] that if s_j is a root of multiplicity m of $f_0(s) := f(s) = \det(s A - B)$, then it must be a root of multiplicity greater than or equal to $m - 1$ of $f_1(s)$, and

$$e_1(s) = \frac{f(s)}{f_1(s)} = \gamma_1 \prod_{j=1}^{k} (s - s_j)^{m_{1j}},$$

where $m_{1j} = 1$. In other words, it does not matter whether the characteristic polynomial $f(s)$ has multiple roots. What is important is that the roots of $e_1(s)$ all be distinct. Put another way, ϕ and ψ defined as above can be transformed into $\sum_{i=1}^{n} (x_i')^2$ and $\sum_{i=1}^{n} s_i (x_i')^2$, respectively. Weierstrass had penetrated beneath the generic reasoning to uncover the essential structure governing the algebraic behavior.

---

[50] For the setup that follows, with slightly different notation, see Hawkins, 1977b, p. 120. Hawkins has somewhat modernized and streamlined Weierstrass's original notation.
[51] See Weierstrass, 1868.
[52] Weierstrass, 1858, p. 214 and Hawkins, 1977b, p. 132.

A decade later, he developed this into what he termed a theory of elementary divisors in the more general context of bilinear forms.[53] Now letting $\phi = \sum_{i,j=1}^{n} a_{ij}x_i y_j$ and $\psi = \sum_{i,j=1}^{n} b_{ij}x_i y_j$ be two *bilinear* forms (with no special conditions placed on the a_{ij}'s and the b_{ij}'s), he wanted to transform ϕ, ψ into ϕ', ψ', respectively, by means of nonsingular linear transformations T and W defined by $x_i = \sum_{i=1}^{n} t_{ij}x'_{ij}$ and $y_i = \sum_{i=1}^{n} w_{ij}y'_{ij}$, respectively. To do this, he considered the form $p\phi + q\psi$ for variables p and q and its associated determinant $\det(pA + qB)$. He showed that a necessary condition for $p\phi + q\psi$ to be transformed into $p\phi' + q\psi'$ (where $\det(pA' + qB')$ is the associated determinant of the latter) was

$$\det(pA' + qB') = \det(T)\det(W)\det(pA + qB),$$

and a sufficient condition was that $\det(p\phi + q\psi)$ and $\det(p\phi' + q\psi')$ have the same elementary divisors. He demonstrated the latter by establishing that $p\phi + q\psi$ can be put into a *normal form* via a linear change of the original variables x_i, y_i, that is, he showed how to represent $p\phi + q\psi$ by a matrix made up in a very special way of submatrices (or blocks)—one associated with each of the underlying eigenvalues and sized according to the multiplicity of that eigenvalue—and with entries of zeros, ones, and the eigenvalues themselves. In his *Treatise on Substitutions* of 1870, Camille Jordan independently—and as an outgrowth of his group-theoretic work (recall the previous chapter)—arrived at the same sort of block representations of matrices.[54] Today, this special way of representing matrices is called the *Jordan canonical form*.[55]

Although results in a similar vein had been obtained in England by Sylvester, Cayley, and Henry J. S. Smith (1826–1883) between 1851 and 1861—that is, results (although not the canonical form) involving the factorization and meaning of multiple roots of the characteristic polynomial as well as the greatest common divisor of its $(n-1) \times (n-1)$ minors—they had remained disconnected and reflective of the generic

---

[53] See Weierstrass, 1868 and the detailed discussion of it in Hawkins, 1977b, pp. 132–139. Here, we keep, as much as possible, the notation set up in the previous paragraph.

[54] See Jordan, 1870, pp. 114–126.

[55] The theory of similarity of matrices is easily seen to be a special case of the analysis just described, taking A to be the identity matrix. For a modern treatment of elementary divisors (and invariant factors), see, for example, Roman, 2005, pp. 141–179.

Figure 12.4. Georg Frobenius (1849–1917).

reasoning antithetical to the spectral theory evolving in Berlin.[56] It was not, however, these generic results that attracted the rigorizing attentions of Weierstrass's student, Georg Frobenius, in 1878. As Frobenius made clear in the introduction to a lengthy paper entitled, "Ueber lineare Substitutionen und bilineare Formen" (or "On Linear Substitutions and Bilinear Forms"), that appeared that year in Crelle's *Journal*, it was rather those of Cayley and Hermite on the Cayley–Hermite problem in the mid-1850s. "Investigations on the transformation of quadratic forms into themselves have so far been limited to consideration of the general [that is, generic] case," he explained, "while exceptions to which the results are subject in certain special cases have been exhaustively treated only for ternary forms... I have thus attempted to fill in the gaps which occur in the proofs of the formulas that... Cayley... and... Hermite have given."[57]

Like his mentor Weierstrass, Frobenius cast his results in the context of bilinear forms. As he knew from the work of his teachers in Berlin, two such forms $A = \sum_{i,j=1}^{n} a_{ij} x_i y_j$ and $B = \sum_{i,j=1}^{n} b_{ij} x_i y_j$ could be added, and scalar multiplication was a permissible operation. Frobenius took

[56] Hawkins, 1977b, pp. 139–144.
[57] Frobenius, 1878, p. 1, as quoted in Hawkins, 1977a, p. 98 (his translation).

things further and defined multiplication on bilinear forms this way: if P denotes the product of A and B, then

$$P = AB := \sum_{k=1}^{n} \frac{\partial A}{\partial y_k} \frac{\partial B}{\partial x_k}. \tag{12.10}$$

Moreover, considering the partial derivatives in this expression, we have

$$\frac{\partial A}{\partial y_k} = \sum_{i=1}^{n} a_{ik} x_k \quad \text{and} \quad \frac{\partial B}{\partial x_k} = \sum_{\ell=1}^{n} b_{\ell k} y_k,$$

so that substituting these expressions back into (12.10) yields the expression of the product

$$P = \sum_{i,j=1}^{n} \left(\sum_{\ell=1}^{n} a_{i\ell} b_{\ell j} \right) x_i y_j, \tag{12.11}$$

yet another bilinear form. This proved suggestive to Frobenius, given that the multiplication indicated by the sum in parentheses represents the multiplication of the matrices $[a_{ij}]$ and $[b_{ij}]$ associated with the bilinear forms A and B, respectively. It also proved suggestive to both him and others as they pursued a theory of linear associative algebras later in the century. Bilinear forms, matrices, and linear associative algebras were all intimately related (see the next chapter).[58]

Frobenius also noted explicitly that if C and D are two more bilinear forms, the following also hold:

(1) the distributive laws, namely,

$$A(B+C) = AB + AC, \quad (A+B)C = AC + BC,$$
$$(A+B)(C+D) = AC + BC + AD + BD;$$

(2) the laws of scalar multiplication for scalars a and b, namely,

$$(aA)B = A(aB) = a(AB), \quad (aA + bB)C = aAC + bBC; \text{ and}$$

---

[58] For the material in this and the next paragraph, see Frobenius, 1878, pp. 2–5.

(3) the associative law

$$(AB)C = A(BC).$$

As Frobenius remarked, however, the commutative law, that is, $AB = BA$, does *not* hold in general.

After exploring the more general, arithmetic properties of bilinear forms, he next applied them in the context of the Cayley–Hermite problem. If A is a symmetric bilinear form as above, then consider a linear transformation of its variables defined by $x_i = \sum_{j=1}^{n} p_{ij} x_i'$ and $y_i = \sum_{j=1}^{n} q_{ij} y_i'$. This can be used to define two new bilinear forms, namely, $P = \sum_{i,j=1}^{n} p_{ij} x_i y_j$ and $Q = \sum_{i,j=1}^{n} q_{ij} x_i y_j$. The application of the linear transformation to A then generates the product, in Frobenius's sense, $P^t A Q$ of three bilinear forms, where $P^t = \sum_{i,j=1}^{n} p_{ji} x_i y_j$ is the transpose of P. In this setup, the Cayley–Hermite problem can be stated succinctly as that of finding all forms P such that $P^t A P = A$, for A symmetric, that is, where $A^t = A$ and $\det(A) \neq 0$. In his 1878 paper, Frobenius examined further the properties of such substitutions P at the level of elementary divisors in addition to giving a perfectly general proof of the Hamilton–Cayley theorem.[59] In so doing, he not only "filled in the gaps" in the work of his British and French predecessors, but he also beautifully married the spectral theory of the Berlin school, in which he had been so well trained, and the symbolical algebra of linear substitutions (and matrices) as it had developed in the work of Eisenstein, Hermite, Cayley, and others. By the turn of the twentieth century, this marriage had been sanctified in textbook treatments in both German and English and had come to represent the so-called "higher algebra."[60] (For further developments in this vein, see the next chapter.)

THE EVOLUTION OF A THEORY OF INVARIANTS

The phrase "higher algebra" acquired a rather different meaning in the British Isles in the latter half of the nineteenth century. There, it

---

[59] See Frobenius, 1878, pp. 33–42 and 23–29, respectively, and compare Hawkins, 1977a, pp. 99–100.
[60] See Muth, 1899 and Bôcher, 1907.

signified a new area of mathematics—invariant theory—that had arisen particularly at the hands of Cayley and Sylvester and that had developed largely independently on the Continent in the work of Otto Hesse (1811–1874), Siegfried Aronhold (1819–1884), Alfred Clebsch (1833–1872), and Paul Gordan (1837–1912). Cayley's mathematical attentions had been turned in this direction as early as 1844 when he read George Boole's two-part "Exposition of a General Theory of Linear Transformations" of a couple of years earlier. Boole had closed that paper with an invitation to his readers to carry on in the research vein he had begun to mine. As he put it, "to those who may be disposed to engage in the investigation, it will, I believe, present an ample field of research and discovery. It is almost needless to observe that any additional light which may be thrown on the general theory, and especially as respects the properties of the function $\theta(q)$ [that is, the expression $\theta(Q_n)$ in (12.8) above], will tend to facilitate our further progress, and to extend the range of useful applications."[61] Cayley eagerly took up this challenge, writing an introductory letter to Boole on 13 June 1844 in which he expressed "the pleasure afforded me by a paper of yours...'On the theory of linear transformations' and of the interest I take in the subject."[62] In Cayley's view, "the subject...really appears inexhaustible," and he very quickly contributed to its further development.[63]

In an 1845 paper published, like Boole's two-part work, in the *Cambridge Mathematical Journal*, Cayley developed a construct—he called it the *hyperdeterminant*—which, like Boole's $\theta(Q_n)$, generated expressions in the coefficients of a homogeneous form of degree n in m variables that satisfied the invariantive property. However, whereas Boole's method treated homogeneous polynomials of degree n in m variables, Cayley's generalized the setting "by considering...a homogeneous function...of the same order [n], [but] containing n sets of (m) variables, and the variables of each entering linearly."[64] Although the terminology would not be in place for a number of years, Cayley thus envisioned generating *invariants* of *multilinear* forms. As he realized, such invariants could then

[61] Boole, 1842, p. 119.
[62] Arthur Cayley to George Boole, 13 June 1844, Trinity College Cambridge, R.2.88.1, as quoted in Crilly, 2006, p. 86. The next quote is also on this page.
[63] The account of the nineteenth-century history of invariant theory that follows in this section draws from Crilly, 1986; Crilly, 1988; Parshall, 1989; Parshall, 1999; and Parshall, 2011.
[64] Cayley, 1845a, p. 193.

be specialized to yield invariants of homogeneous polynomials of degree n in m unknowns.

Cayley was particularly interested in what he uncovered about the binary quartic form $U = \alpha x^4 + 4\beta x^3 y + 6\gamma x^2 y^2 + 4\delta x y^3 + \epsilon y^4$. His new method produced the invariant[65]

$$I = \alpha\epsilon - 4\beta\delta + 3\gamma^2,$$

whereas Boole's method had generated the discriminant, call it Δ_4, for U.[66] There were at least two *distinct* invariants of the binary quartic, but, as Boole pointed out on learning of this finding, even more was true. There was, in fact, yet another invariant,

$$J = \alpha\gamma\epsilon - \alpha\delta^2 - \epsilon\beta^2 - \gamma^3 + 2\beta\gamma\delta$$

of U, and I, J, and Δ_4 were connected by the polynomial expression

$$I^3 - 27J^2 = \Delta_4. \tag{12.12}$$

Cayley quickly realized the implications of this fact. As he put it in a continuation of his paper published less than a year later in 1846, a new and, as he saw it, rich research program would be "to find all the derivatives [invariants] of any number of functions [homogeneous forms], which have the property of preserving their form unaltered after any linear transformations of the variables."[67] Yet, he continued, as (12.12) suggests, "there remains a question to be resolved, which appears to present very great difficulties, that of determining the *independent* derivatives, and the relation between these and the remaining ones." In other words, Cayley sensed that there exists a minimum generating set for the invariants of any given homogeneous form of degree n in m unknowns and that there may exist algebraic dependence relations—or what Sylvester later called *syzygies*—between them.[68] That these syzygies

---

[65] Cayley, 1845a, p. 207.

[66] Recall from chapter 11 that this notion of the discriminant had arisen in Lagrange's analysis of the solution of the quartic equation.

[67] Cayley, 1846a, p. 104 (his emphasis). The next quote is also on this page.

[68] In the case of the binary quartic U, although he had not yet proven this, $\{I, J\}$ constitutes a minimum generating set for all invariants of U. No syzygy exists between these invariants, that

would be hard to uncover and to understand proved to be a highly prophetic statement, in light of the next fifty years of the history of what became a *theory* of invariants.

By 1850, Cayley had met and befriended Sylvester, and the two men had embarked upon a shared mathematical journey to explore both the geometric and, concomitantly, the purely algebraic ramifications of invariants. Sylvester, who was called to the Bar at the close of 1850 and who earned his living in London as Secretary and actuary at the Equity and Law Life Assurance Society, had reengaged with mathematical research after a troubled period in the latter half of the 1840s. His attentions had been drawn to questions in analytic geometry that involved both determinants and his own determinant-theoretic realization of the discriminant as well as to questions concerning the linear transformation of homogeneous forms f of degree n in m unknowns.[69] The latter work, in particular, had been inspired by an 1844 paper by the German mathematician, Otto Hesse.

Hesse had considered a homogeneous form

$$f = \Sigma a_{\chi\lambda\mu} x_1 x_2 x_3 \;=\; a_{111} x_1^3 + a_{222} x_2^3 + a_{333} x_3^3 + 6 a_{123} x_1 x_2 x_3$$
$$+ 3 a_{223} x_2^2 x_3 + 3 a_{233} x_2 x_3^2 + 3 a_{133} x_1 x_3^2$$
$$+ 3 a_{113} x_1^2 x_3 + 3 a_{122} x_1 x_2^2 + 3 a_{112} x_1^2 x_2, \quad (12.13)$$

of degree three in three variables and had defined the determinant $\phi = \phi(f) := \det\left(\frac{\partial^2 f}{\partial x_i \partial x_j}\right)$ in the course of his study of the critical points of curves in three-space. He showed that if T is a linear transformation such that $\det T = r$, then $\phi' = r^2 \phi$, where $\phi' := \phi(T(f))$.[70] Like Boole's discriminant and the expressions Cayley and Boole discovered associated with the binary quartic form, this expression ϕ, which Sylvester would later term the *Hessian*, was equal to ϕ' up to a power of the determinant of the linear transformation. Yet, whereas the previous examples of invariants involved only the coefficients of the underlying homogeneous

---

is, for no a, $b \in \mathbb{R}$ is it the case that $aI + bJ = 0$. In the case of the binary quintic, however, there is a syzygy among the invariants. Syzygies become more of an issue when the full set of *covariants* (see below) is considered.

[69] On Sylvester's life and work in historical context, see Parshall, 2006.

[70] See Hesse, 1844a, pp. 89–91; we have adopted a somewhat different notation here. For the geometrical results, see Hesse, 1844b.

form f, the Hessian was an expression in both the coefficients *and* the variables of f or what would later be termed a *covariant*. Sylvester foresaw that it should be possible to put all of these results—as well as results being obtained by Charles Hermite in Paris, among others—under "a common point of view."[71] A year later, Cayley had made a fundamental breakthrough toward the realization of that goal.

In a letter written to Sylvester on 5 December 1851, Cayley declared that "every invariant satisfies the partial diff[erentia]l equations

$$\left(a \frac{d}{db} + 2b \frac{d}{dc} + 3c \frac{d}{dd} \cdots + nj \frac{d}{dk} \right) U = 0$$

$$\left(b \frac{d}{db} + 2c \frac{d}{dc} + 3d \frac{d}{dd} \cdots + nk \frac{d}{dk} \right) U = \frac{1}{2} ns U$$

(s the degree of the Invariant) & of course the two equations formed by taking the coeff[icien]ts in a reverse order. This," he concluded, "will constitute the foundation of a new theory of Invariants."[72] Almost immediately, he and Sylvester began to refine and exploit this insight. In a foundational paper "On the Principles of the Calculus of Forms" composed over the course of 1851 and 1852, Sylvester explored the problem of determining a linear transformation that would allow him always to express a homogeneous form in two variables in its simplest, *canonical* form. He found the budding theory of invariants at its core, and in his paper's second installment, he elaborated explicitly on Cayley's announcement of 5 December 1851.

Given a homogeneous form of degree n in two variables (or what he called a binary n-ic form or binary quantic)

$$g = a_0 x^n + a_1 \binom{n}{1} x^{n-1} y + \cdots + a_n \binom{n}{n} y^n, \qquad (12.14)$$

---

[71] Sylvester, 1851b, p. 190.

[72] Parshall, 1998, p. 37. Cayley had, in fact, already hinted at detecting invariants as solutions of a set of partial differential equations in his 1845 paper, but he did not develop the idea further at that time. See Cayley, 1845a, p. 198.

By way of example, consider the binary quartic U and its invariants I and J as written above. It is easy to see that the operator $\left(\alpha \frac{d}{d\beta} + 2\beta \frac{d}{d\gamma} + 3\gamma \frac{d}{d\delta} + 4\delta \frac{d}{d\epsilon} \right)$ applied to both I and J yields zero. Similarly, a quick check confirms that the operator $\left(\beta \frac{d}{d\beta} + 2\gamma \frac{d}{d\gamma} + 3\delta \frac{d}{d\delta} + 4\epsilon \frac{d}{d\epsilon} \right)$ applied to I and J yields $\frac{1}{2} \cdot 4 \cdot 2I$ and $\frac{1}{2} \cdot 4 \cdot 3J$, respectively, where I has degree 2 and J has degree 3.

he defined the following two formal differential operators:

$$\mathfrak{X} = \left(a_0\partial_{a_1} + 2a_1\partial_{a_2} + 3a_2\partial_{a_3} + \cdots + na_{n-1}\partial_{a_n}\right),$$ (12.15)

as in Cayley's formulation, but also

$$\mathfrak{Y} = \left(a_n\partial_{a_{n-1}} + 2a_{n-1}\partial_{a_{n-2}} + 3a_{n-2}\partial_{a_{n-3}} + \cdots + na_1\partial_{a_0}\right).$$ (12.16)

He stated that if I is a homogeneous expression in the coefficients of g, and if $\mathfrak{X}(I) = 0$ and $\mathfrak{Y}(I) = 0$, then I is an invariant of g. Moreover, the partial differential operators $\mathfrak{X} - y\partial_x$ and $\mathfrak{Y} - x\partial_y$ could be used to detect covariants.[73]

In this and other papers in the early 1850s, Sylvester seemingly took the lead in laying out many of the initial ideas of invariant theory, but Cayley, his constant correspondent, talked them through with him behind the scenes in their extensive correspondence and quickly extended and refined those results. While Sylvester's 1853 monographic and freely ranging paper "On a Theory of Syzygetic Relations of Two Rational Integral Functions..." focused on the problem of detecting syzygies, Cayley's more systematic 1854 "Introductory Memoir upon Quantics" and his 1856 "Second Memoir upon Quantics" adapted the differential operators $\mathfrak{X}$ and $\mathfrak{Y}$ to the more general context of covariants and strove to provide a clearer articulation of the evolving theory.[74]

Its first systematic treatment, however, owed to the efforts of the Irish mathematician and third member of what was styled the "invariant trinity," George Salmon (1819–1904).[75] In his 1859 *Lessons Introductory to the Modern Higher Algebra*, Salmon laid out the concepts and defined the research agenda that would shape the British theory as it developed into the 1880s and 1890s.[76] Consider a binary n-ic form g as in (12.14) and a nonsingular linear transformation of it $T : x \mapsto aX + bY, y \mapsto a'X + b'Y$

---

[73] Sylvester, 1852b, pp. 204–214. Cayley and Sylvester's use of $\mathfrak{X}$ and $\mathfrak{Y}$ to detect invariants can today be explained in Lie-algebraic terms. For the details, see Parshall, 1998, pp. 184–186 (notes 93 and 94).

[74] See Sylvester, 1853 and Cayley, 1854b and 1856.

[75] Charles Hermite coined this phrase. See Crilly, 2006, p. 177.

[76] The notation and presentation below, however, draw from the somewhat later and somewhat more modern development found in Elliott, 1895, pp. 27–50.

such that

$$T(g) = A_0 X^n + A_1 \binom{n}{1} X^{n-1} Y + \cdots + A_n \binom{n}{n} Y^n.$$

In general, Cayley, Sylvester, and, later, fellow British invariant theorists Percy MacMahon (1854–1929) and Edwin Bailey Elliott (1851–1937), among others, sought all homogeneous polynomials $K(a_0, a_1, \ldots, a_n; x, y)$ in the coefficients and the variables of (12.14) that satisfy

$$K(A_0, A_1, \ldots, A_n; X, Y) = \Delta^\ell K(a_0, a_1, \ldots, a_n; x, y),$$

where $\Delta = \det T$ and $\ell \in \mathbb{Z}^+$. Such an expression K is called a *covariant*; an *invariant* is a covariant comprised only of coefficients. It is fairly easy to prove that if K is a covariant, then it has a constant degree of homogeneity in x, y, denoted μ and called the *order* of K, and a constant degree of homogeneity in the coefficients $a_0, a_1, \ldots, a_n$, called the *degree* and denoted θ. Moreover, given a monomial of K, its *weight* is defined to be the sum of the products of the subscripts and their corresponding superscripts; for example, the monomial $a_2^2 a_3^4 x^3 y^2$ has weight $(2 \times 2) + (3 \times 4) + (1 \times 3) + (0 \times 2) = 19$, where x is arbitrarily assigned the subscript 1 and y the subscript 0. With these definitions in place, it can be shown that every covariant K of degree θ and order μ has constant weight $\frac{n\theta + \mu}{2}$, where n is the degree (in the modern sense) of (12.14).

Now consider $\mathfrak{X}$ and $\mathfrak{Y}$ as in (12.15) and (12.16) above, and take a perfectly general (homogeneous) polynomial A_0 in the coefficients of (12.14) of degree θ and weight $\frac{n\theta - \mu}{2}$ such that $\mathfrak{X} A_0 = 0$. Setting $A_j = \frac{\mathfrak{Y}(A_{j-1})}{j}$ for $1 \leq j \leq \mu$, define

$$K = A_0 x^\mu + A_1 x^{\mu-1} y + \cdots + A_{\mu-1} x y^{\mu-1} + A_\mu y^\mu, \tag{12.17}$$

a homogeneous polynomial of weight $\frac{n\theta + \mu}{2}$ and order μ. As Cayley argued in his 1854 "Introductory Memoir," K defined in this way is always a covariant of (12.14), and, moreover, *every* covariant of (12.14) of order μ and degree θ can be so expressed. Cayley also concluded that, for given n, θ, and μ, determining the number of linearly independent covariants reduces to determining the number of partitions of $\frac{n\theta-\mu}{2}$ and of $\frac{n\theta-\mu-1}{2}$, with parts taken from the integers $0, 1, 2, \ldots, n$ and repetitions allowed. In other words, Cayley had showed that an important question in invariant theory was nothing more than a problem in partition theory. Charles Hermite proved yet another "counting" result in 1854, the law of reciprocity, namely, for a binary quantic g of degree n, the number of invariants of order p in the coefficients of g equals the number of invariants of order n in the coefficients of a binary quantic of degree p.[77] These and other results ultimately launched nineteenth-century invariant theorists on combinatorial researches to address such issues.[78]

Cayley used much of this evolving theory in his "Second Memoir" of 1856 explicitly to calculate minimum generating sets of covariants for quantics of successive degrees. In particular, he erroneously argued— and Salmon initially perpetuated the error—that a minimum generating set for the general binary quintic form has infinitely many elements. Undaunted, he and Sylvester continued in their search for distinct (that is, both linearly and algebraically independent) covariants for given binary forms with Cayley ultimately writing ten in his series of memoirs on quantics. They were collectors; they were calculators; complete catalogues were their goal. As far as they knew, like the Victorian botanist who sought to catalogue and classify all of the world's plants, they had taken on a problem of infinite proportions, but, in so doing, they created a new area of mathematical research that attracted the attention of fellow mathematicians both at home and abroad.

Contemporaneous with these developments in the British Isles, a new field of mathematics emerged, particularly in Germany, that stemmed from similar observations. As just noted, Hesse, in his analysis of the critical points of third-order plane curves f, had defined the determinant $\phi = \phi(f) := \det\left(\frac{\partial^2 f}{\partial x_i \partial x_j}\right)$ and had showed that it satisfied the invariantive property. At essentially the same time, Gotthold Eisenstein had completed a study in which he had encountered the same expression— although in the more restricted context of a binary cubic form— in his efforts to extend Gauss's number-theoretic work on quadratic forms to cubic forms, and he, too, showed explicitly that it had the

---

[77] Hermite, 1854b.

[78] On Sylvester's combinatorial work, in particular, see Parshall, 1988b, pp. 172–188. See also Wilson, 2011.

invariantive property.[79] By 1849, Siegfried Aronhold, who had taken courses with both Jacobi and Hesse at the University of Königsberg in East Prussia (and not to be confused with the Königsberg in Bavaria that was the birthplace of Regiomontanus), had been attracted by Eisenstein's results as well as by Hesse's 1844 work. He sought to cast the latter in more purely algebraic terms.

In the course of his analysis of the ternary cubic form (12.13), Hesse encountered an equation which he wrote as $G = Ba - Ab = 0$, where A and B are homogeneous forms of degree three in indeterminants a and b (taken over the complex numbers). Hesse's analysis required him to solve this quartic equation, but his solution was laborious and, at least to Aronhold's mind, insufficiently intrinsic. As Aronhold put it in a paper submitted in July of 1849 but published in 1850, "Hesse's presentation, which has a different point of departure, brings out the algebraic problem rather as a corollary to other developments. Therefore, I have considered it necessary to prove systematically the theorems he gives... [so that] they permit the solution of the problem directly."[80] From Aronhold's point of view, understanding G—that is, solving the equation $G = 0$—meant understanding A and B, and he did the latter in invariant-theoretic terms.

In particular, he derived a complicated expression S—homogeneous and of degree four in the coefficients $a_{\chi\lambda\mu}$ of f—that "has the property that its form does not change if it is constructed out of the coefficients of that function into which f changes when one transforms $[f]$ by a linear substitution." For r, the determinant of the linear transformation, and S', the transformed version of S, Aronhold found that $S' = r^4 S$. Similarly, he constructed an even more complicated expression T, homogeneous and of degree six in the coefficients of f, that satisfies $T' = r^6 T$. He then showed that $A = 3Sa^2b + 6Tab^2 + 3S^2b^3$ and $B = a^3 - 3Sab^2 - 2Tb^3$ so that $G = a^4 - 6Sa^2b^2 - 8Tab^3 - 3S^2b^4$. In other words, he expressed G in terms of the invariants S and T of the ternary cubic form (12.13). Dividing G through by b^4 and making the substitution $\lambda = \frac{a}{b}$ reduced G to the fourth-degree polynomial $G = \lambda^4 - 6S\lambda^2 - 8T\lambda - 3S^2$, the roots of which he could find invariant-theoretically using results of Eisenstein.

---

[79] Eisenstein, 1844a, p. 75 and Eisenstein, 1844b, p. 91–92, respectively.
[80] Aronhold, 1850, p. 144 (our translation). For the quote in the next paragraph, see p. 152 (our translation).

In other words, he showed how to express the four roots of the quartic equation $G = 0$ completely in terms of discrete invariants.[81] This was the more intrinsic, algebraic interpretation of $G = 0$ that he had sought.

Aronhold had cast his results of 1850 in terms of a symbolic notation that had determinants as its substructure and that allowed him to express his results more succinctly than would have been possible in terms of the sums and products of coefficients with triple subscripts. For example, he expressed his invariant S as

$$\sum (a_1a_1)^{\chi,\lambda}(a_\chi a_\lambda)^{1,1} = (a_1a_1)^{1,1}(a_1a_1)^{1,1} + (a_1a_1)^{2,2}(a_2a_2)^{1,1}$$
$$+ (a_1a_1)^{3,3}(a_3a_3)^{1,1} + 2(a_1a_1)^{2,3}(a_2a_3)^{1,1}$$
$$+ 2(a_1a_1)^{1,3}(a_1a_3)^{1,1} + 2(a_1a_1)^{1,2}(a_1a_2)^{1,1},$$

$$(12.18)$$

where $(a_\chi a_\lambda)^{1,1} := a_{\chi 33}a_{\lambda 22} + a_{\chi 22}a_{\lambda 33} - 2a_{\chi 23}a_{\lambda 23}$, for $1 \leq \chi$, $\lambda \leq 3$ and for, say, $\chi = 1$ and $\lambda = 2$, $(a_1a_1)^{\chi,\lambda} = (a_1a_1)^{1,2} = a_{113}a_{123} + a_{123}a_{113} - a_{112}a_{133} - a_{133}a_{112}$, all in the coefficients of the ternary cubic form f.[82] By 1858, Aronhold had not only extended the results of his 1850 paper but also familiarized himself with at least some of the invariant-theoretic work—largely on *binary* forms—being done at the hands of Cayley and Sylvester in England. In a hundred-page treatise, he fully explored the behavior as well as the properties of the *ternary* cubic form from an invariant-theoretic point of view, and he did so in the evolving language of invariants, covariants, etc. that his English colleagues had devised.

In his approach, a homogeneous form of degree n in m unknowns $x_1, x_2, \ldots, x_m$ was written symbolically as $(a_1x_1 + a_2x_2 + \cdots + a_mx_m)^n$. For example, the ternary cubic (12.13) would be written as

$$(a_1x_1 + a_2x_2 + a_3x_3)^3 \qquad (12.19)$$

or more simply as a_x^3 or even a^3, and since the a_i's are merely symbols, a_x^3, b_x^3, c_x^3, etc. all represent the ternary cubic. The Cartesian expression (12.13) of the form could be recovered from (12.19) by multiplying it out

---

[81] Aronhold, 1850, pp. 152–156 and Eisenstein, 1844b,c.
[82] Aronhold, 1850, pp. 151–152.

and then equating $a_i a_j a_k$ with the coefficient a_{ijk}. Invariants of (12.13) were then generated by products of symbolic 3×3 determinants. In general, let (rst) denote the 3×3 determinant

$$
\begin{vmatrix}
r_1 & r_2 & r_3 \\
s_1 & s_2 & s_3 \\
t_1 & t_2 & t_3
\end{vmatrix},
$$

for symbols r_i, s_j, t_k. Form the four symbolic determinants $A = (vwp)$, $B = -(uwp)$, $C = (uvp)$, and $D = -(uvw)$—where the negative signs arise when one views these determinants as the "minors" formed from the 4×3 array

$$
\begin{vmatrix}
u_1 & u_2 & u_3 \\
v_1 & v_2 & v_3 \\
w_1 & w_2 & w_3 \\
p_1 & p_2 & p_3
\end{vmatrix},
$$

by striking the appropriate row and appending the sign $(-1)^{1+j}$—and take their product $ABCD$. Finally, identifying the trinomials $u_\chi u_\lambda u_\mu$, $v_\chi v_\lambda v_\mu$, $w_\chi w_\lambda w_\mu$, $p_\chi p_\lambda p_\mu$ with the coefficients $a_{\chi\lambda\mu}$ yields the invariant S in (12.18), which in this new, more general notation would be written simply as $(vwp)(uwp)(uvp)(uvw)$.[83]

Aronhold's work proved extremely suggestive to fellow German, Alfred Clebsch. By 1861, Clebsch, then on the faculty at the Technische Hochschule in Karlsruhe but later at the universities in Giessen and then Göttingen, had developed Aronhold's symbolic notation into a more general form applicable to homogeneous forms regardless of the degree or the number of variables. He had also begun to develop the general theory in those terms and had applied it in the geometrical context in which he was most interested. In particular, he had laid out the theoretical justification within the German approach for the formation of invariants and covariants.[84] In an 1863 paper, entitled "Ueber eine fundamentale Begründung der Invariantentheorie" (or "On a Fundamental Foundation

---

[83] Aronhold, 1858, pp. 106–108.
[84] See Clebsch, 1861a, 117–118 and Clebsch, 1861b, 27–28.

of Invariant Theory"), Aronhold extended all of this into a unified theory for building invariants and covariants from simpler, fundamental ones, that is, like his English colleagues, in terms of a minimum generating set.

Despite Aronhold's research successes especially with the ternary cubic form, calculations involving the ternary n-ic were, as the above might suggest, far from easy. Like the English school of invariant theorists, the German school ultimately made the greatest progress in the case of the *binary* n-ic form. A fundamental theorem in their approach to that simplest, general case was that "every invariant is represented symbolically as the aggregate of products of symbolic determinants of the type (ab); every covariant as the aggregate of products of symbolic determinants (ab) with linear symbolic factors of the type c_x."[85] As we saw, the main theorem on the construction of covariants for the English school was (12.17). The analogous—but functionally different—expression for the German school was thus that every covariant of a binary n-ic could be expressed as a linear combination of symbolic products of the form

$$P = a_x^\alpha b_x^\beta \cdots (ab)^{\alpha_1}(ac)^{\beta_1} \cdots ,$$

where the total number of a's, b's, etc. appearing in the expression is n, the degree of the associated form, or a positive integer multiple of n.[86]

Calculating these symbolic products and determining when two of them are actually the same was nontrivial, and the German school developed an involved calculus for manipulating and interpreting such expressions.[87] They also developed a technique—they called it *Ubereinanderschiebung* or *transvection*—for formally generating covariants. Given two binary forms in x_1 and x_2 (which could be taken to be identical) represented symbolically by $(a_1x_1 + a_2x_2)^n$ and $(b_1y_1 + b_2y_2)^m$, it hinged on the differential operator

$$\Omega = \frac{1}{mn}\frac{\partial^2}{\partial x_1 \partial y_2} - \frac{\partial^2}{\partial y_1 \partial x_2}$$

[85] Clebsch, 1872, p. 32 (our translation). Here c_x denotes $c_1x_1 + c_2x_2$.
[86] Compare Gordan, 1868, p. 324.
[87] For an introduction to their methods, see Clebsch, 1872, pp. 1–43.

and its formal powers Ω^k. Cayley had actually already hit upon this operator in 1846[88] but had fairly quickly abandoned it in favor of (12.17) and its concomitant theory as being more amenable to direct calculation. In hindsight, that decision may be deemed a mistake. It was precisely within that framework that Paul Gordan proved that, for a given binary quantic, the minimum generating set of covariants—that is, the smallest set of algebraically independent covariants that produce all covariants associated with it—is *finite*.[89] His 1868 proof—which thereby contradicted Cayley's 1856 "result" that the minimum generating set of the general binary quintic form was infinite—hinged on carefully established properties of the transvection process.

In the two decades that followed the publication of Gordan's finiteness theorem, the German school worked to extend his result to ternary forms, while the English school continued not only to calculate complete systems of invariants for binary quantics of ever higher degrees (reaching degree ten) but also to vindicate their methods by producing a "British" proof of Gordan's theorem.[90] Indeed, so much invariant-theoretic work was generated in Germany that a new journal, the *Mathematische Annalen* (or *Annals of Mathematics*), was founded in 1869 largely to accommodate the surge of research. (Its inaugural issue contained papers by such invariant-theoretic stalwarts as Cayley, Clebsch, and Gordan.) Still, although progress was made on both sides of the English Channel, neither school succeeded in its quest. The Germans were ultimately unable to adapt their transvection process to extend Gordan's finiteness theorem for binary quantics to homogeneous forms in three or more variables; the British could not push their methods through to an actual *proof* of the finiteness of the minimum generating sets for the binary n-ics. The aim of both schools was explicit calculation, but the tools of neither were strong enough to carry their respective programs significantly further.

That changed in the late 1880s. David Hilbert (1862–1943), as a student at the University of Königsberg, entered the mathematical ranks with a dissertation on a topic in invariant theory for which he earned

---

[88] See Cayley, 1846a, pp. 106–107.
[89] Gordan, 1868.
[90] See Cayley, 1871 on the quintic and Sylvester, 1879 for the calculational results. On the repeated efforts over the course of the 1870s and 1880s to find a British proof of Gordan's theorem, see Parshall, 1989, pp. 180–185 and Sylvester's correspondence in Parshall, 1998, pp. 172–216.

Figure 12.5. David Hilbert (1862–1943).

his degree in 1885.[91] By 1888, he had been attracted to the problem of extending Gordan's theorem to homogeneous forms of degree n in m unknowns. Could it be done? He surprised the international community of invariant theorists that year by answering this question in the affirmative.

Hilbert's insight had been to ask the question somewhat differently. Given a finite set of variables $x_1, x_2, \ldots, x_m$ and an infinite system of forms in terms of them, does a finite set of forms exist such that every form in the infinite system can be written as a linear combination of the forms in the finite set with coefficients that are rational integral functions in $x_1, x_2, \ldots, x_m$? Drawing on ideas of Leopold Kronecker, Hilbert proved that, indeed, such a finite set exists. He did not *construct* it; he provided a mathematical argument that it *exists*.[92] This prompted the famous retort from Gordan that "this is not mathematics. This is theology."[93] By 1893, Hilbert had even provided a constructive argument,

---

[91] On Hilbert's life, see Reid, 1972.

[92] Hilbert, 1888–1889 and Hilbert, 1890. This theorem, here in the special setting of polynomials over a field, is now called the Hilbert basis theorem.

[93] Weyl, 1944, p. 549.

although it used not *modern* algebra in the sense of Salmon's textbook on invariant theory but rather algebra in a new sense that was evolving in work on the concept of algebraic number fields (see the next chapter).[94] Another new era in the history of algebra was dawning, and, as we shall see, it would be ushered in by mathematicians like Hilbert, Richard Dedekind, Emmy Noether (1882–1935), and many others, in the 1890s and into the opening decades of the twentieth century.

[94] Hilbert, 1893.

13

Understanding the Properties of "Numbers"

I f understanding polynomials and polynomial equations defined two intertwining threads of what came to be recognized as the fabric of algebraic research by the end of the nineteenth century, a third thread was spun from the efforts of mathematicians to understand the properties of "numbers." As early as the sixth century BCE, this quest had motivated the Pythagoreans to inquire into the nature of the positive integers and to define concepts such as that of the *perfect* number, that is, a positive integer like $6 = 1 + 2 + 3$ or $28 = 1 + 2 + 4 + 7 + 14$, which satisfies the property that it equals the sum of its proper divisors. Several centuries later, Euclid codified some of the Pythagorean number-theoretic results, especially in Books VII, VIII, and IX of his *Elements*, although, as we saw in chapter 3, his approach was highly geometric.

By the sixteenth century, Girolamo Cardano had encountered—in the course of his work in the *Ars magna* or "great art" of algebra— a new type of "number," the complex number $a + b\sqrt{-1}$, for $a, b \in \mathbb{R}$. There, he acknowledged the "mental tortures involved" in dealing with such expressions, at the same time that he successfully multiplied two complex conjugates (recall chapter 9). Cardano's younger contemporary, Raphael Bombelli, contextualized and developed these ideas—as well as the recently rediscovered indeterminate analysis of Diophantus— in his book on *Algebra* (recall chapters 4 and 9). There, Bombelli had no compunction whatsoever about treating complex solutions of polynomial equations (although he did not treat them consistently), and he gloried in the kind of number-theoretic problems that Diophantus had posed. Pierre de Fermat, a generation later, shared that fascination and famously claimed in the margin of his copy of Diophantus's *Arithmetica* that he could actually prove that there are no nontrivial integer solutions of the equation $x^n + y^n = z^n$, for an integer n greater than 2.

As mathematicians in the eighteenth and nineteenth centuries struggled to understand what Fermat's alleged proof of his so-called "last theorem" might have been, they, as well as others motivated by issues other than Fermat's work, came to extend the notion of "number." And, they did this in much the same spirit that Galois had extended that of "domain of rationality" or field, that is, through the creation and analysis of whole new types of algebraic systems. This freedom to create and explore new systems—and new algebraic constructs like the determinants and matrices that we encountered in the previous chapter—became one of the hallmarks of the *modern* algebra that developed into the twentieth century.

NEW KINDS OF "COMPLEX" NUMBERS

One of the first to consider seriously the matter of Fermat's claim relative to his so-called "last theorem" was Leonhard Euler. As early as 4 August 1753 in a letter to his friend and regular mathematical correspondent, Christian Goldbach (1690–1764), Euler announced that he could prove the theorem's $n = 3$ case,[1] but a published proof materialized only in 1770 in his massive textbook, *Vollständige Anleitung zur Algebra* (or *The Elements of Algebra*). That book culminated with Euler's presentation of "Solutions of Some Questions in Which Cubes Are Required" and, in particular, with his demonstration that "it is impossible to find any two cubes, whose sum, or difference, is a cube."[2] His approach involved a technique near and dear to Fermat's heart, namely, the method of infinite descent. Here is the idea.[3] Suppose that whenever three positive integers a, b, c can be found such that $a^3 + b^3 = c^3$, it is then possible to find other positive integers a', b', c' such that $a'^3 + b'^3 = c'^3$ and $c'^3 < c^3$. Since the process could be carried out on the new triple and so on and so on, it would be possible to find triples of smaller and smaller integers ad infinitum, clearly a contradiction. Thus, the original assumption must be false, namely, it must be *impossible* to find nonzero (positive) integers a, b, c such that $a^3 + b^3 = c^3$.

---

[1] Leonhard Euler to Christian Goldbach, 4 August 1753, as published in Fuss, 1843, p. 618.
[2] Euler, 1770/1984, p. 450.
[3] For the argument, see Euler, 1770/1984, pp. 449–454 and Edwards, 1977, pp. 40–46.

Euler began his argument by noting that he could, without loss of generality, assume that a, b, c are pairwise relatively prime, that is, that the greatest common divisor of any pair of these three integers is one. That being the case, exactly one of the integers must be even and the other two odd. Suppose, for example, that a and b (for $a > b$) are odd (and c is even). Then, $a + b$ and $a - b$ are both even and, so, can be expressed as $2p$ and $2q$, respectively.[4] An elementary analysis of p and q allowed Euler to conclude that p and q have opposite parities (that is, one is even and one is odd) and that they are relatively prime. Therefore, the assumption that $a^3 + b^3 = c^3$ with a and b both odd implies that there are relatively prime positive integers p and q of opposite parity— say, p is even and q is odd—such that $a^3 + b^3 = (p+q)^3 + (p-q)^3 = 2p(p^2 + 3q^2) = c^3$.

Since c^3 is even, it must be divisible by 8, and so $\frac{p}{4}(p^2 + 3q^2) = (\frac{c}{2})^3$ is also a cube. Moreover, since $p^2 + 3q^2$ is odd and so, in particular, is not divisible by 4, $\frac{p}{4}$ must be an integer. As Euler argued, then, $\frac{p}{4}$ and $(p^2 + 3q^2)$ must be relatively prime, unless 3 divides p. Assuming that is not the case, both factors must be cubes (as must $2p$).

Euler then took the bold step of factoring the cube $p^2 + 3q^2$ as

$$p^2 + 3q^2 = (p + q\sqrt{-3})(p - q\sqrt{-3})$$

and operating on the "integers" $\mathbb{Z}[\sqrt{-3}]$ in the same way that one would operate on the "usual" integers $\mathbb{Z}$. In particular, he assumed that these integers form a *unique factorization domain*, namely, a number system in which any nonzero element is either a unit or can be written uniquely (up to order and unit multiples) as the product of a finite number of irreducible elements. He then argued this way. Since p and q have opposite parities and since p and q are relatively prime, the factors $p + q\sqrt{-3}$ and $p - q\sqrt{-3}$ are relatively prime. Thus, each is itself a cube in $\mathbb{Z}[\sqrt{-3}]$, that is,

$$p + q\sqrt{-3} = (t + u\sqrt{-3})^3 \text{ and } p - q\sqrt{-3} = (t - u\sqrt{-3})^3.$$

[4] The other case, in which a and c are even and b is odd, is handled similarly.

Expanding the cubes gives

$$p = t^3 - 9tu^2 = t(t - 3u)(t + 3u) \text{ and } q = 3t^2u - 3u^3 = 3u(t - u)(t + u).$$

Necessarily, t and u are relatively prime of opposite parity (and assume, for convenience, that they are both positive integers).

The ingredients for the infinite descent are now in place. Since

$$2p = 2t(t - 3u)(t + 3u) \tag{13.1}$$

is a cube and neither p nor t is divisible by 3, the three factors on the right-hand side of (13.1) are pairwise relatively prime. Thus, each is a cube, say, $2t = d^3$, $t - 3u = e^3$, and $t + 3u = f^3$. Adding the last two of these equations and incorporating the first yields $d^3 (= 2t) = e^3 + f^3$. Noting that $d^3 e^3 f^3 = 2p$, a cube less than c^3 by construction, Euler concluded that $d < c$, which gives the desired contradiction as required for the infinite descent. The second case, in which 3 divides p, goes through similarly.[5] Although it takes more work actually to complete the proof,[6] this much of the argument should suffice to give its general flavor.

Although Euler's proof contained the flawed assumption about the unique factorization of "integers" of the form $p + q\sqrt{-3}$ (for p, $q \in \mathbb{Z}$),[7] his argument is, in fact, salvageable. This is perhaps not surprising since, in modern terms, the "integers" $\mathbb{Z}[\sqrt{-3}]$ form a subring of the domain $\mathbb{Z}[\zeta]$, where $\zeta = \frac{-1+\sqrt{-3}}{2}$ is a primitive cube root of unity, and $\mathbb{Z}[\zeta]$ *is* a unique factorization domain.[8] Its flaw aside, Euler's argument is noteworthy for its fearless definition and usage of a brand-new type of "integer" as well as for its inherent assumption that the "regular" integers can be understood more fully by extending them to some other realm of "number." The latter idea would prove especially potent as number theory was increasingly "arithmetized" over the course of the nineteenth century.

Perhaps the mathematician most influential in setting number theory on that new course was Carl Friedrich Gauss in his *Disquisitiones arithmeti-*

[5] See Edwards, 1977, p. 42.

[6] For the rest of the details, see Edwards, 1977, pp. 52–54.

[7] That they are not is made clear by the fact that, for example, $4 = 2^2 = (1 + \sqrt{-3})(1 - \sqrt{-3})$, two distinct factorizations of 4 as a product of irreducibles.

[8] See Edwards, 1977, pp. 52–54 for a proof using Eulerian methods that plugs the gap.

cae of 1801. As we saw in the previous two chapters, Gauss introduced, among many other things, a rich theory of quadratic forms there that others later in the century worked to extend to a theory of ternary and higher forms. Gauss opened his text, however, with an exposition of the basic, but key, notion of congruence. "If a number *a* divides the difference of the numbers *b* and *c*, *b* and *c* are said to be *congruent relative to a*; if not, *b* and *c* are *noncongruent*. The number *a* is called the *modulus*. If the numbers *b* and *c* are congruent, each of them is called a *residue* of the other. If they are noncongruent, they are called *nonresidues*."[9] In other words, and using Gauss's notation, for *a* a nonzero integer, $b \equiv c \pmod{a}$ if and only if *a* divides $b - c$.

Gauss proceeded to show that this new binary operation enjoyed many of the properties of the familiar binary operations of addition and multiplication. In particular, he demonstrated that if $k \in \mathbb{Z}$, if $p, q, r, \ldots \in \mathbb{Z}^+$, and if $a \equiv b \pmod{n}$, $c \equiv d \pmod{n}$, $e \equiv f \pmod{n}$, etc. for any (positive) integer *n*, then[10]

(1) $a + c + e + \cdots \equiv b + d + f + \cdots \pmod{n}$;

(2) $ace \cdots \equiv bdf \cdots \pmod{n}$;

(3) $a - c \equiv b - d \pmod{n}$;

(4) $ka \equiv kb \pmod{n}$; and

(5) $Aa^p + Ba^q + Ca^r + \cdots \equiv Ab^p + Bb^q + Cb^r + \cdots \pmod{n}$, for $A, B, C, \ldots \in \mathbb{Z}$.

He next established the effectiveness of this notion of congruence—and its associated modular arithmetic—by employing it in tackling a variety of number-theoretic questions.

Perhaps one of the questions that intrigued him the most in this vein was the so-called law of quadratic reciprocity, a result conjectured by both Euler and Adrien-Marie Legendre (1752–1833) but first proved by Gauss in print in the *Disquisitiones*.[11] At issue was the general question, "Given a number, to assign all numbers of which it is a [quadratic] residue or a

---

[9] Gauss, 1801/1966, p. 1 (his emphases).
[10] Gauss, 1801/1966, pp. 2–3.
[11] See Gauss, 1801/1966, §6, pp. 63–107, especially pp. 82–104. For the next two quotations, see pp. 72 and 105, respectively.

[quadratic] nonresidue." What does this mean? If a and m are integers (with $|m| > 2$), then a is a *quadratic residue* $(\mathrm{mod}\, m)$, if there is an integer x such that $x^2 \equiv a$ $(\mathrm{mod}\, m)$. Consider two distinct odd primes p and q, and suppose it is known that p is a quadratic residue $(\mathrm{mod}\, q)$. Is q a quadratic residue $(\mathrm{mod}\, p)$? In what Gauss judged an "excellent tract" on "Recherches d'analyse indéterminée" (or "Researches on Indeterminate Analysis") published in 1785, Legendre had answered this question in the affirmative, if *at least* one of the primes p or q is congruent to 1 $(\mathrm{mod}\, 4)$. If, however, *both* p and q are congruent to 3 $(\mathrm{mod}\, 4)$, then, he claimed, if p is a quadratic residue $(\mathrm{mod}\, q)$, then q is a quadratic nonresidue $(\mathrm{mod}\, p)$, and vice versa.[12] As Gauss realized, however, Legendre's proof was incomplete. What Gauss succeeded in proving in full generality—and in eight (!) different ways over the course of his lifetime— was that if $p \equiv 1$ $(\mathrm{mod}\, 4)$, then p is a quadratic residue (nonresidue) modulo q if and only if q is a quadratic residue (nonresidue) modulo p, while if $p \equiv 3$ $(\mathrm{mod}\, 4)$, then $-p$ has the same properties. As he described it almost two decades after publishing its proof, this "fundamental theorem on quadratic residues is one of the most beautiful truths of higher Arithmetic."[13]

By the late 1820s, Gauss was at work extending this "beautiful truth" to the setting of biquadratic residues, that is, congruences $x^4 \equiv p$ $(\mathrm{mod}\, q)$; in other words, he sought conditions under which such a congruence is solvable.[14] As Euler had recognized in his work on Fermat's last theorem, answering questions about the integers may require going outside of the integers proper. Gauss used this idea in fundamental and far-reaching ways to tackle a law of biquadratic reciprocity. In particular, he defined new "numbers" of the form $a + b\sqrt{-1}$, for $a, b \in \mathbb{Z}$ (that is, a subset of the usual complex numbers in which the coefficients of the real and imaginary parts are restricted to the integers), and he showed that they behave just like the "regular" (from now on we call them the *rational*) integers $\mathbb{Z}$. Since they are a subset of the complex numbers, they inherit all of the usual binary operations and their properties from $\mathbb{C}$. Now, Euler had recognized all of these properties for his new "numbers" $p + q\sqrt{-3}$, too, but, as we saw, he also mistakenly assumed unique factorization.

[12] Legendre, 1785.
[13] Gauss, 1818, p. 496 (our translation).
[14] See Gauss, 1828 and Gauss, 1832.

In his 1832 study of "numbers" $a + b\sqrt{-1}$, for a, $b \in \mathbb{Z}$—now called the *Gaussian integers*—Gauss was more methodical.[15] As he did in the *Disquisitiones* relative to establishing a priori the properties of congruence and the fundamentals of modular arithmetic, so in his study of biquadratic residues, Gauss clearly established the properties of the new number system that guided his work. In particular, he extended the notion of a unit—that is, ± 1 in the rational integers $\mathbb{Z}$—to the Gaussian integers, where the units are ± 1 and $\pm\sqrt{-1} := \pm i$. A Gaussian integer is composite if it can be written as the product of two other Gaussian integers neither of which is a unit, and prime otherwise. Moreover, for a Gaussian integer g, he defined the norm $N(g)$ of g to be $a^2 + b^2 = (a + bi)(a - bi)$. It was thus easy to see that the norm satisfied the property that $N(g)N(h) = N(gh)$, for Gaussian integers g and h.

How, then, do Gaussian integers behave, and how does their behavior differ from that of the rational integers $\mathbb{Z}$? Consider, for example, 17. In $\mathbb{Z}$, it is a prime of the form $4n + 1$, but in the Gaussian integers, it is composite, since $17 = (1 + 4i)(1 - 4i)$. In fact, since it can be demonstrated that any prime of the form $4n + 1$ can be written as the sum of two squares, it is clear that every prime in $\mathbb{Z}$ of the form $4n + 1 = c^2 + d^2 = (c + di)(c - di)$ is composite in the Gaussian integers. On the other hand, Gauss proved that rational primes of the form $p = 4n + 3$ are both rational *and* Gaussian primes.

With this machinery in place, Gauss was able to show in his work of 1832 that a Gaussian integer $a + bi$, for which $ab \neq 0$, is prime if its norm is a rational prime, and composite otherwise. This led him to conclude that the Gaussian primes were of the form $1 + i$, $1 - i$, $4n + 3$ (when $4n + 3$ is a rational prime), or the conjugate factors $c + di$ and $c - di$ of primes of the form $4n + 1$, together with their unit multiples (which Gauss called their "associates"). With this understanding of primes and given that he needed to grapple with the factorization of Gaussian integers in his work toward a law of biquadratic reciprocity, it was natural for him to ask whether an analogue of the fundamental theorem of arithmetic held in his new system, that is, whether, given a Gaussian integer, it can always be factored *uniquely* into the product of powers of Gaussian primes. He showed not only that unique factorization holds

---

[15] Gauss, 1832, pp. 540ff. and compare Kline, 1972, p. 817 and Bashmakova and Smirnova, 2000, pp. 132–133.

but also that it was possible to find the greatest common divisor of two Gaussian integers using an analogue of the Euclidean algorithm. For all intents and purposes, then, it was possible to establish an *arithmetic* for the Gaussian integers fully analogous to that for the rational integers $\mathbb{Z}$. That is, theorems that hold in $\mathbb{Z}$ like Fermat's "little theorem"—namely, if p is a prime rational integer and if a is an arbitrary integer relatively prime to p, then $a^{p-1} \equiv 1 \pmod{p}$—have natural analogues in the Gaussian integer setting.[16] As Gauss recognized, he had found "the natural source of a general theory" through an "extension of the domain of arithmetic."

Still, he only stated the law of biquadratic reciprocity toward the end of the second installment of his paper,[17] reserving the actual proof for what was to have been the paper's third part. When that follow-up failed to materialize, first Jacobi in the course of his lectures in Königsberg in 1836–1837 and then Gauss's student, Gotthold Eisenstein, in print in 1844, proved the result.[18] All three mathematicians, Gauss, Jacobi, and Eisenstein, also worked on and proved a law of cubic reciprocity, with Gauss, in particular, conceiving of it in terms of "numbers," this time of the form $a + b\rho$, where $a, b \in \mathbb{Z}$ and $\rho \neq 1$ is a cube root of unity, as defined above.

NEW ARITHMETICS FOR NEW "COMPLEX" NUMBERS

In the case both of Euler's attempt to prove Fermat's last theorem for $n = 3$ and of Gauss's work on the law of biquadratic reciprocity, the issue of factorization—and so that of divisibility—was key. Over the course of the nineteenth century, these notions were developed further as mathematicians continued to push for a proof of Fermat's last theorem in particular cases as well as in general. Among those who participated in this effort were the largely self-taught French mathematician, Sophie Germain (1776–1831), who believed that she had proven the theorem in the early 1800s in the case of even exponents $2p$ for rational primes p

[16] In the context of the Gaussian integers, Fermat's little theorem states that if $p = a + bi$ is a Gaussian prime and if k is a Gaussian integer not divisible by p, then $k^{N(p)-1} \equiv 1 \pmod{p}$, where $N(p)$ is the norm of p. Compare Gauss, 1832, p. 560. For the quotation that follows, see p. 540 (our translation).

[17] Gauss, 1832, p. 576.

[18] See Kline, 1972, p. 818 and Eisenstein, 1844d.

Figure 13.1. Sophie Germain (1776–1831).

of the form $8m \pm 3$ ($m \in \mathbb{Z}^+$); Adrien-Marie Legendre and Peter Lejeune Dirichlet, who proved it in the $n = 5$ case in 1825; Dirichlet, who succeeded in a proof of the $n = 14$ case seven years later in 1832; and Gabriel Lamé (1795–1870), who cracked the $n = 7$ case. Then, on 1 March 1847, Lamé famously and erroneously claimed on the very public floor of the Paris Academy of Sciences to have a proof of the theorem in general.[19]

The idea that Lamé propounded may be summed up this way. For p an odd rational prime, factor the polynomial $x^p + y^p \in \mathbb{C}[x, y]$ over the complex numbers. Clearly,

$$\left(\frac{x}{y}\right)^p - 1 = \prod_{i=0}^{p-1} \left(\frac{x}{y} - \alpha^i\right), \tag{13.2}$$

for $\alpha = e^{\frac{2\pi i}{p}}$, a primitive pth root of unity. Replacing $\frac{x}{y}$ by $-\frac{x}{y}$ in (13.2), making the necessary adjustments of signs, and multiplying through both

[19] Edwards, 1977, pp. 59–80. The account below of Lamé's work and the reaction to it draws from pp. 76–80. For more on Sophie Germain's work toward a proof of Fermat's last theorem, in particular, see Laubenbacher and Pengelley, 2010.

sides of the resulting equation by y^p then gives the factorization

$$x^p + y^p = (x + y)(x + \alpha y)(x + \alpha^2 y) \cdots (x + \alpha^{p-1} y). \qquad (13.3)$$

Two cases next presented themselves. First, if, for integer values x and y, the factors on the right-hand side of (13.3) are all relatively prime, then, Lamé contended, if $x^p + y^p = z^p$, for $z \in \mathbb{Z}$, each of the factors is a pth power in $\mathbb{Z}[\alpha]$. From there, just like Euler in the $n = 3$ case, he could generate the infinite descent argument which would demonstrate unsolvability. If, however, the factors are *not* all relatively prime, then he would find their common divisor, divide them all through by it so that they *are* relatively prime, and proceed as in the first case. The problem with this strategy, as Joseph Liouville noted from the floor of the Académie during the discussion of Lamé's presentation, was that it would only work if one knew that complex "numbers" of the form $x + \alpha^i y$, for $x, y \in \mathbb{Z}$—or, more generally, of the form

$$f(\alpha) := a_0 + a_1 \alpha + a_2 \alpha^2 + \cdots + a_{p-1} \alpha^{p-1}, \qquad (13.4)$$

for α a (primitive) pth root of unity (p still a prime) and $a_i \in \mathbb{Z}$, that is, the so-called *cyclotomic integers* denoted $\mathbb{Z}(\alpha)$—can be uniquely factored into a product of primes. In other words, this would only work if the fundamental theorem of arithmetic held in this *new* and more general number system. (Liouville had put his finger on exactly the same issue that had left a gap in Euler's argument regarding Fermat's last theorem in the $n = 3$ case.) In order to establish *that*—if, indeed, it were the case—one would need to develop a full *arithmetic* of cyclotomic integers, which would include defining notions like unit, prime, norm, irreducibility, etc. in this new number system.

Lamé's claim and Liouville's questioning of it sparked a period of high drama in France's highest scientific body that did not fail to capture the attention of mathematicians outside the French capital. On 24 May 1847, Liouville read a letter before the assembled *savants* that he had received unexpectedly from the German mathematician, Ernst Kummer. There, Kummer noted that he had (1) proven, in a paper published three years earlier, that complex "numbers" of precisely the type (13.4) that came up in Lamé's argument do *not*, in general, enjoy the property of unique factorization (it first fails when $p = 23$) and (2) shown, in an 1846 paper, how to salvage factorization by introducing yet another new kind of "number," a collection that he termed the "ideal complex numbers."[20] He had apparently come to these results through his work on higher reciprocity laws, but he also recognized their import in research toward a proof of Fermat's last theorem. Kummer's intervention effectively shifted the focal point of investigations of the latter open problem from France to Germany.

One of Kummer's key ideas was to distinguish between the notions of "irreducible" and "prime." A cyclotomic integer is *irreducible* if it cannot be factored into two other cyclotomic integers, neither of which is a unit,[21] whereas it is *prime* if, whenever it divides a product of two cyclotomic integers, then it must divide one of them. In the rational integers $\mathbb{Z}$, these two notions are equivalent, but, as Kummer realized in the context of his watershed example of the cyclotomic integers associated with $\alpha \neq 1$, a 23rd root of unity, although all cyclotomic prime integers are irreducible, there are irreducible cyclotomic integers that are not prime. Since reducibility—that is, factorability—is at the heart of the matter, Kummer needed to get a handle on when cyclotomic integers were factorable. Like Gauss, he did this via the notion of the norm.

Keeping the notation above, Kummer defined—in his work in the 1840s—the norm of a cyclotomic integer $f(\alpha)$ to be the product

$$N(f(\alpha)) := f(\alpha)f(\alpha^2)\cdots f(\alpha^{p-1}), \qquad (13.5)$$

where $f(\alpha^i)$ is the cyclotomic integer gotten by replacing α by α^i in (13.4). He then proved, as in the case of the Gaussian integers, that $N(f(\alpha)) \in \mathbb{Z}$ is always greater than or equal to zero, and, for two cyclotomic integers $f(\alpha)$ and $g(\alpha)$, $N(f(\alpha)g(\alpha)) = N(f(\alpha))N(g(\alpha))$.[22] Then, just as Gauss exploited the factorability of the norm in the Gaussian integers to determine when a Gaussian integer was composite or prime, so Kummer employed the factorability of the norm he defined for the cyclotomic integers to the same end: if the norm of a cyclotomic integer was prime, then so was the cyclotomic integer.

---

[20] See Kummer, 1844 and Kummer, 1846, respectively.
[21] A unit $g(\alpha) \in \mathbb{Z}[\alpha]$ is a number such that there exists an $h(\alpha) \in \mathbb{Z}[\alpha]$ with $g(\alpha)h(\alpha) = 1$.
[22] Kummer, 1844, p. 187.

As he highlighted in 1846, one remarkable thing about cyclotomic integers is that "even if $f(\alpha)$ cannot be decomposed in any way into complex factors, it therefore does not yet have the true nature of a complex prime number, because it lacks the usual, first and most important property of prime numbers: namely, that the product of two prime numbers is not divisible by any prime number different from the two." "Many such numbers $f(\alpha)$," Kummer continued, "even if they are not divisible into complex factors, still have the nature of composite numbers; the factors, however, are then not true but *ideal complex numbers*."[23] He thus had to determine how the cyclotomic integers decomposed into numbers of this—yet another new—*ideal* kind.

His strategy? It was exactly the same as Euler's; factor the "integer" by going beyond it into some larger mathematical realm. But what does "factorability" mean there? This was precisely the question that Kummer explored in two papers, one "Zur Theorie der complexen Zahlen" (or "On the Theory of Complex Numbers") and published initially in 1846, and the other "Über die Zerlegung der aus Wurzeln der Einheit gebildeten complexen Zahlen in ihre Primfactoren" (or "On the Decomposition into Prime Factors of Complex Numbers Formed from Roots of Unity") published a year later.[24] While the details behind Kummer's methods are involved,[25] it is nevertheless possible to get a sense of them through a brief consideration of his analysis of the $p = 23$ case, that is, the smallest case in which unique factorization fails.[26]

Let $\alpha \neq 1$ be a 23rd root of unity, and consider the collection $f(\alpha) \in \mathbb{Z}[\alpha]$ of cyclotomic integers with respect to α. Kummer's machinery allowed him to determine that for no $f(\alpha) \in \mathbb{Z}[\alpha]$ is it the case that $N(f(\alpha)) = 47$ or 139. In fact, neither 47 nor 139 has prime factors in $\mathbb{Z}[\alpha]$. However, the product $47 \cdot 139$ *is* the norm of $g(\alpha) := 1 - \alpha + \alpha^{21} \in \mathbb{Z}[\alpha]$. These facts about 47 and 139 considered as elements first in $\mathbb{Z}[\alpha]$ and then as elements in $\mathbb{Z}$—together with the usual properties of divisibility in $\mathbb{Z}$—allowed Kummer to conclude that $g(\alpha)$ is irreducible but not prime in $\mathbb{Z}[\alpha]$, since $N(g(\alpha))$ can be written in at least two ways as the product of irreducibles in $\mathbb{Z}[\alpha]$. Specifically, one factorization is in terms of

[23] Kummer, 1846, p. 319 (our translation; his emphasis).
[24] Kummer, 1846 and Kummer, 1847.
[25] See Edwards, 1977, pp. 76–151 for an extensive treatment and explanation.
[26] For an exposition of his proof of the failure of unique factorization when $p = 23$, see Edwards, 1977, pp. 104–106. The next paragraph captures the highlights of that proof.

twenty-two irreducible factors each of norm $47 \cdot 139$, and the other is in terms of eleven irreducible factors with norm 47^2 and eleven irreducible factors with norm 139^2.

But what about factorization in terms of *ideal* prime factors? Given the factorization, on the one hand (call it the left-hand side), into twenty-two irreducible factors each of norm $47 \cdot 139$ and, on the other (call it the right-hand side), in terms of eleven irreducible factors with norm 47^2 and eleven irreducible factors with norm 139^2, it should be the case that each of the twenty-two irreducible factors on the left should be divisible by two *ideal prime* factors on the right, one a factor of 47 and the other a factor of 139. What do those ideal prime factors look like, then? Kummer defined them, not explicitly, but rather in terms of divisibility.

Suppose $\mathfrak{p}$ is the *ideal* prime factor of 47 that divides both $g(\alpha)$ and the product ϕ of the twenty-one conjugates of $g(\alpha)$, namely, $\prod_{i=2}^{22} g(\alpha^i) = \frac{N(g(\alpha))}{g(\alpha)}$ (compare equation (13.5) above). Then, ϕ will be divisible by all of the ideal prime factors of 47 except $\mathfrak{p}$, and $\psi\phi$ will be divisible by 47 if and only if ψ is divisible by $\mathfrak{p}$. In other words, and this was Kummer's definition, an integer $\psi \in \mathbb{Z}[\alpha]$ is *divisible* by the ideal prime factor $\mathfrak{p}$ if $\psi\phi$ is divisible by 47, and similarly, ψ is divisible m times by $\mathfrak{p}$ if $\psi\phi^m$ is divisible by 47^m. The ideal prime factor is thus not something that is written down explicitly but rather something that is detected via its behavior in the context of divisibility relative to a particular prime. As historian of mathematics Harold Edwards put it, "the kernel of Kummer's theory of ideal complex numbers is the simple observation that the test for divisibility by the hypothetical factor of 47 ... is perfectly meaningful even though there is no actual factor of 47 for which it tests. One can choose to regard it as a test for divisibility by an *ideal* prime factor of 47 and this, in a nutshell, is the idea behind Kummer's theory."[27]

Here, we have considered this definition only in the particular context of $\mathbb{Z}[\alpha]$ for $\alpha \neq 1$, a 23rd root of unity. In the late 1840s and early 1850s, however, Kummer also applied it in the setting of any arbitrary $\mathbb{Z}[\zeta]$, where $\zeta \neq 1$ is a pth root of unity, by interpreting, for a given prime p, an arbitrary ideal complex number to be the formal product of its prime ideal complex numbers. In so doing, he was able to prove Fermat's last theorem for all primes $p < 100$ with the exceptions of $p = 37, 59$,

---

[27] Edwards, 1977, p. 106 (his emphasis).

and 67.[28] His methods—because they were dependent on the prime, that is, because they determined ideal complex prime factors a prime at a time—were thus termed "local."

Now, in the setting of the "usual" integers, $\mathbb{Z}$ is the *ring* of integers in the *field* $\mathbb{Q}$, where, loosely speaking, a ring is a domain in which addition (and so subtraction) as well as multiplication—with all of their "usual" properties—hold, and a field is a domain in which addition, subtraction, multiplication, and division—again, with all of their "usual" properties—make sense.[29] Similarly, the ring of cyclotomic integers $\mathbb{Z}[\zeta]$ is the ring of algebraic integers (see (13.6) below) in the field $\mathbb{Q}(\zeta)$, called the cyclotomic field. As we have just seen, this setting is special, in that the root of unity ζ has very particular properties associated with the fact that ζ ($\neq 1$) satisfies the equation $0 = x^{p-1} + x^{p-2} + \cdots + x^2 + x + 1$ over $\mathbb{C}$. But, what if the situation were more general? That is, can this notion of integers in a field be interpreted in less or differently structured settings?

At essentially the same time that Kummer was doing his work on the cyclotomic integers, Gotthold Eisenstein, Peter Lejeune Dirichlet, and Charles Hermite also came to define a new type of "integer"—of which Kummer's cyclotomic integers were a special case—during the course of their respective number-theoretic research. They considered monic polynomial equations of the form

$$f(x) = x^n + a_1 x^{n-1} + \cdots + a_{n-1} x + a_n = 0, \tag{13.6}$$

with coefficients $a_i \in \mathbb{Z}$, and they defined the *algebraic integers* to be the roots of such equations. As Eisenstein proved in 1850, the collection of these new, even more general "numbers" is closed under both addition and multiplication,[30] that is, if η and θ are algebraic integers, then so are $\eta + \theta$ and $\eta\theta$. By 1871 and in the tenth supplement to his second edition of Dirichlet's *Vorlesungen über Zahlentheorie* (or *Lectures on Number Theory*), Richard Dedekind had also arrived at the algebraic integers, but through his efforts (1) to generalize Kummer's results in the context of cyclotomic fields to arbitrary algebraic number fields $\mathbb{Q}(\theta)$, where θ is a root of an

---

[28] See Edwards, 1977, pp. 181–244. On Kummer's looseness relative to an actual definition of ideal complex numbers, see Edwards, 1980, p. 342.

[29] For more on these concepts, see chapter 14. For a brief overview of the development of ring theory, see Kleiner, 1998.

[30] Eisenstein, 1850, pp. 236–237.

equation of the form (13.6) but with *rational* coefficients and (2) to set those results on a firmer mathematical foundation.

In all, Dedekind gave four different versions of his ideas. The first, in 1871, represented his most direct response to Kummer's work and presented, as Dedekind himself expressed it, Kummer's approach "clothed in a different garment."[31] The second, which appeared serially in French in the *Bulletin des sciences mathématiques et astronomiques* (or *Bulletin of the Mathematical and Astronomical Sciences*) in 1876 and 1877 and then in book form a year later as *Sur la théorie des nombres entiers algébriques* (or *On the Theory of Algebraic Numbers*), aimed at introducing what he thought were his neglected ideas to a broader and more general mathematical audience. It also presented those ideas with fundamentally different, less Kummerian, emphases.[32] Finally, the third and fourth versions supplemented the third (1879) and fourth (1894) editions that Dedekind published of Dirichlet's lectures on number theory and continued to reflect the new orientation of the second version. Out of what sort of cloth was this new, Dedekindean garment cut?

In his efforts to generalize Kummer's work on cyclotomic integers in a cyclotomic field to the setting of algebraic integers in an algebraic number field, Dedekind encountered fundamental stumbling blocks, one of which was the very nature of the "integers" in an algebraic number field. First of all, Dedekind defined his terms:[33] "A number θ is called an *algebraic number* if it satisfies an equation

$$\theta^n + a_1\theta^{n-1} + a_2\theta^{n-2} + \cdots + a_{n-1}\theta + a_n = 0$$

with finite degree n and rational coefficients $a_1, a_2, \ldots, a_{n-1}, a_n$. It is called an *algebraic integer*, or simply an *integer*, when it satisfies an equation of the form above in which all the coefficients $a_1, a_2, \ldots, a_{n-1}, a_n$ are rational integers" in $\mathbb{Z}$. This definition of an algebraic integer was just the same as the one adopted by Eisenstein and others, although Dedekind very self-consciously understood it within the broader setting of an algebraic number field, proving, for example, that the sum and product of two algebraic numbers was an algebraic number. But what

---

[31] Dirichlet, 1871, pp. 380–497 and compare Edwards, 1980, pp. 342–349. For the quote, see Dirichlet, 1871, p. 451 (our translation) and compare Edwards, 1980, p. 351.
[32] Dedekind, 1877/1996.
[33] Dedekind, 1877/1996, pp. 53–54 (Stillwell's translation).

about the integers in this kind of field? As we have seen, the cyclotomic integers are all of the form $a_0 + a_1\alpha + a_2\alpha^2 + \cdots + a_{p-1}\alpha^{p-1}$, for $a_i \in \mathbb{Z}$, $\alpha \neq 1$ a pth root of unity, and $p \in \mathbb{Z}$ a prime. Moreover, by definition, α satisfies the equation $x^p - 1 = 0$, and so is an algebraic integer. Thus, *cyclotomic* integers are *algebraic* integers. The "integers" in an *arbitrary* algebraic number field, however, are not quite so "nice." For example, $\frac{1}{2} + \frac{\sqrt{3}}{2}i$ is a root of the monic polynomial equation $x^2 - x + 1 = 0$, and so, by definition, is an *algebraic* integer. Getting a handle on just what the algebraic integers look like, then, was essential in order to generalize Kummer's theory, as was developing an arithmetic of them.

Critical to the latter goal was coming up with a meaningful analogue of Kummer's ideal prime factors (and his notion of ideal complex numbers) in the more general context of algebraic number fields and thereby understanding a concept of divisibility there. Dedekind's resolution of this problem focused on what he deemed Kummer's "entirely legitimate"—but extrinsic and hence, to Dedekind's lights, philosophically unsatisfactory—definition of an ideal prime factor in terms of a divisibility test not in general but based on the selection of a particular number (47 in the definition as given above).[34] Dedekind sought an intrinsic definition and, in formulating it, he introduced not only a fresh way of approaching algebraic questions but also a number of what would prove to be far-reaching concepts.

Dedekind opened the general exposition of his new ideas in section 159 of the tenth supplement of 1871. As he put it, 'In that we are trying to introduce the reader to these new ideas, we position ourselves at a somewhat higher standpoint and begin therefrom to introduce a concept which seems better suited to serve as a foundation for the higher algebra and for the parts of number theory connected with it."[35] That "higher standpoint" hinged on the notion of a *field*—or *Körper* in Dedekind's terminology—namely, the concept that he defined as a "system of infinitely many real or complex numbers, which is so closed and complete that the addition, subtraction, multiplication, and division of any two of these numbers always yields a number of the same system." This was precisely the notion that had arisen in another guise

---

[34] See the discussion in Edwards, 1980, pp. 342–343. For the quote, see Dedekind, 1877/1996, p. 57 (Stillwell's translation)

[35] Dirichlet, 1871, p. 424 (our translation). The next quotation is also from this page (our translation).

in the work of Galois (recall the discussion in chapter 11). Although this definition is cast specifically in terms of the real and complex numbers, its insistence on the four binary operations and closure under them— as well as Dedekind's formulation of the notions of subfield, basis, and degree—foreshadowed not only a fuller axiomatic definition of a field but also the components critical to field theory in its twentieth-century development at the hands of Ernst Steinitz (1871–1928) and others (see the next chapter). Dedekind proceeded to show that the collection of all algebraic numbers is, in fact, a field, and he linked his notion of a field with that of Galois.[36] It was within this concept of an algebraic number field that Dedekind explored the properties of the collection of algebraic integers and, in particular, their arithmetic.

As in the case of Kummer's analysis of the cyclotomic integers, Dedekind needed to define the "usual" arithmetic concepts. In his setup, "an algebraic integer α is said to be *divisible* by an algebraic integer $\beta, \ldots$ if the quotient $\frac{\alpha}{\beta}$ is also an algebraic integer."[37] In other words (and this is how Dedekind put it in 1877), the algebraic integer "α will be said to be *divisible* by an [algebraic] integer β if $\alpha = \beta\gamma$, where γ is likewise an [algebraic] integer."[38] Since Dedekind wanted to understand divisibility in as general and as meaningful a context as possible, he recognized that he could restrict his analysis to those algebraic integers that satisfy an *irreducible* polynomial equation of degree n with rational integer coefficients. That would allow him best to define the concepts of irreducibility and prime algebraic integer.[39]

Having established all of these analogues, Dedekind next defined two new structures in terms of which he could probe the algebraic integers. The first was the notion of a *module*, which he defined (unlike our modern definition) as "a system $\mathfrak{a}$ of real or complex numbers α, whose sums and differences also belong to the system $\mathfrak{a}$,"[40] while the second was that of an *ideal*. Considering $\mathfrak{d}$, the algebraic integers inside the field of algebraic numbers (he termed $\mathfrak{d}$ an *Ordnung* or *order*),[41] Dedekind defined an ideal to be "a system $\mathfrak{a}$ of infinitely many elements of $\mathfrak{d} \ldots$ if it satisfies both of

[36] Dirichlet, 1871, pp. 436–438 and 428, respectively.
[37] Dirichlet, 1871, p. 437 (his emphasis; our translation).
[38] Dedekind, 1877/1996, p. 54.
[39] Dirichlet, 1871, pp. 438–442.
[40] Dirichlet, 1871, p. 442 (our translation).
[41] Dirichlet, 1871, p. 471.

these conditions:

> I. The sum and difference of every two numbers in $\mathfrak{a}$ is also a
> number in $\mathfrak{a}$ [in other words, it is a module in Dedekind's
> sense].
> II. Every product of a number in $\mathfrak{a}$ and a number in $\mathfrak{d}$ is also a
> number in $\mathfrak{a}$.[42]

Dedekind proceeded to sketch a theory of ideals—that is, a theory of these infinite *sets* of algebraic integers—in which he interpreted definitions of "divisibility" set-theoretically.

Consider two ideals $\mathfrak{a}$ and $\mathfrak{b}$ of algebraic integers. For Dedekind, an algebraic integer α is *divisible by an ideal* $\mathfrak{a}$, if $\alpha \in \mathfrak{a}$; if $\mathfrak{a} \subset \mathfrak{b}$, then $\mathfrak{a}$ is divisible by $\mathfrak{b}$; the intersection $\mathfrak{a} \cap \mathfrak{b}$ is the *least common multiple* of $\mathfrak{a}$ and $\mathfrak{b}$; the *greatest common divisor* of $\mathfrak{a}$ and $\mathfrak{b}$ is the set of algebraic integers $\alpha + \beta$, where $\alpha \in \mathfrak{a}$ and $\beta \in \mathfrak{b}$; $\mathfrak{a}$ is a prime ideal if its only divisors are itself and $\mathfrak{d}$, that is, the full ring of algebraic integers; for a *prime ideal* $\mathfrak{a}$, if $\alpha\beta \equiv 0 \pmod{\mathfrak{a}}$, then at least one of the algebraic integers α or β is divisible by $\mathfrak{a}$; and, finally, the *product* of two ideals $\mathfrak{a}$ and $\mathfrak{b}$ is the set $\mathfrak{a}\mathfrak{b} := \{\sum_{i=1}^{m} \alpha_i \beta_i,$ where $\alpha_i \in \mathfrak{a}$ and $\beta_i \in \mathfrak{b}$, for $1 \leq i \leq m\}$.[43] The full elaboration of this arithmetic of ideals allowed Dedekind to provide the intrinsic and explicit description of the prime factors in the algebraic integers by essentially translating the discussion into purely set-theoretic terms. In so doing, he replaced Kummer's "entirely legitimate" but philosophically problematic notion of ideal prime factors by equivalent and explicitly definable sets.[44] This represented a major shift in algebraic thinking that would increasingly come to hold sway and that would come to characterize the so-called *modern* algebra in the early twentieth century.[45]

---

[42] Dirichlet, 1871, p. 452.

[43] Dirichlet, 1871, pp. 452–462. The modern notion of a *Dedekind domain* would develop from this context. An *integral domain* is a commutative ring which has no zero divisors. A Dedekind domain is an integral domain in which every nonzero proper ideal factors into a product of prime ideals.

[44] Edwards, 1980, pp. 343–346. Six years later, when Dedekind published the second version of his theory, the notion of the product of two ideals had come to play an even more critical role. See Dedekind, 1877/1996 and the discussion in Edwards, 1980, pp. 350–352.

[45] It should be noted here that both the Russian mathematician, Egor Zolotarev (1847–1878) and Kummer's student, Leopold Kronecker, extended Kummer's work on divisibility to the context of algebraic number fields. Their methods, although important, ultimately proved to

By 1882, Dedekind and his collaborator, Heinrich Weber, had also succeeded in transferring many of these ideas into the context of fields of algebraic *functions* and thereby in setting the theory of Riemann surfaces on a new, *algebraic* geometric course. This intersection of research in number theory and in geometry, in fact, marked a key turning point in the history of the latter field. In the words of the French mathematician, Jean Dieudonné, "Dedekind and Weber propose[d] to give algebraic proofs of all of Riemann's algebraic theorems. But their remarkable originality (which in all the history of algebraic geometry is only scarcely surpassed by that of Riemann) leads them to introduce a series of ideas that will become fundamental in the modern era."[46] Those ideas were reflective of the structural algebra that arose especially in the final decade of the nineteenth century and the opening decades of the twentieth (see the next chapter) and that helped to steer algebraic geometry onto a new algebraic course, especially under the guidance of Baertel van der Waerden (1903–1996), André Weil (1906–1998), and Oscar Zariski (1899–1986) in the 1930s and 1940s.[47]

WHAT IS ALGEBRA?: THE BRITISH DEBATE

While continental mathematicians, especially in the 1840s, explored the implications of both Fermat's famous claim and Gauss's seminal work, mathematicians in the British Isles engaged in a different set of questions that affected the development of nineteenth-century algebraic thought. From the beginning of the eighteenth century and in the wake of the infamous Newton–Leibniz controversy over priority relative to the conception of the calculus, British mathematicians had doggedly adhered to Newton's geometrical approach to the mathematical underpinnings of Newtonian physics and had trained generations of British students in that tradition, particularly at Cambridge. Because the geometry

---

be less far reaching than those of Dedekind. On Kronecker's work, in particular, see Edwards, 1980, pp. 353–360. Neumann, 2007, pp. 84–87 gives a brief treatment of some of Zolotarev's ideas.

[46] Dieudonné, 1985, p. 29, as quoted in Stillwell, 2012, p. 27. A discussion of these developments would take us too far afield, but this, Stillwell's introduction to Dedekind and Weber 1882/2012, gives a nice historical account of the work of Dedekind and Weber in context.

[47] See Schappacher, 2007 and Slembek, 2007 for more on these developments as well as Dieudonné, 1985 for a full historical overview.

of three-dimensional space set the standard, the British—much like Cardano—were loath to accept mathematical concepts—such as negative and imaginary numbers—that ran counter to geometric intuition and interpretation. They were thus loath to accept the analytic (that is, algebraic) techniques that were being developed on the continent by mathematicians like Euler and Lagrange.

Indicative of this skepticism, Frances Maseres (1731–1824), a prominent civil servant and Fellow of Clare Hall, Cambridge, published in 1758 a *Dissertation on the Use of the Negative Sign in Algebra* with the revealing subtitle, *Containing a Demonstration of the Rules Usually Given Concerning It; And Shewing How Quadratic and Cubic Equations May Be Explained, without the Consideration of Negative Roots*. For Maseres, while the negative sign was legitimate, it was not so, as he put it, "in any other light than as the mark of the subtraction of a lesser quantity from a greater."[48] He thus aimed in his book to show how to effect the solutions of quadratic and cubic equations through an examination of each possible configuration of + and − terms.

To get the flavor of his thinking, consider his treatment of the quadratic equations. "Now, all affected quadratic equations are evidently reducible to one of these three forms;

$$xx + px = r$$

$$xx - px = r \quad \text{and}$$

$$px - xx = r;$$

in all of which, x denotes the unknown, and p and r the known quantities. . . . The first and second of these equations are always possible, whatever be the magnitude of r; the last, only when r is not greater than $\frac{pp}{4}$."[49] For Maseres, it only made sense to subtract a given number from a number of magnitude greater than it, and it did not make sense to consider, say, −5 divorced from some number from which it was to be subtracted.

Maseres's younger contemporary, William Frend (1757–1841), continued this battle against the negative numbers in his 1796 textbook, *The Principles of Algebra*. There, Frend denounced those algebraists "who

---

[48] Maseres, 1758, p. i.
[49] Maseres, 1758, p. 20.

talk of a number less than nothing, of multiplying a negative number into a negative number and thus producing a positive number, of a number being imaginary. Hence. . . they talk of solving an equation, which requires two impossible roots to make it solvable: they can find out some impossible numbers, which, being multiplied together produce unity. This is all jargon, at which common sense recoils."[50] Fundamentally at issue here was the question, what is algebra?, and while Maseres and Frend agreed that it had to conform to "common sense," others were prepared to take a different philosophical stance.[51]

At Cambridge in 1812, a group of students spearheaded by John Herschel (1792–1871), Charles Babbage (1791–1871), and George Peacock (1791–1858) founded the so-called Cambridge Analytical Society in an effort, once and for all, to replace what they viewed as the antiquated geometrical slant of the post-Newtonian era in Britain with the more modern analytical techniques issuing from the continent. Although the society itself was short-lived, its members went on to translate Sylvestre Lacroix's *Traité élémentaire de calcul différentiel et de calcul intégral* of 1802 into English (as *An Elementary Treatise on the Differential and Integral Calculus*) in 1816 and to publish *A Collection of Examples of the Applications of the Differential and Integral Calculus* in 1820 so as to introduce the continental approach to the Cambridge student body.[52] By the 1820s and 1830s, as these same students, now graduates, established themselves within the Cambridge educational sphere, they pressed for a modernization and reorientation in an analytic direction of the Cambridge Tripos examination, the exit examination in mathematics required of all who hoped to attain a Cambridge degree with honors.[53]

In particular, Peacock, a Fellow and Tutor at Trinity College, Cambridge, published a *Treatise on Algebra* in 1830, in which he sought better to understand what algebra was and what it could be expected to achieve. His analysis turned on the distinction that he drew between what he termed "arithmetical" and "symbolical" algebra. For him, arithmetical algebra—essentially the rules of arithmetic of the nonnegative integers—was "the science, whose operations and the general consequences of

---

[50] Frend, 1796, pp. x–xi.
[51] For more on Frend's views and his place in the nineteenth-century debate on the nature of algebra, see Pycior, 1982.
[52] Lacroix, 1802/1816 and Peacock, 1820.
[53] On these developments, see Enros, 1983 and Fisch, 1994, among many other sources.

them should serve as the guides to the assumptions which become the foundation of symbolical algebra."[54] The arithmetic of the positive whole numbers thus determined the rules of arithmetical algebra, and arithmetical algebra, in turn, fundamentally informed the behavior of symbolical algebra. As Peacock articulated it a couple of years later in 1834, "For in as much as symbolical algebra though arbitrary in the authority of its principles, is not arbitrary in their application, being required to include arithmetical algebra as well as other sciences, it is evident that their rules must be identical with each other, as far as those sciences proceed together in common."[55] Peacock termed this the "principle of the permanence of equivalent forms," and in light of it, he conceived of symbolical algebra as a generalization of both the symbols and the operations of arithmetical algebra.

The publication of Peacock's *Treatise* did not go unnoticed by his mathematical contemporaries. From his home at the Dunsink Observatory outside Dublin, William Rowan Hamilton (1805–1865) entered into the algebraic debate in November of 1833 when he read the opening sections of his "Theory of Conjugate Functions, or Algebraic Couples; With a Preliminary and Elementary Essay on Algebra as the Science of Pure Time" at a meeting of the Royal Irish Academy. There, in addition to the representation of the complex numbers as what he termed "couples" (or what we would call ordered pairs, see below), Hamilton laid out his philosophical views on algebra (which had been partially inspired by his reading of Immanuel Kant's *Critique of Pure Reason*) as well as his thoughts on the ramifications of that philosophy. To his way of thinking, "the study of algebra may be pursued in three very different schools, the Practical, the Philological or the Theoretical, according as Algebra itself is accounted as an Instrument, or a Language, or a Contemplation."[56] "The Practical person," he continued, "seeks a Rule which he may apply, the Philological person seeks a Formula which he may write, the Theoretical person seeks a Theorem on which he may meditate." Hamilton placed himself in the "Theoretical" camp. He sought to make algebra a science based on proven theorems and clearly thought that was the proper way

---

[54] Peacock, 1830, p. xi, as quoted in Fisch, 1999, pp. 166–167. For what follows in this section, compare also Parshall, 2011, pp. 339–345.

[55] Peacock, 1834, p. 195 and Fisch, 1999, p. 168.

[56] Hamilton, 1831, p. 293. The next quotation is also on this page. For Hamilton's biography, see Hankins, 1980.

Figure 13.2. William Rowan Hamilton (1805–1865).

for algebra to develop and to be used. He regarded Peacock, on the other hand, as a "Practical" sort, who sought rules.

Augustus De Morgan (1806–1871), the professor of mathematics at University College London, had formulated yet another philosophical stance by the early 1840s. He used the forum of the Cambridge Philosophical Society to express his views "On the Foundations of Algebra" first in November of 1839 and then in a second installment in November of 1841. For him, algebra should be set up in terms both of the operations of addition, subtraction, multiplication, and division and of a system of axioms that these operations obey. Although he did not give an explicit formulation of the axiom that we would term "associativity," De Morgan's "algebra" otherwise satisfied all of the axioms that today define a field. Moreover, he held that "the first step to logical algebra is the separation of. . . its laws of operation from the explanation of the symbols operated upon or with. . . .The literal symbols a, b, c, &c. have no necessary relation except that whatever any one of them may mean in any one part of a process, it means the same in every other part of the same process." De Morgan's algebra, then, fully admitted both negative and imaginary numbers, since its symbols were a priori independent of any meaning other than what the axioms dictated. His symbols, unlike Peacock's, did not even necessarily stand for numbers, but rather could, in principle, represent anything as long as the axioms were correctly followed. Meaning and interpretation would follow later. By 1847, De Morgan had carried these ideas into the realm of logic, which he cast and analyzed in terms of algebraic symbols and operations.[57]

The axioms, however, while important for De Morgan, were actually of only secondary interest to him. They provided a certain structure, but algebra, and gaining algebraic insights, involved much more than axioms in his view. He captured his sense of what algebra is via the metaphor of the jigsaw puzzle. "A person who puts one of these together by the backs of the pieces, and thereby is guided only by their forms, and not by their meanings," he explained, "may be compared to one who makes the transformations of algebra by the defined laws of operation only: while one who looks at the fronts, and converts his general knowledge of the countries painted on them into one of a more particular kind by help of the forms of the pieces, more resembles the investigator and the mathematician."[58] De Morgan's algebra of the 1840s was thus a "symbolical algebra," although in a different sense from Peacock's, since its axiomatic structure made it formally independent of a principle of the permanence of forms.

This debate about the nature of algebra—as well as the reform of the Cambridge curriculum at the hands of Peacock, Babbage, and others—fundamentally informed the mathematical worldview of a cadre of Cambridge graduates in the 1830s and 1840s. Two of those graduates, Duncan Gregory (1813–1844) and Robert Ellis (1817–1859), founded the *Cambridge Mathematical Journal* in 1837 primarily as a publication venue for young and aspiring mathematicians interested in pursuing the mathematical ramifications of a symbolical algebra. They took as one of their points of departure an analysis of the operations involved in the calculus. The idea was to separate the symbols of operation from the operands and then to study and understand those symbols algebraically.

---

[57] See De Morgan, 1847. A discussion of the interplay between algebra and logic would take us too far afield. For a nice overview of the work of De Morgan, George Boole, and other British contributors, see Grattan-Guinness, 2011.

[58] De Morgan, 1842, pp. 289–290. De Morgan's views on algebra changed over time. For an analysis of his evolving thought, see Pycior, 1983.

Manipulation and simplification of the symbols then allowed for simpler solutions to specific analytic problems.

Consider, for example, Duncan Gregory's sense of the separation of symbols in the context of the solution of simultaneous ordinary differential equations in a paper in the first volume of the *Cambridge Mathematical Journal* in 1837–1839.[59] "Let us take," he wrote,

$$\frac{dx}{dt} + ay = 0, \qquad \frac{dy}{dt} + bx = 0.$$

We have to separate $\frac{d}{dt}$ from the variable, and eliminate one of the variables y or x, as would be done if $\frac{d}{dt}$ were an ordinary constant. This will be done if we multiply the first equation by $\frac{d}{dt}$ and the second by a, and subtract, then we obtain

$$\left(\frac{d^2}{dt^2} - ab \right) x = 0. \ldots$$

Having now eliminated y, we may integrate the equation in x at once. The result is $x = c_1 e^{(ab)^{\frac{1}{2}}t} + c_2 e^{-(ab)^{\frac{1}{2}}t}$.

Although hardly earth-shattering, these sorts of techniques generated a new area of mathematics—the calculus of operations—that became a cottage industry of British mathematics in the 1840s and 1850s.[60]

Much more significant was the work in an algebraic vein that William Rowan Hamilton presented in the context of his philosophical thoughts on algebra as the science of pure time. At issue were the complex numbers and their representation in the form $a + bi$, for a, $b \in \mathbb{R}$. In Hamilton's view, this addition of a real and an imaginary part was tantamount to adding apples and oranges and, so, was philosophically unsatisfactory. By 1831, he had hit upon a representation that he found not only more philosophically pleasing but also suggestive of further research.

Writing the complex number $a + bi$ as the ordered pair—he called it the "algebraic couple" or "number-couple"—(a, b), Hamilton identified the complex numbers with points in the real plane, defined the usual

[59] Gregory, 1837–1839, p. 191.
[60] For more on this British work, see Koppelman, 1971–1972.

operations of addition, subtraction, multiplication, and division in terms of this notation, and proceeded to explore the algebraic properties of the complex numbers in this new form. As he put it, "it is easy to see the reasonableness of the following definitions, and even their necessity:"

$$(b_1, b_2) + (a_1, a_2) = (b_1 + a_1, b_2 + a_2);$$

$$(b_1, b_2) - (a_1, a_2) = (b_1 - a_1, b_2 - a_2);$$

$$(b_1, b_2)(a_1, a_2) = \cdots = (b_1 a_1 - b_2 a_2, b_2 a_1 + b_1 a_2);$$

$$\frac{(b_1, b_2)}{(a_1, a_2)} = \left(\frac{b_1 a_1 + b_2 a_2}{a_1{}^2 + a_2{}^2}, \frac{b_2 a_1 - b_1 a_2}{a_1{}^2 + a_2{}^2} \right).$$

Were these definitions even altogether arbitrary, they would at least not contradict each other, nor the earlier principles of Algebra, and it would be possible to draw legitimate conclusions, by rigorous mathematical reasoning, from premises thus arbitrarily assumed: but... these definitions are really *not arbitrarily chosen*, and... though others might have been assumed, no others would be equally proper.[61]

If points in the real plane thus corresponded to ordered pairs with "proper" definitions of the usual binary operations, then could the same thing be effected for three-space? In other words, was it possible to represent points in three-space by ordered triples with suitably defined binary operations and thereby to analyze space in a new algebraic way? This question perplexed Hamilton for a decade, as he sought in vain to define what—in terminology that would be developed only later—would be called a three-dimensional (division) *algebra* over the real numbers (see below).

In 1843, Hamilton realized that while he could not devise a suitable system of triples, he *could* do so for 4-tuples, that is, "numbers"—he called them *quaternions*—of the form

$$(w, x, y, z) := w + xi + yj + zk, \tag{13.7}$$

---

[61] Hamilton, 1831, p. 403 (his emphasis).

where w, x, y, $z \in \mathbb{R}$ and where i, j, and k satisfy the relations

$$ij = -ji = k, \ \ jk = -kj = i, \ \ ki = -ik = j, \ \text{and } i^2 = j^2 = k^2 = -1.$$
(13.8)

As Hamilton noted,[62] addition and subtraction were componentwise, that is, for two quaternions $Q = (w, x, y, z)$ and $Q' = (w', x', y', z')$, we have

$$Q \pm Q' = (w \pm w') + (x \pm x')i + (y \pm y')j + (z \pm z')k,$$

while multiplication is defined—using the relations in (13.8)—as

$$QQ' = (ww' - xx' - yy' - zz') + (wx' + xw' + yz' - zy')i$$
$$+ (wy' + yw' + zx' - xz')j + (wz' + zw' + xy' - yx')k.$$

Here, the first three relations in (13.8) indicate that multiplication in the quaternions is *not* commutative, while the fourth shows the analogy between the quaternions and the "usual" complex numbers, in which $i^2 = -1$. Hamilton acknowledged that "it must, at first sight, seem strange and almost unallowable, to define that the product of two imaginary factors in one order differs (in sign) from the product of the factors in the opposite order $(ji = -ij)$."[63] Yet, he continued, "it will, ... it is hoped, be allowed, that in entering on the discussion of a new system of imaginaries, it may be found necessary or convenient to surrender *some* of the expectations suggested by the previous study of products of real quantities." "Whether the choice of the system of definitional equations... has been a judicious, or at least a happy one," he concluded, "will probably be judged by... trying whether those equations conduct to results of sufficient consistency and elegance."

Hamilton had *created* a "new system of imaginaries" and with them operational rules that no longer satisfied all of the "usual" arithmetic properties. While he was willing to "surrender" commutativity, his friend, John Graves (1806–1870), was nervous. In a letter to Hamilton on 31 October 1843, Graves allowed that "I have not yet a clear view as to

---

[62] See Hamilton, 1844, pp. 11–12. This was just the first of numerous installments of this paper on quaternions by Hamilton in this journal.

[63] Hamilton, 1844, p. 11. The quotes that follow in this paragraph are also on this page (his emphasis).

the extent to which we are at liberty arbitrarily to create imaginaries, and to endow them with supernatural properties."[64] Graves and Arthur Cayley had both nevertheless exercised that freedom almost immediately on learning of Hamilton's discovery. They independently discovered the *octonions*—a system of 8-tuples with real entries defined analogously to the quaternions—the multiplication of which is both noncommutative and, as Hamilton later discovered, nonassociative.[65]

Hamilton pushed this line of thought even further in 1844. He defined the *biquaternions* (we would call them the "complex quaternions") as in (13.7) and (13.8), except that w, x, y, $z \in \mathbb{C}$ rather than $\mathbb{R}$. In so doing, he had to "surrender" even more than commutativity. Consider the biquaternions

$$(-\sqrt{-1}, 1, 0, 0) = -\sqrt{-1} + i \quad \text{and} \quad (\sqrt{-1}, 1, 0, 0) = \sqrt{-1} + i.$$

A quick check confirms that $(-\sqrt{-1}, 1, 0, 0)(\sqrt{-1}, 1, 0, 0) = (-\sqrt{-1} + i)(\sqrt{-1} + i) = 0$. The biquaternions thus contain zero divisors, that is, nonzero "numbers," the product of which is zero.[66] In Britain, De Morgan and various of his contemporaries—among them, Thomas Kirkman (1806–1895), James Cockle (1818–1895), and John Graves's brother, Charles (1812–1899)—pursued other new algebraic structures and their properties throughout the 1840s inspired by Hamilton's path-breaking willingness to open up and explore new mathematical terrains.[67]

AN "ALGEBRA" OF VECTORS

Hamilton's discovery of the quaternions did more than spark a purely algebraic search for new types of "number" systems. It also suggested new possibilities for representing physical properties analytically, that is, algebraically. Hamilton sensed these possibilities almost immediately. Writing in one of his personal journals on the day in 1843 when he

---

[64] As quoted in Hankins, 1980, p. 300.

[65] Graves laid out his discovery in a letter to Hamilton on 26 December 1843. Cayley published his result in Cayley, 1845b.

[66] See Hamilton, 1853, pp. 633ff. for Hamilton's exploration of the biquaternions. Because they contain zero divisors, the biquaternions are not a division algebra.

[67] For further references, see Flood, Rice, and Wilson, 2011, p. 448 (note 20).

announced his discovery of the quaternions to his scientific colleagues at the Royal Irish Academy, he noted that relative to the quaternion $v + xi + yj + zk$, "xyz may determine *direction* and *intensity*; while v may determine the *quantity* of some agent such as electricity. x, y, z are *electrically polarized, v electrically unpolarized*....The Calculus of Quaternions may turn out to be a CALCULUS OF POLARITIES."[68]

Hamilton paved the way notationally for this kind of an interpretation by conceiving of his quaternions in two discrete parts: a real or *scalar* part and an imaginary or *vector* part. Although this addition of two dissimilar entities was precisely what he had found philosophically unsatisfactory about the complex numbers some twelve years earlier, Hamilton's philosophical stance had clearly changed by the early 1840s.[69] As he explained,

> The algebraically *real* part may receive... all values contained on the one *scale* of progression of number from negative to positive infinity; we shall call it therefore the *scalar part*, or simply the *scalar* of the quaternion, and shall form its symbol by prefixing, to the symbol of the quaternion, the characteristic Scal., or simply S., where no confusion seems likely to arise from using this last abbreviation. On the other hand, the algebraically *imaginary* part, being geometrically constructed by a straight line, or radius vector, which has, in general, for each determined quaternion, a determined length and determined direction in space, may be called the *vector part*, or simply the *vector* of the quaternion; and may be denoted by prefixing the characteristic Vect., or V.[70]

He continued by developing the notation explicitly, saying that "we may therefore say that *a quaternion is in general the sum of its own scalar and vector parts*, and may write

$$Q = \text{Scal}.Q + \text{Vect}.Q = S.Q + V.Q$$

---

[68] Graves, 1882–1889/1975, 2:439–440, as quoted in Crowe, 1967, p. 45 (note 36) (Hamilton's emphases).

[69] For more on this shift, see Hankins, 1980, pp. 310–311.

[70] See Hamilton, 1846, pp. 26–27 (his emphases), as quoted in Crowe, 1967, pp. 31–32 for this and the quote that follows.

or simply

$$Q = SQ + VQ."$$

Moreover, given two quaternions $\alpha = xi + yj + zk$ and $\alpha' = x'i + y'j + z'k$ with scalar parts each zero, Hamilton defined two operations on their vector parts this way: $S.\alpha\alpha' = -(xx' + yy' + zz')$ and $V.\alpha\alpha' = (yz' - zy')i + (zx' - xz')j + (xy' - yx')k$. Today, we would call these the (negative of the) scalar *dot product* and the vector *cross product*, respectively. For Hamilton, then, vectors represented *yet another new* type of algebraic entity on which (at least two) binary operations could be meaningfully and usefully defined. This realization marked the beginning of Hamilton's version of vector analysis—termed quaternionics by his contemporaries—and he explored its ramifications both algebraic and physical for the remainder of his life.

Two young Scottish mathematical physicists, Peter Guthrie Tait (1831–1901) and James Clerk Maxwell (1831–1879), took up the quaternionic banner, advocating the explicit application of quaternions in physics. By the 1880s, however, first Josiah Willard Gibbs (1839–1903) in the United States and then Oliver Heaviside (1850–1925) in England independently recognized that a simpler mathematical formulation sufficed for capturing the key aspects of physical concepts. Instead of the full algebra of quaternions, vectors and their associated dot and cross products were really all that was necessary. Both Gibbs and Heaviside introduced their versions of vector analysis beginning in the 1880s, and both ultimately published book-length treatments, Heaviside in 1894 in the first of his three-volume *Electromagnetic Theory* and Gibbs in his *Vector Analysis* seven years later in 1901.[71]

Hamilton had actually not been the first to formulate a geometrical algebra applicable to three-dimensional space. The German mathematician, August Ferdinand Möbius (1790–1868), in his *Der barycentrische Calcul* (or *Barycentric Calculus*) of 1827 and his younger Italian contemporary, Giusto Bellavitis (1803–1880), in a paper published in Italian in 1833, both devised vectorial systems for the mathematical interpretation of three-space before Hamilton announced his discovery

---

[71] Heaviside, 1894 and Gibbs, 1901. For more details on this work, see Crowe, 1967, pp. 150–162 (on Gibbs) and pp. 162–176 (on Heaviside).

of the quaternions.[72] In the title of an 1843 paper published in the *Cambridge Mathematical Journal*, moreover, Arthur Cayley had introduced the terminology "geometry of n dimensions," but his was an algebraic analysis of solving systems of linear equations in n unknowns in terms of determinants as opposed to an actual geometrical algebra of n dimensions.[73] The latter was devised in 1844 by the German secondary school teacher, Hermann Grassmann (1809–1877), in his philosophically cast, largely unread, yet highly original book, *Die lineale Ausdehnungslehre* (or *Linear Extension Theory*). Eighteen years later in 1862, Grassmann published a new version of his book in a more mathematical style that he hoped, largely in vain, would allow for a wider audience for his ideas.[74]

Grassmann had been inspired in this mathematical work by his efforts in 1840 more adequately to mathematize the theory of tides in terms of what he called his "principles of the methods of *geometrical analysis*."[75] While his 1840 work had merely applied the new theory in a particular context, his 1844 publication of it in *Die lineale Ausdehnungslehre* represented his first full philosophical and mathematical treatment of it. As he explained in his preface,

> It had for a long time been evident to me that geometry can in no way be viewed, like arithmetic..., as a branch of mathematics; instead geometry relates to something already given in nature, namely, space. I also had realized that there must be a branch of mathematics which yields in a purely abstract way laws similar to those in geometry, which appears to bound space. By means of the new analysis it appeared possible to form such a purely abstract branch of mathematics; indeed this new analysis, *developed without the assumption of any principles established outside of its own domain and proceeding purely by abstraction*, was itself this science.[76]

As Grassmann saw it, "the essential advantages which were attained through this conception were [first] in relation to the form—now all

[72] Möbius, 1827 and Bellavitis, 1835. For a brief discussion of their work in historical context, see Crowe, 1967, pp. 48–54 and Dorier, 1995, pp. 234–236.

[73] See Cayley, 1843, and compare the brief discussion in Crilly, 2006, pp. 81–83.

[74] Compare Grassmann, 1844 and Grassmann, 1862.

[75] Grassmann, 1894–1911, 3(1):18 as quoted and translated in Crowe, 1967, p. 60 (Grassmann's emphasis).

[76] See Grassmann, 1844, pp. ix–x, as quoted in Crowe, 1967, p. 64 for this and the next quote (Crowe's translations; our emphases) .

principles which express views of space are entirely omitted, and consequently the beginnings of the science were as direct as those of arithmetic—and [second] in relation to the content—*the limitation to three dimensions is omitted.*" He proceeded to give a very abstract analysis of binary operations—later styled "a nearly impassable barrier for most mathematicians of the time"—in which he cast the usual concepts of associativity, commutativity, and distributivity.[77]

With this abstract machinery in place, Grassmann defined a vector as a straight line segment with a given length and direction. Addition of two vectors is achieved by joining the beginning point of the second vector to the endpoint of the first. Subtraction is the addition of the first vector and the negative of the second, that is, the vector of the same length as the second but in the opposite direction. Grassmann showed that addition and subtraction so defined obeyed the abstract laws of addition and subtraction that he had established. More generally, although he presented his ideas in a narrative style rather than in this more compact notation, Grassmann considered what was tantamount to a set of linearly independent quantities $e_1, e_2, \ldots, e_m$ and considered a vector to be a linear combination $\sum_{i=1}^m \alpha_i e_i$, where $\alpha_i \in \mathbb{R}$. Given two vectors $\sum_{i=1}^m \alpha_i e_i$ and $\sum_{i=1}^m \beta_i e_i$, then, their sum and difference were expressible as $\sum_{i=1}^m (\alpha_i + \beta_i) e_i$ and $\sum_{i=1}^m (\alpha_i - \beta_i) e_i$, respectively.[78]

Grassmann also realized that products, while not difficult to define, came in several flavors, given his abstract setup. They thus depended on the possible meanings of the product $[e_i e_j]$, specifically, all of the products $[e_i e_j]$ could be independent, or $[e_i e_j] = [e_j e_i]$, or $[e_i e_j] = -[e_j e_i]$, or $[e_i e_j] = 0$, for $1 \leq i, j \leq m$.[79] His systematic analysis of these various scenarios included his versions of the scalar dot product and the vector cross product as well as many of the concepts that we view as fundamental in linear algebra today—linear dependence and independence of vectors, the definition of a vector space, the notion of a basis of a vector space, and the idea of the dimension of a vector space—all in the general context of n dimensions. (For Grassmann, vector spaces were always over the base field of real numbers.) In particular, he proved the formula for the dimension

---

[77] Crowe, 1967, p. 69.

[78] Grassmann, 1844, pp. 15–32.

[79] Grassmann, 1844, pp. 51–61.

of the sum and intersection of two subspaces of a vector space, namely (in modern notation),

$$\dim(E + F) = \dim E + \dim F - \dim(E \cap F),$$

for E and F subspaces of some vector space.[80] He also explored some of the terrain that now comprises the theory of multilinear algebra, defining, among others, the notions of the exterior product, that is, the product that satisfies

$$[e_i e_j] = -[e_j e_i] \tag{13.9}$$

and of the exterior algebra of a vector space.[81]

Even after their publication in a second version in 1862, however, Grassmann's ideas entered the wider mathematical consciousness only slowly. In Germany, Hermann Hankel (1839–1873) incorporated Grassmann's work in his 1867 study on the *Theorie der complexen Zahlensysteme* (or *Theory of Complex Number Systems*), but his premature death cut his direct influence short.[82] Similarly, the geometer and invariant theorist, Alfred Clebsch, championed the German schoolteacher's work in the early 1870s, introducing it to the students in his courses at Göttingen, but he, too, died prematurely. One of those students, however, Felix Klein, not only highlighted Grassmann's work in his influential *Erlanger Programm* of 1872 but also played a key role in arranging for the publication, beginning in 1894, of Grassmann's collected works under the general editorship of Friedrich Engel (1861–1941). It was only with this publication that Grassmann's contributions began to be more fully appreciated by the wider mathematical community.[83]

Despite the delayed reception of the *Ausdehnungslehre*, some of the ideas it contained had independently made their way into the mathematical mainstream via different routes and at the hands of other

---

[80] Grassmann, 1844, pp. 183–185.

[81] Grassmann, 1844, pp. 147–181. A more detailed discussion of Grassmann's work would take us too far afield. The interested reader should consult, among other possible sources, Lewis, 1977; Fearnley-Sander, 1979; Flament, 1992; and Petsche, Lewis, Liesen, and Russ, 2011.

[82] See Hankel, 1867.

[83] On the reception of Grassmann's work in Germany, see Tobies, 1996 and Rowe, 1996. The other authors who contributed to the third section of Schubring, 1996 broach the issue of reception from a variety of different perspectives.

mathematicians. As we saw in the previous chapter, the problem of solving n linear equations in n unknowns or, more generally, the problem of solving m linear equations in n unknowns (for m not necessarily equal to n), had actively occupied Leibniz, Cramer, Euler, Cauchy, Cayley, Sylvester, and others beginning at least at the close of the seventeenth century. In his analysis of Cramer's paradox, for example, Euler had recognized that some systems of equations had a unique solution, while some had infinitely many solutions. With the development of the notions of determinants, matrices, and linear transformations as well as of the theories attendant upon them, it became clear that n-tuples—the same basic constructs as the vectors at the foundation of Grassmann's "extension theory"—were fundamental. The solutions of systems of equations in n unknowns were n-tuples. The rows of both the matrix representing and the determinant associated with such a system were n-tuples comprised of the coefficients of each of the equations. Linear transformations could be represented by matrices as could a system of linear equations. Given these commonalities, it should come as little surprise that individual analyses of these various underlying algebraic structures should have yielded some of the same findings and resulted in new ideas. One mathematician who pushed these lines of research was the Berliner mathematician, Georg Frobenius.

As discussed in the previous chapter, Frobenius was one of the key figures in the development of the theory of matrices, and consequently, of what would become known as linear algebra. In the course of his work in this vein in the 1870s, he gave the modern definitions not only of the notion of the *linear independence* of n-tuples in the context of his analysis of solving systems of n linear equations in n unknowns but also that of the *rank* of a determinant, a concept that proved to be key to the resolution of questions such as Cramer's paradox. For Frobenius, given a homogeneous system of linear equations in n unknowns, solutions $(x_{11}, x_{12}, \ldots, x_{1n})$, $(x_{21}, x_{22}, \ldots, x_{2n})$, \ldots, $(x_{k1}, x_{k2}, \ldots, x_{kn})$ are linearly independent provided, if $c_1 x_{1j} + c_2 x_{2j} + \cdots + c_j x_{kj} = 0$, for $j = 1, 2, \ldots, n$, then $c_1 = c_2 = \cdots = c_k = 0$.[84] On the other hand, he defined the rank of a determinant this way: "If in a determinant all minors of order $m + 1$ vanish, but not all of those of order m are zero, then

---

[84] Frobenius, 1877, p. 236. This 1877 paper by Frobenius contains a wealth of novel ideas. For an historical analysis of its contents, see Hawkins, 2005.

I call m the rank of the determinant."[85] As Frobenius recognized, if a homogeneous system of p linear equations in n unknowns has rank m, then it has $n - m$ linearly independent solutions of n-tuples. Frobenius's older English contemporary, Charles Dodgson (1832–1898) (perhaps better known as Lewis Carroll), had come close to this realization in his 1867 *Elementary Treatise on Determinants with Their Application to Simultaneous Linear Equations and Algebraical Geometry*.[86] Such results were, in a sense, in the air in the closing decades of the nineteenth century, but they would not be united into what is today recognized as the material constitutive of a first course in linear algebra—the union of results on the solution of systems of linear equations, determinants, matrices, linear transformations, and vector spaces—until the 1930s (see the next chapter).

A THEORY OF ALGEBRAS, PLURAL

Also in the air was that spirit of algebraic freedom that Hamilton had loosed with his work on the *noncommutative* system of new "numbers," the quaternions. As we saw, that four-dimensional number system over the real numbers was soon joined in the algebraic pantheon by the eight-dimensional octonions, while De Morgan and others isolated and studied systems of other dimensions over $\mathbb{R}$. By 1870, the American mathematician, Benjamin Peirce (1809–1880), had based a classification of systems of dimension 1 through 6 over the *complex* numbers on a general theory of what he termed "linear associative algebra." The Harvard mathematician presented these ideas to his scientific colleagues—physicists, astronomers, chemists, naturalists, and others, but very few mathematicians—at a meeting in 1870 of the National Academy of Sciences, where they were met with general befuddlement. Peirce also had 100 copies of the text lithographed and distributed to various friends both at home and abroad, but his work did not really begin to enter into the more general mathematical milieu until 1881 when James Joseph Sylvester, from his position as professor of mathematics at the Johns

---

[85] Frobenius, 1879, p. 1 (our translation).
[86] See Dodgson, 1867, p. 50.

Figure 13.3. James Joseph Sylvester (1814–1897).

Hopkins University, published it posthumously in his *American Journal of Mathematics*.[87]

Not unlike Hamilton (whose work on quaternions he knew) and Grassmann (whose work he did not know), Peirce cast his new ideas in broadly philosophical terms, carrying on the debate of his predecessors, Peacock, Hamilton, De Morgan, and others. He opened by defining mathematics as "the science which draws necessary conclusions," elaborating that

> this definition of mathematics is wider than that which is ordinarily given, and by which its range is limited to quantitative research. The ordinary definition, like those of other sciences, is objective; whereas this is subjective. Recent investigations, of which quaternions is the most noteworthy instance, make it manifest that the old definition is too restricted. The sphere of mathematics is here extended, in accordance with all the derivation of its name, to all

---

[87] Sylvester had left England for this post in 1876 and remained at Hopkins until December 1883 when he returned to England to assume the Savilian Professorship of Geometry at Oxford. For more on this aspect of his biography, see Parshall, 2006, pp. 223–277.

demonstrative research, so as to include all knowledge capable of dogmatic teaching.[88]

Algebra had an important role to play within this sweeping purview. To Peirce's way of thinking, "where there is a great diversity of physical appearance, there is often a close resemblance in the process of deduction. It is important, therefore, to separate the intellectual work from the external form. Symbols must be adopted, and mathematics treated by such symbols is called *algebra*. Algebra, then, is formal mathematics," which Peirce understood within a tripartite framework.[89] For him, "the symbols of an algebra, with the laws of combinations, constitute its *language*; the methods of using the symbols in the drawing of inferences is its *art*; and their interpretation is its *scientific application*."

In his 1870 work, Peirce aimed to elaborate that "language" by means of its "alphabet"—that is, its letters—its vocabulary—that is, "its signs and the elementary combinations of its letters"—and its grammar—that is, "the rules of composition by which the letters and signs are united into a complete and consistent system."[90] Since he sought a complete classification of algebras over the complex numbers of successive dimensions from 1 to 6, his alphabets consisted of the basis elements (or basis vectors) of those algebras and so had—case by case—one, two, three, four, five, or six letters. Peirce likened his work to that of the first botanist to explore the flora of a particular region. "This artificial division of algebras is cold and uninstructive like the artificial Linnean system of botany," he offered. "But it is useful in a preliminary investigation of algebras, until sufficient variety is obtained to afford the material for a natural classification." Like the early invariant theorists we encountered in the previous chapter, Peirce was on a systematic hunt.

He set out on his quest by first establishing the preliminary concepts that made up his vocabulary, next determining the axioms of operation that formed his grammar, and finally using those to work out the internal structure of his algebras. Since, for example, Hamilton's biquaternions were among the algebras he isolated in his search, algebras, for Peirce,

---

[88] Peirce, 1881, p. 97. Peirce's work is treated in the historical context of the development of the structure theory of algebras in Parshall, 1985, pp. 250–258. See also Pycior, 1979 and Grattan-Guinness, 1997.

[89] Peirce, 1881, p. 97 (his emphasis). The next quote (with his emphases) is also on this page.

[90] Peirce, 1881, p. 99. The quote that follows in this paragraph is also on this page.

could contain zero divisors (he called them "nilfactors"), but even more was true. His vocabulary included *nilpotents,* that is, elements which vanish when raised to some power, and *idempotents* or nonzero elements which when raised to the square or any higher power give the element back again.[91] Peirce was unabashed in his acceptance of these sorts of elements. "However incapable of interpretation the nilfactorial and nilpotent expressions may appear," he stated categorically, "they are obviously an essential element of the calculus of linear algebras. Unwillingness to accept them has retarded the progress of discovery and the investigation of quantitative algebras. But the idempotent basis seems to be equally essential to actual interpretation." Algebra for Peirce was thus even farther removed from the algebra ruled by Peacock's principle of the permanence of equivalent forms than Hamilton had dared to venture.

As for the grammar that governed the behavior of the elements in an algebra, the two binary operations of addition and multiplication were linked by the usual left and right distributive laws, but multiplication, while associative, was not necessarily commutative. There was also a scalar multiplication, that is, a multiplication of elements in the ground field (of complex numbers) by elements in the algebra that behaved nicely. In his words, "an algebra in which every expression is reducible to the form of an algebraic sum of terms, each of which consists of a single *letter* with a quantitative coefficient, is called a *linear algebra.*"[92] Elements in the algebras Peirce considered were thus linear combinations of the form $C = \sum_{i=1}^{n} c_i a_i$, where $c_i \in \mathbb{C}$, and where the a_i's were the basis elements.

With all of these basic concepts and operations in place, Peirce proceeded to prove general theorems about algebras. He first showed that "in every linear associative algebra, there is at least one [nonzero] idempotent or one nilpotent element," critically using the notion of linear independence.[93] Then, considering an arbitrary algebra that contains at least one idempotent element, call it e_1, he established what has since been dubbed the *Peirce decomposition* of the algebra, namely, every element in an algebra containing an idempotent can be written as a sum of elements, one from each of the following sets: $\{b \mid e_1 b = b \text{ and } b e_1 = b\}$, $\{b \mid e_1 b = b \text{ and } b e_1 = 0\}$, $\{b \mid e_1 b = 0 \text{ and }$

---

[91] Peirce, 1881, p. 104. For the quotation that follows, see p. 118. Peirce did not assume a priori that his algebras contained the idempotent 1.

[92] Peirce, 1881, p. 107 (his emphases).

[93] Peirce, 1881, p. 109.

Figure 13.4. Joseph H. M. Wedderburn (1882–1948).

$be_1 = b\}$, and $\{b \mid e_1 b = 0 \text{ and } be_1 = 0\}$.[94] In the modern notation that the Scottish mathematician, Joseph Henry Maclagan Wedderburn (1882–1948), would essentially use to cast and significantly extend Peirce's groundbreaking work, this says that an algebra A containing at least one idempotent e_1 can be realized as

$$A = B \oplus e_1 B_1 \oplus B_2 e_1 \oplus e_1 A e_1,$$

where $B_1 = \{x \in A \mid xe_1 = 0\}$, $B_2 = \{x \in A \mid e_1 x = 0\}$, and $B = B_1 \cap B_2$.[95] (For example, if $M_2(F)$ denotes the two-by-two matrices over the field F, then the matrix E_{11} with a 1 in the $(1, 1)$ position and zeros everywhere else is an idempotent in $M_2(F)$. It is easy to see that, for b, c, $d \in F$, B_1 is the set of all matrices in $M_2(F)$ of the form $\left(\begin{smallmatrix} 0 & b \\ 0 & d \end{smallmatrix}\right)$, the matrices in B_2 are of the form $\left(\begin{smallmatrix} 0 & 0 \\ c & d \end{smallmatrix}\right)$, and those in B are of the form $\left(\begin{smallmatrix} 0 & 0 \\ 0 & d \end{smallmatrix}\right)$. The Peirce

---

[94] Peirce, 1881, pp. 109–111.
[95] Wedderburn, 1907, pp. 91–92. Wedderburn, however, used a simple + instead of $\oplus$ to denote the direct sum.

decomposition thus yields

$$M_2(F) = \left\{\left(\begin{smallmatrix} 0 & 0 \\ 0 & d \end{smallmatrix}\right)\right\} \oplus \left\{\left(\begin{smallmatrix} 0 & b \\ 0 & 0 \end{smallmatrix}\right)\right\} \oplus \left\{\left(\begin{smallmatrix} 0 & 0 \\ c & 0 \end{smallmatrix}\right)\right\} \oplus \left\{\left(\begin{smallmatrix} a & 0 \\ 0 & 0 \end{smallmatrix}\right)\right\},$$

where a is also in F.)

Peirce next considered the other class of algebras—namely, those with at least one nonzero nilpotent element—and demonstrated first that "in... an algebra which has no idempotent expression, all the expressions are nilpotent" and then that, therefore, the successive nonzero powers of a nilpotent element are linearly independent.[96] As he expressed it in a letter dated 8 November 1870 to his friend, George Bancroft (1800–1891), then the American Ambassador to Germany, these propositions "give the key to all the research," and that research resulted in the classification of over 150 algebras of dimension 1 to 6 over $\mathbb{C}$.[97]

Peirce's research immediately resonated with his son, the mathematician and philosopher Charles Sanders Peirce (1839–1914). When Charles edited his father's paper for publication in Sylvester's *American Journal* in 1881, he incorporated the many new insights he had had since as early as 1870. In particular, he added footnotes that made explicit the connections between his research in logic and his father's algebras, connections that also linked both men's work to matrices. Moreover, he proved, independently of Frobenius, the major theorem that the only (finite-dimensional) division algebras over the real numbers are the real numbers themselves, the complex numbers, and the quaternions.[98] Three years earlier, the English mathematician, William Kingdon Clifford (1845–1879), had published a paper in the inaugural volume of the *American Journal* in which he explicitly interpreted Hamilton's quaternions in terms of elements satisfying Grassmann's exterior product (13.9). Moreover, influenced by both Grassmann's *Ausdehnungslehre* and Benjamin Peirce's 1870 paper on "Linear Associative Algebra," Clifford

---

[96] Peirce, 1881, pp. 113 and 115, respectively.

[97] Ginsburg, 1934, p. 282 and Pycior, 1979, p. 546.

[98] See B. Peirce, 1881, pp. 225–229 for C. S. Peirce's proof of this theorem, and compare Frobenius, 1878, pp. 402–405. A linear associative algebra is a *division algebra* if and only if it has a multiplicative identity $1 \neq 0$ and every nonzero element a in it has a multiplicative inverse, that is, an element b such that $ab = ba = 1$. For more on the work of the Peirces, see Hawkins, 1972, p. 246 and Parshall, 1985, pp. 258–260. On the place of this work in the establishment of the first research-oriented mathematics department in the United States, at the Johns Hopkins University, see Parshall, 1988b and Parshall and Rowe, 1994, pp. 130–134.

had shown how to form a $4m$-dimensional algebra over $\mathbb{R}$ by taking what would today be called the tensor product of m copies of the quaternions.[99]

By 1882, Sylvester himself had entered into this mathematical dialogue, aware of the research of the Peirces and Clifford as well as of Cayley's earlier work on matrices (recall the previous chapter). Sylvester determined a specific system of *matrices* that satisfy a set of relations analogous to those satisfied by the *quaternions*. Taking $\theta = \sqrt{-1}$, the matrices

$$I = \begin{pmatrix} 1 & 0 \\ 0 & 1 \end{pmatrix}, \quad u = \begin{pmatrix} 0 & \theta \\ \theta & 0 \end{pmatrix}, \quad v = \begin{pmatrix} 0 & 1 \\ -1 & 0 \end{pmatrix}, \quad \text{and } w = uv = \begin{pmatrix} -\theta & 0 \\ 0 & \theta \end{pmatrix}$$

satisfy the relations in (13.8) above (with u corresponding to i, v to j, and w to k) and so form a basis of 2×2 *matrices* for the quaternions viewed as an algebra over $\mathbb{R}$. He was positively delighted when he realized that he could extend this construction to 3×3 matrices. This time taking $\rho = \sqrt[3]{-1}$, he set

$$U = \begin{pmatrix} 0 & 0 & 1 \\ \rho & 0 & 0 \\ 0 & \rho^2 & 0 \end{pmatrix} \quad \text{and } V = \begin{pmatrix} 0 & 0 & 1 \\ \rho^2 & 0 & 0 \\ 0 & \rho & 0 \end{pmatrix}$$

and showed that the matrices

$$I, \ U, \ V, \ U^2, \ UV, \ V^2, \ U^2V, \ UV^2 \text{ and } U^2V^2$$

form a basis of a nine-dimensional algebra over $\mathbb{R}$. He called this new matrix algebra the *nonions* and published his results in the United States as well as in France in an effort to make the ideas known to the broadest possible mathematical audience.[100] Both the Peirces and Sylvester came to recognize that linear associative algebras and matrices were intimately related.

---

[99] Clifford, 1878, pp. 356–357.
[100] See Sylvester, 1882 and Sylvester, 1883. The Peirces had apparently discovered the matrix interpretations of both the quaternions and the nonions before Sylvester, but they had not published their results. Sylvester discovered them independently and so has publication priority. See Taber, 1889, p. 354. On Sylvester's strategy for publicizing his ideas, see Parshall and Seneta, 1997.

Sylvester pursued this line of thought in his lectures at the Johns Hopkins University in 1882 and 1883. By 1884, he had published what was to have been the first in an installment of "Lectures on the Principles of the Universal Algebra," in which he aimed to codify a general theory of linear associative (matrix) algebras. Following Peirce, he defined the three operations of addition, multiplication, and scalar multiplication and laid out the elementary properties, such as the associativity of multiplication, that they satisfied. He then proceeded to tease out some of the algebra's general, underlying theory, focusing on the characteristic equation and its roots (recall the discussion of Cayley's paper on matrices in the previous chapter), the notions of spanning or nonspanning sets of matrices for a matrix algebra of a given dimension over its base field, and the idea of the matrix unit representation of a matrix, that is, the representation of an $n \times n$ matrix $A = [a_{ij}]$ as the linear combination $\sum_{i,j=1}^{n} a_{ij} E_{ij}$, where E_{ij} denotes the $n \times n$ basis matrix with a 1 in the ijth place and zeros elsewhere. In particular, he distinguished between what are now termed the characteristic and minimum polynomials of a given matrix, in light of the fact that the characteristic roots of a particular characteristic equation need not be distinct. His approach to all of this, however, so heavily dependent on explicit calculations with small square matrices, did not lend itself to developing the latter distinction further.[101]

That would require the techniques—albeit in the guise of linear transformations and bilinear forms—that had been honed on the continent in the 1870s and 1880s by Weierstrass and, especially, by Frobenius (recall the previous chapter), and of which Sylvester was unaware. In fact, Weierstrass had already been thinking along these lines as early as the winter semester of the 1861–1862 academic year. As he explained in a letter dated 19–27 June 1883 to his student, Hermann Amandus Schwarz (1843–1921), and published in the *Nachrichten* (or *Reports*) of Göttingen University a year later, he had been inspired by a remark that Gauss had made in the abstract to the 1832 installment of his paper on the theory of biquadratic residues. There, in recognition of his work particularly on the Gaussian integers, Gauss had raised the following question: "Why can the relation between things which present

---

[101] See Sylvester, 1884. Compare the discussion of Sylvester's matrix-theoretic work in Parshall, 1985, pp. 243–250 in the context of the history of the theory of algebras and in Parshall and Rowe, 1994, pp. 135–138 in that of the establishment of research-level mathematics in the United States in the final quarter of the nineteenth century.

a multiplicity of more than two dimensions not furnish still further kinds of quantities permissible in the general arithmetic"?[102] Although Gauss never pursued this line of thought, Weierstrass did, as he reported to Schwarz. He saw that, for elements $a = a_1e_1 + a_2e_2 + \cdots + a_ne_n$ in an n-dimensional algebra A over $\mathbb{R}$ with basis elements $e_1, e_2, \ldots, e_n$, he could define addition as $a + b = \sum_{i=1}^{n}(a_i + b_i)e_i$ and multiplication of the basis elements by $e_ie_j = \sum_{k=1}^{n} \eta_{ijk}e_k$, where the structure constants η_{ijk} are in $\mathbb{R}$. Thus, the multiplication of arbitrary elements a and b in A behaved this way: $ab = \sum_{i,j=1}^{n}(a_ib_j)e_ie_j = \sum_{i,j,k=1}^{n}(\eta_{ijk}a_ib_j)e_k$. For Weierstrass, though, the multiplication associated with an algebra was commutative, reflective of Gauss's number-theoretic setting.[103]

Another algebraist steeped in number theory, Richard Dedekind, responded almost immediately to Weierstrass's ideas. In an 1885 paper, he noted that if the relations $e_re_s = e_se_r$ and $(e_re_s)e_t = (e_re_t)e_s$ hold among the basis elements, then the structure constants satisfy $\eta_{trs} = \eta_{tsr}$ and $\sum_{i,j=1}^{n} \eta_{jti}\eta_{irs} = \sum_{i,j=1}^{n} \eta_{jsi}\eta_{irt}$.[104] This line of commutative algebraic thought, interwoven with the developments sketched above on algebraic number and function fields, developed powerfully over the rest of the course of the nineteenth and into the twentieth century. Especially at the hands of David Hilbert, Emmy Noether, and others, commutative algebra came to play a transformative role in ushering in a new era of *algebraic geometry*. In particular, key results that Hilbert published in 1893 not only rendered the invariant theory that had developed over the course of the last half of the nineteenth century largely obsolete (recall the close of the previous chapter) but also established key links like the so-called *Nullstellensatz* (or "zero locus theorem") between the geometric notion of the set of zeros of a polynomial in n variables and the algebraic construct of ideals in a polynomial ring over an algebraically closed field.[105]

As we have seen, however, unlike the number systems that arose in the work of Dedekind, Hilbert, and Noether, those that concerned Hamilton and Sylvester—termed *hypercomplex number systems* or *linear associative algebras* or just *algebras* for short—were *not* commutative. The quickly rising star of French mathematics, Henri Poincaré (1854–1912),

---

[102] See Gauss, 1870–1929, 2:178 (our translation).
[103] Weierstrass, 1884, pp. 312–313.
[104] Dedekind, 1885, pp. 142–143.
[105] See Hilbert, 1893.

recognized new and potent connections to that body of research in the noncommutative setting. In a short 1884 note published, like many of Sylvester's matrix-theoretic results, in the *Comptes rendus* (or *Proceedings*) of the Paris Academy of Sciences, he wrote primarily to his fellow French mathematicians, acknowledging that "the remarkable works of M. Sylvester on matrices have recently attracted attention once again to the [hyper]complex numbers analogous to Hamilton's quaternions." He was quick to point out, however, that "the problem of [hyper]complex numbers reduces easily to the following: To find all continuous groups of linear substitutions in n variables whose coefficients are linear functions of n arbitrary parameters."[106] Poincaré thus connected the theory of continuous transformation groups that Sophus Lie had been developing since the 1870s to the emergent theory of hypercomplex number systems, that is, the linear associative algebras (over $\mathbb{R}$ and $\mathbb{C}$) explored by Hamilton, the Peirces, and Sylvester. Relative to the algebras analogous to the quaternions that Sylvester was studying, Poincaré saw more than algebras of matrices. He recognized that every element x in such an algebra defined a linear transformation, namely, left translation by x, and that the theory of what would later be called Lie groups could be applied to those linear transformations. To restate in modern terminology the problem that Poincaré laid out in his 1884 *Comptes rendus* note, then, it was a matter of finding all subgroups of the Lie group $GL_n(\mathbb{C})$.[107]

The latter connection generated quite a bit of new research in Europe in the 1880s and 1890s particularly at the hands of the German mathematicians, Eduard Study (1862–1930), Friedrich Schur (1856–1932), and Georg Scheffers (1866–1945), the Latvian mathematician, Theodor Molien (1861–1941), and the French mathematician, Élie Cartan (1869–1951).[108] As a doctoral student of Lie, Scheffers, in particular, pushed

---

[106] Poincaré, 1884, p. 740 (our translation). Thomas Hawkins first pointed out the importance of Poincaré's note in Hawkins, 1972, p. 249.

[107] Compare Parshall, 1985, p. 262. A discussion of Lie groups and their historical development would take us beyond the mathematical prerequisites we have assumed for this book. The interested reader should consult Hawkins, 2000 and Borel, 2001. The Lie group $GL_n(\mathbb{C})$ is the collection of $n \times n$ matrices over the complex numbers with nonzero determinant.

[108] What follows in this chapter's closing paragraphs is a very cursory overview of the development of the theory of algebras. To do the theory justice would, once again, take us beyond the mathematical prerequisites assumed. For more on Study's results, see Happel, 1980. For details on the work of Scheffers, Molien, and Cartan, consult Hawkins, 1972 and Parshall, 1985. For a mathematical overview of the theory of hypercomplex number systems, see Kantor and Solodovnikov, 1989.

this research agenda in a series of papers published between 1889 and 1893, in which he began laying the groundwork for a structure theory of hypercomplex number systems (over $\mathbb{C}$) and classifying such systems of small dimensions (up to and including eight).[109] Lie theory also inspired Molien in his research on hypercomplex numbers, but he focused on developing general structure theorems that applied to n-dimensional algebras over $\mathbb{C}$. In the version of his doctoral thesis published in the *Mathematische Annalen* and entitled "Ueber Systeme höherer complexer Zahlen" (or "On Systems of Higher Complex Numbers"), Molien defined the key concept of the *simple* algebra as well as a notion equivalent to that of the two-sided ideal.[110] Making critical use of the characteristic equation and its properties, he succeeded in showing that all simple algebras over $\mathbb{C}$ are, in fact, matrix algebras. He thereby linked the emergent theory of algebras with that of matrices (recall the previous chapter).

By 1898, Cartan, basing his work on that of Scheffers and independently of that of Molien, had established the correspondence between hypercomplex number systems and bilinear groups and showed how to translate the structural properties of the former into those of the latter.[111] His paper, entitled "Sur les groupes bilinéaires et les systèmes de nombres complexes" (or "On Bilinear Groups and Systems of Complex Numbers"), makes clear his starting point. More than exploring these connections, however, he developed an intrinsic and free-standing theory of algebras (over $\mathbb{R}$ and $\mathbb{C}$), that is, a theory, unlike that of Scheffers and Molien, that did not depend on the local Lie group structure. Like Molien, Cartan relied heavily on notions such as the simple algebra and the two-sided ideal and used the characteristic equation to uncover an algebra's internal structure. Like Peirce, but independently of his work, Cartan also exploited the existence of idempotents in (nonnilpotent) algebras and successfully showed how to decompose such algebras by means of what would today be termed pairwise orthogonal idempotents, that is, idempotents e_i and e_j satisfying $e_i e_j = e_j e_i = 0$. With this

---

[109] See Scheffers, 1889a,b; 1891; 1893.

[110] See Molien, 1893. Recall from above that Dedekind had defined the notion of an ideal in a ring in 1871. A *two-sided ideal* $\mathfrak{A}$ in an algebra A is a linear subset of elements such that $\mathfrak{A}A \subseteq \mathfrak{A}$ and $A\mathfrak{A} \subseteq \mathfrak{A}$. An algebra A is *simple* if the only proper (two-sided) ideal of A is the zero ideal and if A is not a zero algebra of order 1.

[111] See Cartan, 1898. For Cartan, a bilinear group is, in modern terms, a Lie subgroup of $GL_n(\mathbb{C})$.

machinery in place, he succeeded in proving, again independently, a number of key results, among them, that all simple algebras over $\mathbb{C}$ are matrix algebras.

This line of research reached a local climax in 1907 with the groundbreaking paper "On Hypercomplex Numbers" by Joseph Wedderburn. There, Wedderburn succeeded in establishing a theory of algebras *over arbitrary fields* (as opposed just to over $\mathbb{R}$ and $\mathbb{C}$) based on structures—particularly, idempotent and nilpotent elements, subalgebras, and two-sided ideals—intrinsic to the algebras themselves. In so doing, he obviated the need for those more complicated, extrinsic arguments of his predecessors that involved the characteristic equation. In particular, Wedderburn focused on the notion of nilpotence and showed that, given an algebra A, all nilpotent two-sided ideals in A—that is, all two-sided ideals comprised of elements $a \in A$ with the property that $a^n = 0$, for some integer n—are contained in a *maximal* nilpotent ideal N. The beauty of this result is that, in some sense, the maximal nilpotent ideal (later called the *radical*) of an algebra contains all of the algebra's "badly behaved" elements. Thus, the algebra A/N—that is, the algebra A factored out by the maximal nilpotent ideal N—is "nice," specifically, its maximal nilpotent ideal is the zero ideal or, in other words, it is *semisimple*. Moreover, it can be written as a direct sum of simple subalgebras. In order to understand an arbitrary semisimple algebra, then, it is enough to understand its simple components. One of the main results in Wedderburn's paper did just that. He succeeded in demonstrating that any simple algebra can be expressed as a full matrix algebra over a division algebra, that is, as $M_n(D)$.[112]

Wedderburn's emphasis both on uncovering the general, underlying structure of a particular mathematical object and on the development of methods intrinsic to a given mathematical setting was, by the close of the nineteenth century, increasingly becoming the hallmark of algebraic research. This, together with the contemporaneous evolution of meaningful sets of axioms for algebraic entities, would come to characterize the *modern* algebra of the twentieth century, a topic to which we now turn briefly in our final chapter.

---

[112] Wedderburn, 1907, p. 99. This theorem is today termed the Wedderburn–Artin theorem after both Wedderburn and Emil Artin (1898–1962), who extended Wedderburn's original result to rings satisfying the so-called descending chain condition in 1927. See Artin, 1927 and the next chapter.

14

The Emergence of Modern Algebra

A s the previous three chapters have documented, by the first decade of the twentieth century, the topography of algebra had changed significantly. Galois's ideas on groups, which had developed in the specific context of finding the roots of algebraic equations, had not only entered the algebraic mainstream but had developed over the course of the last half of the nineteenth century—thanks to the work of Arthur Cayley, Camille Jordan, Ludwig Sylow, and Heinrich Weber, among many others—into an independent and freestanding theory of groups. Weber, in particular, had also abstracted out and explicitly expressed a viable set of group axioms by the 1890s that applied even in the setting of infinite groups. At the same time, the group concept had surfaced beyond its primary point of origin in the theory of permutations—in geometry, in number theory, and elsewhere. By 1900, the exploration of the properties and internal structures of groups per se had come to define an important branch of algebraic research that was recognized as having ever-growing applications, both mathematical and physical.

Not dissimilarly, a concrete problem—that of linearly transforming homogeneous polynomials in n unknowns—had arisen in contexts as diverse as celestial and terrestrial mechanics, analytic geometry, and number theory. It had spurred the development of a theory of determinants, a spectral theory of matrices and linear transformations, and a theory of invariants. As eighteenth- and nineteenth-century mathematicians came to realize, they could reduce the problem of simultaneously solving systems of linear equations to one of analyzing a new construct, the determinant, formed in a specific way from the equation's coefficients. From there, it was only a small step to studying determinants in and of themselves and to developing a full-blown theory of their properties and behavior. The same kind of square array of coefficients—or matrix— that characterized the determinant had also cropped up in the context of effecting a linear transformation of the variables of a homogeneous polynomial. Whether treating actual square arrays as algebraic objects to be manipulated, as Arthur Cayley did, or taking a more analytical point of view, as Karl Weierstrass and Georg Frobenius did, approaches had converged by the 1890s into a spectral theory that would play a key role in the establishment of so-called linear algebra over the course of the first half of the twentieth century. Contemporaneously, the behavior of linear transformations of a homogeneous polynomial of given degree had led mathematicians on an inductive search for expressions in the polynomial's coefficients and variables that essentially remain fixed under linear transformation. Their findings coalesced into yet another new field of algebraic research, invariant theory, that thrived until the 1890s and experienced a resurgence in the closing decades of the twentieth century.

All of this work was done concurrently with a major nineteenth-century push better to understand the properties of numbers and, particularly, to find a proof of Fermat's infamous "last theorem." The latter quest resulted in a vast extension of the notion of "integer" as well as in a new body of algebraic thought that developed in work such as that of Ernst Kummer and Richard Dedekind. Their research critically extended such basic notions as prime number and divisibility to more general and abstract algebraic definitions and structures. In a different vein, British mathematicians probed the very meaning of algebra and asked questions such as, what should the operational rules of algebra be?, how should they be determined?, and what do they legitimately operate on? Given that the integers fundamentally served as their model, their assumptions were severely shaken when William Rowan Hamilton discovered the *noncommutative* system of quaternions in 1843 and when others followed suit in devising additional new "number" systems that did not satisfy the "usual" rules. By the opening decade of the twentieth century, this line of thought had resulted in a theory of algebras—that is, in yet another new, fundamental algebraic structure—at the hands of mathematicians such as Benjamin Peirce in the United States, Theodor Molien in Latvia, Élie Cartan in France, and Joseph Wedderburn first in Scotland and then in the United States. Hamilton also independently joined mathematicians like August Möbius and Hermann Grassmann not only in recognizing, around mid-century, the possibility of formulating a geometrical algebra applicable to three-dimensional space but also in exploring the algebraic properties of what were termed vectors.

As this overview makes clear, another key characteristic of the evolving topography of algebra was the increasingly international dialogue
between mathematicians in Great Britain, on the continent, in the
United States, and elsewhere. These men (and some women) were united
through research journals as well as through abstracting journals such
as the *Jahrbuch über die Fortschritte der Mathematik*.[1] By the 1890s, they
were also united formally through International Congresses of Mathematicians, with the so-called "zeroth" congress being held in Chicago in
1893 and more formal congresses taking place in Zürich in 1897 and in
Paris in 1900.[2]

The 253 mathematicians in attendance in Paris, for example, had the
opportunity to hear a bold lecture on "Mathematical Problems," in which
the thirty-eight-year-old David Hilbert laid out twenty-three problems
that he felt merited the concerted efforts of what was increasingly
viewed as an *international* mathematical community. Of those, at least six
stressed the importance of solid work on the foundations of mathematics.
Hilbert made his position on this point clear. "I think," he said, "that
whenever, from the side of the theory of knowledge or in geometry,
or from the theories of natural or physical science, mathematical ideas
come up, the problem arises for mathematical science to investigate
the principles underlying these ideas and so to establish them upon
a simple and complete system of axioms, that the exactness of the
new ideas and their applicability to deduction shall be in no respect
inferior to those of the old arithmetic concepts."[3] Moreover, he had
distinct ideas as to what should constitute a "complete system of axioms."
"When we are engaged in investigating the foundations of a science,"
he explained,

> we must set up a system of axioms which contains an exact and
> complete description of the relations subsisting between the ele-
> mentary ideas of that science. The axioms so set up are at the same
> time the definitions of those elementary ideas; and no statement
> within the realm of the science whose foundation we are testing is
> held to be correct unless it can be derived from those axioms by a
> finite number of logical steps. Upon close examination, the question

---

[1] For more on this theme of internationalization in mathematics, see Parshall, 1995, Parshall,
2009, and the essays collected in Parshall and Rice, 2002.

[2] On the history of the International Congresses of Mathematicians, see Lehto, 1998.

[3] Hilbert, 1902, p. 442. For the next quote, see p. 447 (his emphasis).

arises: Whether, in any way, certain statements of single axioms depend upon one another, and whether the axioms may not therefore contain certain parts in common, which must be isolated if one wishes to arrive at a system of axioms that shall be altogether independent of one another.

As we have seen, work toward the axiomatization of specific algebraic entities actually preceded Hilbert's pronouncements in Paris. Mathematicians like Weber, influenced by the set-theoretic ideas of Georg Cantor (1845–1918), had already begun to consider mathematical objects as sets of elements with well-defined axiom systems for governing how two or more elements could be combined. Nevertheless, Hilbert set an agenda which algebraists—and the term is not ahistorical by the early twentieth century—very quickly took to heart.

REALIZING NEW ALGEBRAIC STRUCTURES AXIOMATICALLY

Twelve years before Hilbert took the podium at the International Congress of Mathematicians in Paris, the Italian mathematician Giuseppe Peano (1858–1932), who was in Paris to hear Hilbert speak in 1900, had already anticipated his German colleague's call. In his 1888 book, *Calcolo geometrico secondo l'Ausdehnungslehre di H. Grassmann* (or *Geometric Calculus According to the Ausdehnungslehre of H. Grassmann*), Peano had sought, like other later nineteenth-century mathematicians (recall the previous chapter), to bring the ideas that Grassmann had presented before a broader audience. In closing his work, however, Peano had departed from Grassmann's presentation to give a set of axioms for what he termed a "linear system" but what we would call a vector space over the real numbers today. In his words,

There exist systems of objects for which the following definitions are given:

(1) There is an *equivalence* between two objects of the system, i.e., a proposition denoted by $\mathbf{a} = \mathbf{b}$...
(2) There is a *sum* of two objects $\mathbf{a}$ and $\mathbf{b}$. That is, there is defined an object denoted by $\mathbf{a} + \mathbf{b}$, which also belongs to the given

system and satisfies the conditions:

$$(\mathbf{a} = \mathbf{b}) \quad \text{[implies]} \qquad \mathbf{a} + \mathbf{c} = \mathbf{b} + \mathbf{c},$$
$$\mathbf{a} + \mathbf{b} \;=\; \mathbf{b} + \mathbf{a},$$
$$\mathbf{a} + (\mathbf{b} + \mathbf{c}) \;=\; (\mathbf{a} + \mathbf{b}) + \mathbf{c}.$$

(3) Letting **a** be an object of the system and m be a positive integer, we mean by $m\mathbf{a}$ the sum of m objects equal to **a**. It is easy to see that if **a**, **b**, ... are objects of the system and m, n, ... are positive integers, then

$$(\mathbf{a} = \mathbf{b}) \quad \text{[implies]} \qquad m\mathbf{a} = m\mathbf{b};$$
$$m(\mathbf{a} + \mathbf{b}) \;=\; m\mathbf{a} + m\mathbf{b};$$
$$(m + n)\mathbf{a} \;=\; m\mathbf{a} + n\mathbf{a};$$
$$m(n\mathbf{a}) \;=\; (mn)\mathbf{a};$$
$$1\mathbf{a} \;=\; \mathbf{a}.$$

We assume that a meaning is assigned to $m\mathbf{a}$ for any real number m in such a way that the previous equations are still satisfied. The object $m\mathbf{a}$ is said to be the *product* of the [real] number m by the object **a**.

(4) Finally, we assume that there exists an object of the system, which we... denote by **0**, such that, for any object **a**, the product of the number 0 by the object **a** is always the object **0**, i.e.,

$$0\mathbf{a} = \mathbf{0}.$$

If we let $\mathbf{a} - \mathbf{b}$ mean $\mathbf{a} + (-1)\mathbf{b}$, then it follows that

$$\mathbf{a} - \mathbf{a} = \mathbf{0}, \;\; \mathbf{a} + \mathbf{0} = \mathbf{a}.$$

DEF: Systems of objects for which definitions (1)–(4) are introduced in such a way as to satisfy the given conditions are called *linear systems*.[4]

---

[4] Peano, 1888, pp. 141–142 as quoted in Moore, 1995, pp. 267–268 (Moore's translation; Peano's emphases). For the examples that follow, see Peano, 1888, pp. 142–154.

Peano proceeded to give concrete examples of structures that satisfied these axioms, among them, the real numbers, the complex numbers, and vectors in a plane or in space, but, more interestingly, the collection of polynomial functions of a real variable and the set of all linear transformations from a linear system A to a linear system B. As Peano clearly realized, examples of vector spaces had been arising in diverse contexts in the mathematical literature over the course of the last half of the nineteenth century.

Unfortunately, however, Peano did not ultimately achieve the desired result of more widely broadcasting Grassmannian ideas. Only three of his countrymen, Cesare Burali-Forti (1861–1931), Roberto Marcolongo (1862–1943), and Salvatore Pincherle (1853–1936), adopted his axiomatic system, and their work had little impact outside of Italy. In fact, a set of axioms for a vector space would only become more widely adopted thanks first to the work of Hermann Weyl (1885–1955) in setting up the general theory of relativity in terms of real vector spaces and to the work in analysis of Hans Hahn (1879–1934), Stefan Banach (1892–1945), and Norbert Wiener (1894–1964), among others.[5] Their axioms, however, effectively differed little from those found in Peano's 1888 text.[6]

Nor was the notion of a vector space the only new algebraic structure in "need" of an adequate set of axioms around the turn of the twentieth century. Dedekind's work in algebraic number theory had generated several new concepts, especially the notions of module, ring, and ideal, while his results with Weber on algebraic function fields as well as earlier developments particularly at the hands of Galois, Serret, and Jordan had highlighted fields as essential to algebraic inquiry. As we saw in the previous chapter, as early as 1871 in his second edition of Dirichlet's *Vorlesungen über Zahlentheorie*, Dedekind had understood the algebraic integers as what he termed a ring living inside the field of algebraic numbers. He then initially defined a module—solely in that specific context—as a set of real or complex numbers closed under addition and subtraction and an ideal as a subset of a ring of algebraic integers closed under addition, subtraction, and multiplication by any algebraic integer

---

[5] See, for example, Weyl, 1918; Hahn, 1922; and Banach, 1922.

[6] See Moore, 1995, pp. 269–271 on the role of the Italians and pp. 273–288 on the impetus from analysis for the development of vector space axioms. Moore provides a fuller discussion of the relevant literature.

in the ring. Over the course of the next two decades as he continued to re-edit Dirichlet's work, Dedekind developed and refined these ideas, although always in the context of algebraic number theory.[7]

In 1897, Hilbert elaborated on them further in the massive report on "Die Theorie der algebraischen Zahlkörper" (or "The Theory of Algebraic Number Fields") that he wrote for the Deutsche Mathematiker-Vereinigung (or German Mathematical Society).[8] In this so-called "Zahlbericht" (or "Report on Numbers"), Hilbert recognized a hierarchy among the structures that Dedekind had examined but, again, only within the framework of algebraic number theory and not in some more general setting. An algebraic number field was a structure closed under addition, subtraction, multiplication, and division; the ring (Hilbert actually called it a *Zahlring* or more simply a *Ring* as opposed to Dedekind's terminology *Ordnung*) of algebraic integers within it was closed only under the first three of these operations; a module was a set of algebraic integers closed only under the first two. Within this architectural context, the notion of an ideal in some sense straddled those of ring and module. Although not closed under multiplication like a ring, an ideal did have a multiplicative structure relative to the entire field of algebraic numbers and, so, had more structure than a module.[9]

By 1914, the German mathematician, Adolf Fraenkel (1891–1965), had recognized that the basic notion of a ring had actually arisen in contexts other than that of algebraic number theory. In particular, he appreciated the fact that a structure very like Dedekind's *Ordnung* and Hilbert's *Zahlring*—as well as like what Kurt Hensel (1861–1941) had defined in the context of his 1908 number-theoretic work on so-called p-adic numbers—could be found in the theory of hypercomplex number systems as well as in the theory of matrices. Referring explicitly to the prior work not only of Dedekind, Hilbert, and Hensel in number theory

---

[7] For more on the evolution of his thought, see Corry, 1996/2004, especially pp. 92–120/93–120.

[8] The Deutsche Mathematiker-Vereinigung had been founded in 1890 to encourage interactions between German-speaking mathematicians within and outside the main mathematical centers such as Göttingen, Berlin, and Leipzig. National mathematical societies were largely a phenomenon of the latter half of the nineteenth century with key societies being established in England (1865), France (1872), Italy (1884), the United States (1888), and elsewhere, in addition to Germany. On the professionalization of mathematics internationally, see the essays collected in Parshall and Rice, 2002.

[9] See Hilbert, 1897, especially pp. 177–188 and the discussions in Moore, 1995, p. 290 and Corry, 1996/2004, pp. 147–154/147–154.

but also that of Weierstrass and others in the spectral theory of matrices and in the theory of hypercomplex numbers, Fraenkel axiomatized a ring in terms of a set of ten axioms. As he expressed it in a narrative footnote, the elements in a ring $\Re$ will (1) form a group under addition as well as multiplication "as much as is possible, that is, with the exception of zero divisors," (2) satisfy the distributive law, and (3) have the property that "if a and b are any two elements in $\Re$, then there exists a regular element $\alpha_{a,b}$ [that is, an element that is not a zero divisor] in $\Re$ such that $a \cdot b = \alpha_{a,b} \cdot b \cdot a$ and a regular element $\beta_{a,b}$ such that $a \cdot b = b \cdot a \cdot \beta_{a,b}$. The multiplication in the ring thus need not be commutative," he continued, "but $a \cdot b$ and $b \cdot a$ differ from one another at most by a regular factor either in front or behind."[10] More structured, that is, less general, than the now-standard formulation of a ring that Emmy Noether would give in the 1920s (see the next section), this axiomatization reflected the influence of Hensel's work, in which, in particular, every nonzero element in a "ring" had a multiplicative inverse. In maintaining this criterion, Fraenkel created a notion of a "ring" that thus did not include what had, in some sense, been the canonical number-theoretic example of the integers $\mathbb{Z}$. Nevertheless, his represents the first general axiomatization of a ring in the literature, and Fraenkel used it to explore some of the structural properties with which such objects were endowed.[11]

Fraenkel had been inspired in his axiomatic work to some extent by a 1910 paper, entitled "Algebraische Theorie der Körper" (or "Algebraic Theory of Fields"), by his fellow countryman, Ernst Steinitz (1871–1928). There, Steinitz had sought to unite under a common framework the diverse instances of the notion of a field that had proliferated especially over the course of the last quarter of the nineteenth century and to develop a fuller theory of this construct. A decade and a half earlier in the same paper on Galois theory in which he had given a set of axioms for a group, Heinrich Weber had actually already laid out a set of axioms of a field, too. As we saw in chapter 11, those two notions—group and field—were intimately interconnected in the work of Galois. Moreover, as we saw in chapter 13, Weber and Dedekind had worked with fields in the very different contexts of algebraic number fields and algebraic function fields. Weber was thus in an ideal position to consider the commonalities

[10] Fraenkel, 1914, pp. 144–145 (the quote is on p. 144; our translation).
[11] For more on Fraenkel's work, see Corry, 1996/2004, pp. 202–214/201–213.

of the field notion as it had arisen in all three of these settings and to begin to establish Galois theory as a whole on an axiomatic foundation.

Weber's "field" was a natural extension of that of a commutative group. It was a set of elements on which *two* operations—addition *and* multiplication—were defined and were both commutative. Moreover, it was a commutative group under addition, was closed under multiplication, and satisfied the following additional axioms: for elements a, b, and c in the set,[12]

(1) $a(-b) = -ab$ (which implies that $(-a)(-b) = ab$);

(2) $a(b + c) = ab + ac$ (which implies, when one sets $c = -b$, that $a \cdot 0 = 0$);

(3) if $ab = ac$, then $b = c$ unless $a = 0$;

(4) for given b and c, there is a unique element a such that $ab = c$ except when $b = 0$;

(5) the product of two elements equals zero when at least one of the elements is zero; and

(6) there is a multiplicative identity, called 1, and every nonzero element has a multiplicative inverse with respect to 1.

In formulating this axiomatic system, however, Weber had strongly in mind his three case studies. While his field axioms—like his group axioms—applied to the case of both finite and infinite fields, he did not explicitly consider what would come to be called the *characteristic* of the field. (The "usual" fields like the rationals, the real numbers, and the complex numbers, have characteristic zero, that is, it is not the case that for some positive integer n, $na = 0$ for all elements a in the field. Finite fields with p elements have characteristic p.)

Steinitz, however, saw beyond the "usual" examples of a field. Like Fraenkel after him, he was inspired by the example of Hensel's p-adic numbers, "a field which," in Steinitz's words, "counts neither as the field of functions nor as the field of numbers in the usual sense of the word."[13]

---

[12] Weber, 1893, pp. 526–528.

[13] Steinitz, 1910, p. 167 (note ***) and Corry, 1996/2004, p. 194/193 (Corry's translation). For the following two quotes, see Steinitz, 1910, p. 5 and Corry, 1996/2004, p. 195/193–194 (Corry's translations; Steinitz's emphasis).

In a mathematical universe thus enlarged by Hensel's research, Steinitz aimed "*to advance an overview of all the possible types of fields and to establish the basic elements of their interrelations.*" For him, the abstract notion of a field was paramount, "whereas," as he acknowledged, "Weber's aim [had been] a general treatment of Galois theory."

In order to realize his goal, Steinitz began by laying out his own set of field axioms. A set of elements on which the two binary operations of addition and multiplication were defined and closed was thus a field to Steinitz's way of thinking, provided it satisfied the following seven axioms:

1) The associative law of addition:.

$$(a+b)+c = a+(b+c),$$

2) the commutative law of addition:.

$$a+b = b+a,$$

3) the associative law of multiplication:.

$$(a \cdot b) \cdot c = a \cdot (b \cdot c),$$

4) the commutative law of multiplication:.

$$a \cdot b = b \cdot a,$$

5) the distributive law:.

$$a(b+c) = ab+ac,$$

6) the law... of subtraction: For a, b elements of the system, there is one and only one element x, for which $a+x = b$.—This element is denoted $b-a$ and is called the *difference* of a and b, so that in general $a+(b-a) = b$,.

7) the law... of division: The system contains besides 0 at least one element, and if a does not equal 0 [and] b is a given element of the system, then there is one and only one

element x for which $a \cdot x = b$.—We call x the *quotient* of b and a and denote $x = \frac{b}{a}$ or $b : a$.[14]

With this setup, Steinitz immediately deduced several other elementary properties of fields. From axioms (1), (2), and (6), for example, he proved the existence of an additive identity 0 before invoking it in axiom (7), and from axiom (7), he established not only the existence of a multiplicative identity 1 but also the property that "a product of [nonzero] elements equals 0, only if at least one factor vanishes." His axiomatic system thus very closely resembles that found in modern algebra textbooks today.

Here, however, it was not so much the axiomatization that marked the novelty of Steinitz's work. Rather, it was the program he laid out for field-theoretic research as a whole. In emphasizing the role of the characteristic of a field, in particular, Steinitz provided not only a certain categorization of fields—both finite and infinite—but also a philosophy for how to go about studying them.[15] As Leo Corry so aptly put it, Steinitz's approach to a *theory* of fields hinged on this research sequence: "First, it is necessary to consider the simplest possible fields. Then, one must study the methods through which from a given field, new ones can be obtained by extension. One must then find out which properties are preserved when passing from the simpler fields to their extensions."[16] "Here," Corry continued, "we find an illuminating example of the kinds of procedures that would later come to be central for any structural research in mathematics, explicitly formulated and effectively realized for the first time." It was precisely this newly emergent algebraic ethos that Fraenkel adopted in his axiomatic approach to a *theory* of rings four years later. It was also in much this same spirit that a system of axioms for the new notion of an algebra (recall the previous chapter) developed at the hands first of the American mathematician, Leonard Eugene Dickson (1874–1954), then of his friend Joseph Wedderburn, and finally of the German mathematician, Emmy Noether.[17] The work of Noether and her students, moreover, proved fundamental in orienting algebraic research toward a quest for structure.

---

[14] Steinitz, 1910, pp. 173–174 (our translation; his emphases). The next quote (our translation) is also on pp. 173–174.

[15] See section four of Steinitz, 1910, pp. 179–181 for the initial discussion of characteristic.

[16] Corry, 1996/2004, p. 195/194. The next quote is also on this page.

[17] See Moore, 1995, pp. 291–293. The pertinent references may be found in his bibliography.

THE STRUCTURAL APPROACH TO ALGEBRA

Emmy Noether had come to mathematics at a time when women were officially debarred from attending university in Germany. Daughter of the noted algebraic geometer and professor at the University of Erlangen, Max Noether (1844–1921), Emmy Noether had grown up among mathematicians and in a university setting but had initially pursued an educational course deemed culturally suitable for one of her sex, that is, she qualified herself to teach modern languages in "institutions for the education and instruction of females."[18] This vocational preparation aside, Noether studied mathematics—from 1900 to 1902 and then in the winter semester of 1903–1904—as an auditor and at the discretion of selected professors at her hometown University of Erlangen and at Göttingen University, respectively. At the time, there was no other way for a German woman to attend courses at either institution.[19] Things changed at the University of Erlangen in 1904 and throughout the country four years later, when women were officially allowed to attend and to earn degrees from German universities. Noether enrolled at Erlangen in 1904 and earned a doctorate there under the invariant theorist, Paul Gordan, in December of 1907.

Her degree in hand, however, Noether had nowhere to go within the German system. Even as holders of the doctoral degree, women were allowed neither to hold faculty positions nor even to lecture. Noether thus remained in Erlangen until 1915 as an unofficial assistant to her father, while pursuing her own algebraic research agenda. In 1915 and at the invitation of her former professors, Felix Klein and David Hilbert, she finally had the opportunity to make the move to Göttingen. There, she lectured in courses, although in physics, officially listed under Hilbert's

---

[18] Dick, 1981, p. 12 (for the quote) and pp. 10–27 on Noether's early life. This is the main biography of Noether in English. See Brewer and Smith, 1981 and Srinivasan and Sally, 1983 for overviews of Noether's life but especially of her work.

[19] German universities had been more liberal with respect to foreign women. Sonya Kovalevskaya (1850–1891), for example, had been allowed, under entirely exceptional circumstances, to take a doctorate in absentia at Göttingen in 1874 after having studied with Leo Königsberger (1837–1921) in Heidelberg and Karl Weierstrass in Berlin. Beginning in 1894, however, foreign women had been officially allowed to take German degrees. See Parshall and Rowe, 1994, pp. 239–253 on American women who pursued mathematical studies in Germany during the decade of the 1890s.

Figure 14.1. Emmy Noether (1882–1935), courtesy of Drs. Emiliana and Monica Noether.

name, beginning in the winter semester of 1916–1917,[20] while Hilbert and Klein fought with the university authorities to allow her to obtain the *Habilitation* or credentials that would allow her to lecture in her own right. They won that final battle in 1919 in a Germany defeated in World War I and under the new parliamentary representative democratic government of the Weimar Republic that had replaced the old imperial regime. Noether soon began to attract young students from both Germany and abroad to classes as well as to more informal series of lectures and seminars.

[20] Hilbert was very interested in physics at this time, particularly in Einstein's new ideas, so that Noether's initial lectures were on physics rather than pure mathematics. On Hilbert and physics, see Corry, 2004.

During the course of the calendar years 1921 and 1922, for example, Noether had her first opportunity to present the fruits of her then-current research. In particular, she gave four-hour lectures on each of the following topics: "higher algebra, finiteness theorems, [the] theory of fields, [the] elementary theory of numbers, and algebraic number fields,"[21] in which she undoubtedly discussed some of the groundbreaking ideas that she had expounded in her paper dated October 1920 on "Idealtheorie in Ringbereichen" (or "Ideal Theory in Domains of Rings"). As she stated in opening that paper, "the aim of the present work is to translate the factorization theorems of the rational integer numbers and of the ideals in algebraic number fields into ideals of arbitrary integral domains and domains of general rings."[22] Drawing on Fraenkel's formulation of a ring, Noether made the key observation that the abstract notion of ideals in rings could be seen not only to lay at the heart of prior work on factorization in the contexts of algebraic number fields and of polynomials (in work of Hilbert, Francis Macaulay (1862–1937), and Emmanuel Lasker (1868–1941)) but also to free that work from its reliance on the underlying field of either real or complex numbers. In her paper, she built up that theory—not surprisingly, in the context of commutative rings given her motivating examples—in structural terms.

Noether began by presenting what are essentially the modern axioms for a (commutative) ring:

> The underlying domain Σ is a (commutative) *ring* in abstract definition; that is, Σ consists of a system of elements $a, b, c, \ldots, f, g, h, \ldots, \ldots$ in which from two ring elements a and b through two operations (means of combination), *addition* and *multiplication*, a unique third element will be obtained called the sum $a + b$ and the product $a \cdot b$. The ring and the otherwise completely arbitrary operations must satisfy the following axioms:
>
> 1. *The associative law of addition*: $(a + b) + c = a + (b + c)$.
>
> 2. *The commutative law of addition*: $a + b = b + a$.

---

[21] Dick, 1981, p. 37.
[22] Noether, 1921, p. 25 or Corry, 1996/2004, p. 227/226 (his translation).

3. *The associative law of multiplication*: $(a \cdot b) \cdot c = a \cdot (b \cdot c)$.

4. *The commutative law of multiplication*: $a \cdot b = b \cdot a$.

5. *The distributive law*: $a \cdot (b + c) = a \cdot b + a \cdot c$.

6. *The law of absolute and unique subtraction.*

Σ contains a unique element x such that the equation $a + x = b$ is satisfied. (One writes $x = b - a$.)[23]

As she immediately noted, it follows from these axioms that a ring has a zero element, that is, an additive identity, although it need not have a multiplicative identity 1 and the product of two nonzero elements may vanish.

With the notion of a ring thus established, Noether proceeded to define an ideal in terms somewhat different from those employed by Dedekind in 1871. In her words,

By an *ideal* $\mathfrak{M}$ in Σ, we will understand a system of elements that satisfies both of these two conditions:

1. $\mathfrak{M}$ *contains in addition to* f, $a \cdot f$, *where a is an arbitrary element in* Σ.

2. $\mathfrak{M}$ *contains in addition to* f *and* g, the difference $f - g$ as well as nf for each integer n.

Since, for Noether, ideals in general rings were to play the role of factors in the setting of the rational numbers as well as of ideals, in Dedekind's sense, in algebraic number fields, she needed to define a general notion of divisibility. "If f is an element of $\mathfrak{M}$," she explained, "we generally denote that by $f \equiv 0(\mathfrak{M})$ and say that f *is divisible by* $\mathfrak{M}$. If every element of $\mathfrak{N}$ is also an element of $\mathfrak{M}$ and divisible by $\mathfrak{M}$, we say that $\mathfrak{N}$ *is divisible by* $\mathfrak{M}$ and denote that by $\mathfrak{N} \equiv 0(\mathfrak{M})$. $\mathfrak{M}$ is called a *proper divisor* of $\mathfrak{N}$, if it contains elements different from $\mathfrak{N}$ and is also not, vice versa, divisible by $\mathfrak{N}$. From $\mathfrak{N} \equiv 0(\mathfrak{M})$; $\mathfrak{M} \equiv 0(\mathfrak{N})$ it follows that $\mathfrak{N} = \mathfrak{M}$."

Another key ingredient to Noether's approach was the fact that ideals were finitely generated, that is, they had a finite basis of elements $f_1, f_2, \ldots, f_\varrho$ such that every element f in the ideal could be written as $a_1 f_1 + a_2 f_2 + \cdots + a_\varrho f_\varrho$ for elements $a_i \in \Sigma$. This assumption allowed her

---

[23] Noether, 1921, p. 29 (her emphases; our translation). For the quotes in the next paragraph, see p. 30 (her emphases; our translation).

immediately to prove one of the hallmarks of her approach to ring theory, namely, the so-called ascending chain condition on ideals or, as she termed it, the "theorem on finite chains": "If $\mathfrak{M}, \mathfrak{M}_1, \mathfrak{M}_2, \ldots, \mathfrak{M}_v, \ldots$ is a countably infinite system of ideals in Σ, such that each is divisible by that which follows, then there exists a finite index n such that all the ideals $\mathfrak{M}_n = \mathfrak{M}_{n+1} = \cdots$. In other words, *if we construct a simply ordered chain of ideals $\mathfrak{M}, \mathfrak{M}_1, \mathfrak{M}_2, \ldots, \mathfrak{M}_v, \ldots$ in such a way that each ideal is a proper divisor of the one immediately preceding it, then the chain is finite.*"[24] Noether used this structural notion—based on clearly and crisply articulated axioms—to show how to decompose ideals and, in so doing, to establish the factorization theories of her predecessors and contemporaries in the general context of commutative rings.[25]

Noether effected a similarly spirited unification during the course of her lectures in the winter semester of 1927–1928 on "Hyperkomplexe Grössen und Darstellungstheorie" (or "Hypercomplex Numbers and Representation Theory"). Representation theory had developed beginning in the closing decade of the nineteenth century in the work of Frobenius and, in the context of finite groups, concerned analyzing abstract groups in terms of groups of linear transformations. In particular, a representation of a finite group G is a map (a so-called group homomorphism) from G into the general linear group $GL_n(\mathbb{C})$ that assigns to every element g in G a nonsingular $n \times n$ matrix over the complex numbers in such a way that the group multiplication is preserved. As this definition makes clear, then, representation theory was intimately related to the theory of matrices over $\mathbb{C}$, which was, in turn, a special case of an algebra over a given base field. Noether recognized that by developing a noncommutative theory of ideals in a ring that satisfies both the ascending and the descending chain conditions, she could bring representation theory and the theory of algebras together in the general context of *non*commutative rings. She did precisely this in a paper published in 1929 and based on the notes that the young Dutch mathematician, Baertel van der Waerden, had taken during the course of her 1927–1928 lectures.[26]

---

[24] Noether, 1921, pp. 30–31 (her emphasis; our translation).

[25] For more details on her work in this vein, see Corry, 1996/2004, pp. 229–239/228–237. Today, rings with the ascending chain condition are termed *Noetherian rings* in Noether's honor.

[26] See Noether, 1929. A discussion of representation theory and its historical development especially at the hands of Georg Frobenius, William Burnside, and Issai Schur (1875–1941) in

Figure 14.2. Baertel van der Waerden (1903–1996).

Van der Waerden had earned his PhD in mathematics at the University of Amsterdam in 1926 and had spent the last half of the 1926–1927 and all of the 1927–1928 academic years in Germany. In Göttingen, he worked primarily with Noether, having already spent time with her in the winter semester of 1924–1925; in Hamburg, he talked with Emil Artin and the mathematicians in his circle. All of these mathematicians embraced and espoused what came to be understood as a structural approach to algebra, and van der Waerden drew critically from their point of view.

In the case of Noether, he plumbed the depths of Steinitz's 1910 paper on fields with her and came to appreciate its emphasis not only on an axiomatic formulation of the field concept but also on a systematic,

---

the years around the turn of the twentieth century would take us beyond the mathematical prerequisites we have assumed for this book. The interested reader should consult Curtis, 1999. In particular, Curtis gives a detailed discussion of the results in Noether, 1929 on pp. 214–223. For the statement of the descending chain condition, replace the phrase "one immediately preceding it" by "one immediately following it" in the theorem on finite chains above.

from-the-ground-up analysis of fields in relation to their subfields and to their extension fields. Steinitz had understood a *theory* of fields in terms of a structural hierarchy and of the interrelations between well-defined constructs at the different levels. Also from Noether, van der Waerden gained firsthand experience in her lectures on hypercomplex numbers and representation theory of the fruitfulness of an abstract and structural approach that allowed seemingly disparate theories to be united under one common rubric. In fact, as Noether freely acknowledged in the published work, "we [van der Waerden and I] undertook the treatment for the published version together. I am also indebted to B. L. van der Waerden for a series of critical remarks."[27]

In the mathematical seminar in Hamburg, on the other hand, van der Waerden witnessed how Artin and his collaborator, Otto Schreier (1901–1929), conceived the abstract and axiomatic notion of a *real field* as an abstract field having the additional property that -1 (that is, the negative with respect to addition of the field's multiplicative identity element 1) cannot be written as a sum of squares of other elements in the field. Pushing this purely abstract formulation, they effected an algebrization of the real numbers that hinged neither on the notion of continuity nor on similar analytic concepts. They showed, in other words, the essential, algebraic nature of the real numbers at the same time that they gave a purely algebraic characterization of the differences between the fields of rational numbers, algebraic numbers, real numbers, and complex numbers.[28] Algebra was thus liberated from the underlying properties of the real (and complex) numbers that had been assumed throughout the course of the nineteenth century.

By 1930–1931 and in the context of his two-volume textbook entitled *Moderne Algebra*, van der Waerden had internalized in a very novel way these algebraic developments in which he had been such a critical participant. He conceived of algebra as a structured and hierarchical whole, which he graphically illustrated in the organizational plan or *Leitfaden* that followed his book's table of contents (figure 14.3). In his words, this schematic gave an "overview of the chapters of both volumes" of his book as well as "their logical dependence."[29]

---

[27] Noether, 1929, p. 641 (our translation).

[28] Corry, 1996/2004, p. 49/47–48. For much more on real fields and their historical significance, see Sinaceur, 1991.

[29] Van der Waerden, 1930–1931, 1:viii (our translation).

Organizational Plan*

Overview of the chapters in both volumes and their logical dependence.

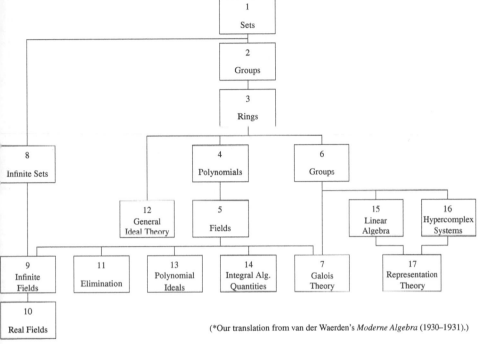

Figure 14.3.

(*Our translation from van der Waerden's *Moderne Algebra* (1930–1931).)

Working axiomatically from the least to the more structured, van der Waerden opened the book with a chapter on sets and followed that with one on groups and one on rings and fields. He also included additional chapters on other structures—such as ideals and hypercomplex number systems—as well as theories—like Galois theory and representation theory—that arose in the context of the basic domains. In particular, he devoted an entire chapter to so-called "linear algebra," an area which, as we have seen, had been emerging in diverse contexts within the mathematical literature over the course of the latter half of the nineteenth century but which received its first unified and axiomatic treatment in *Moderne Algebra* as the study of modules over a ring together with their homomorphisms (that is, linear transformations). Moreover, for the first time in the literature, van der Waerden not only presented axiomatizations of the key algebraic domains but also systematically

organized his study of each of them in terms of specific concepts—
"isomorphism, homomorphism, residue classes, composition series, direct products, etc."—and analyzed how each behaved with respect to the
same kinds of problems.[30] Just like Noether in her unification of the
theories of the factorization of algebraic integers and polynomials via
the theory of ideals in commutative rings, van der Waerden asked and
answered questions about factorization in various contexts, defining as
he went along necessary analogues of notions such as "prime" in the
diverse contexts he considered. He also asked, for each algebraic domain,
what its meaningful subdomains were and precisely how they related
to the original domain. Moreover, he constructed from given domains
new entities, such as quotients or extensions, and determined what
properties those new constructs shared with the original domain from
which they sprang. Finally, he explored whether it was possible, given
an axiomatization of a particular domain, to understand that domain
completely in simpler terms as one can do for a vector space, say, by
realizing that, up to isomorphism, vector spaces are determined by their
field of coefficients and by their dimension over that field.

What was so different about this approach? In contrast to the "image"
of algebra that Weber had conveyed in his *Lehrbuch der Algebra* of 1895
(recall the discussion at the end of chapter 11) in which the theory
was built on a foundation of the numbers characterizing the roots of
polynomial equations, van der Waerden put forth a "structural image"
of algebra in *Moderne Algebra* characterized by "the recurrent use of
certain kinds of concepts, the focusing on certain kinds of problems."
For van der Waerden, the number systems—$\mathbb{Z}$, $\mathbb{Q}$, $\mathbb{R}$, $\mathbb{C}$—and their
properties were no longer in any way privileged. Algebraic domains were
defined axiomatically and their structures were then explored in terms
of specific kinds of questions. Within this kind of a framework, number
systems merely represented particular examples of abstract algebraic
constructs. It is the algebra in this book that first fully meets Mahoney's
definition of algebra from chapter 1. Indeed, in *Moderne Algebra* there
is no "ontological commitment." All concepts were presented using an
operative symbolism in terms of which the relations between objects were

---

[30] Corry, 1996/2004, p. 47/45. The discussion that follows draws from Corry, 1996/2004,
pp. 46–53/45–52. For the quote in the next paragraph, see p. 52/51.

central. This structural image of algebra still guides the way in which we approach algebra decades later.

Van der Waerden's has proved to be a very potent conceptualization. It marked a critical shift from a notion of algebra as a means for finding the roots, whether real or complex, of an algebraic equation that, as we have seen, had held sway at least since the ninth century. The publication in 1930–1931 of his compelling presentation in *Moderne Algebra* thus marked a turning point in the history of algebra. In defining a common language that would come to be shared by algebraists internationally, it marked, at least symbolically, the "moment" when algebra became *modern*.

References

Abel, Niels Henrik. 1826. "Beweis der Unmöglichkeit algebraische Gleichungen von höheren Graden als dem vierten allgemein auflösen." *Journal für die reine und angewandte Mathematik* 1, 65–84 (or Abel, 1881, 1:7).

———. 1828. "Sur la résolution algébrique des équations," in Abel, 1881, 2:217–243.

———. 1829. "Mémoire sur une classe particulière d'équations résolubles algébriquement." *Journal für die reine und angewandte Mathematik* 4, 131–156 (or Abel, 1881, 1:478–507).

———. 1881. *Oeuvres complètes*. 2 Vols. Ed. Ludwig Sylow and Sophus Lie. Christiana: Imprimerie de Grondahl & Søn.

Al-Bīrūnī, Abū Rayḥān. 1910. *Alberuni's India*. 2 Vols. Trans. Edward C. Sachau. London: Kegan Paul, Trench, Trübner.

Alexander, Amir. 2010. *Duel at Dawn: Heroes, Martyrs, and the Rise of Modern Mathematics*. Cambridge, MA: Harvard University Press.

Amir-Moéz, Ali R. 1963. "A Paper of Omar Khayyam." *Scripta Mathematica* 26, 323–337.

Anbouba, Adel. 1975. "Al Samaw'al." In Gillispie, 1970–1990, 12:91–95.

Apollonius of Perga. 1998. *Conics: Books I–III*. Trans. R. Catesby Taliaferro. New Rev. Ed. Santa Fe: Green Lion Press.

Archimedes. 1953. *The Works of Archimedes with the Method of Archimedes*. Ed. with Introduction by Thomas L. Heath. New York: Dover.

Aronhold, Siegfried. 1850. "Zur Theorie der homogenen Functionen dritten Grades von drei Variabeln." *Journal für die reine und angewandte Mathematik* 39, 140–159.

———. 1858. "Theorie der homogenen Functionen dritten Grades von drei Veränderlichen." *Journal für die reine und angewandte Mathematik* 55, 97–191.

Artin, Emil. 1927. "Zur Theorie der hypercomplexen Zahlen." *Abhandlungen aus dem Mathematischen Seminar der Hamburgischen Universität* 5, 251–260.

Ayoub, Raymond G. 1980. "Paolo Ruffini's Contributions to the Quintic." *Archive for History of Exact Sciences* 23, 253–277.

Bachmann, Paul. 1881. "Ueber Galois' Theorie der algebraischen Gleichungen." *Mathematische Annalen* 18, 449–468.

Bajri, Sanaa Ahmed. 2011. "Arithmetic, Induction, and the Algebra of Polynomials: Al-Samaw'al and His 'Splendid Book of Algebra.'" Master's Thesis. University of Christchurch, New Zealand.

Baltus, Christopher. 2004. "D'Alembert's Proof of the Fundamental Theorem of Algebra." *Historia Mathematica* 31, 414–428.

Banach, Stefan. 1922. "Sur les opérations dans les ensembles abstraites et leur application aux équations intégrales." *Fundamenta Mathematicae* 3, 133–181.

Bashmakova, Isabella and Smirnova, Galina. 2000. *The Beginnings and Evolution of Algebra*. Washington: Mathematical Association of America.

Belhoste, Bruno. 1991. *Augustin-Louis Cauchy: A Biography*. Trans. Frank Ragland. New York: Springer.

Belhoste, Bruno and Lützen, Jesper. 1984. "Joseph Liouville et le Collège de France." *Revue d'histoire des sciences* 37, 255–304.

Bell, Eric T. 1937. *Men of Mathematics*. New York: Simon and Schuster.

Bellavitis, Giusto. 1835. "Saggio di applicazioni di un nuovo metodo di geometria analitica (Calcolo delle equipollenze)." *Annali delle scienze del Regno Lombardo-Veneto. Padova* 5, 244–259.

Berggren, J. Lennart. 1986. *Episodes in the Mathematics of Medieval Islam*. New York: Springer.

———. 2007. "Mathematics in Medieval Islam." In Katz, Ed. 2007, 515–675.

Bezout, Étienne. 1765. "Mémoire sur la résolution générale des équations de tous les degrés." *Mémoires de l'Académie Royale des Sciences de Paris*, 533–552.

Bhāskara II. 2000. *Bīja-gaṇita*. Unpublished translation by David Pingree. Parts of this are in Plofker, 2007.

Biermann, Kurt-R. 1988. *Die Mathematik und ihre Dozenten an der Berliner Universität, 1810–1933*. Berlin: Akademie.

Birembaut, Arthur. 1970. "Bravais, Auguste." In Gillispie, Ed. 1970–1990, 2:430-432.

Birkeland, Bent. 1996. "Ludvig Sylow's Lectures on Algebraic Functions and Substitutions, Cristiania (Oslo), 1862." *Historia Mathematica* 23, 182–199.

Black, Ellen. 1986. *The Early Theory of Equations: On Their Nature and Constitution*. Annapolis: Golden Hind Press.

Bôcher, Maxime. 1907. *Introduction to the Higher Algebra*. New York: Macmillan.

Boi, Luciano; Flament, Dominique; and Salanskis, Jean-Michel, Ed. 1992. *1830–1930: A Century of Geometry*. Berlin: Springer.

Bombelli, Rafael. 1579. *L'Algebra*. Bologna. G. Rossi.

———. 1966. *L'Algebra*. Milan: Feltrinelli Editore.

Boole, George. 1841. "Exposition of a General Theory of Linear Transformations. Part 1." *Cambridge Mathematical Journal* 3, 1–20.

———. 1842. "Exposition of a General Theory of Linear Transformations. Part 2." *Cambridge Mathematical Journal* 3, 106–119.

Borel, Armand. 2001. *Essays in the History of Lie Groups and Algebraic Groups*. AMS/LMS Series in the History of Mathematics. Vol. 21. Providence: American Mathematical Society and London: London Mathematical Society.

Bos, Henk. 2001. *Redefining Geometrical Exactness: Descartes' Transformation of the Early Modern Concept of Construction*. New York: Springer.

Bottazzini, Umberto. 1994. *Va' Pensiero: Immagini della matematica nell'Italia dell'Ottocento*. Bologna: Il Mulino.

Boyer, Carl. 1966. "Colin Maclaurin and Cramer's Rule." *Scripta Mathematica* 27, 377–379.

Brewer, James W. and Smith, Martha K., Ed. 1981. *Emmy Noether: A Tribute to Her Life and Work*. New York: Marcel Dekker.

Brown, Joseph E. 1978. "The Science of Weights." In Lindberg, Ed. 1978, 179–205.

Brown, Nancy Marie. 2010. *The Abacus and the Cross: The Story of the Pope Who Brought the Light of Science to the Dark Ages*. New York: Basic Books.

Bühler, Curt F. 1948. "The Statistics of Scientific Incunabula." *Isis* 39, 163–168.

Bühler, Walter Kaufmann. 1981. *Gauss: A Biographical Study*. Berlin: Springer.

Burnett, Charles. 2001. "The Coherence of the Arabic-Latin Translation Program in Toledo in the Twelfth Century." *Science in Context* 14, 249–288.

Burnside, William. 1897. *Theory of Groups of Finite Order*. Cambridge: Cambridge University Press.

Cajori, Florian. 1928. *A History of Mathematical Notations*. Vol. 1. Chicago: Open Court.

Cardano, Girolamo. 1545/1968. *The Great Art or The Rules of Algebra*. Trans. and Ed. T. Richard Witmer. Cambridge, MA: MIT Press.

———. 1575/2002. *The Book of My Life*. Intro. by Anthony Grafton. New York: New York Review Book Classics.

Cartan, Élie. 1898. "Sur les groupes bilinéaires et les systèmes de nombres complexes." *Annales de la Faculté des Sciences de Toulouse* 12B, pp. B1–B99 (or Cartan, 1952–1955, 1, pt. 2:7–105.)

———. 1952–1955. *Oeuvres complètes*. 3 Vols. in 6 Pts. Paris: Gauthier-Villars.

Cauchy, Augustin. 1815a. "Mémoire sur le nombre des valeurs qu'une fonction peut acquérir, lorsqu'on y permute de toutes les manières possibles les quantités qu'elle renferme." *Journal de l'École Polytechnique* 10, 1–28 (or Cauchy, 1882–1974, Ser. 2, 1:64–90).

———. 1815b. "Mémoire sur les fonctions qui ne peuvent obtenir que deux valeurs égales et de signes contraires par suite des transpositions opérées entre les variables qu'elles renferment." *Journal de l'École Polytechnique* 10, 29–112 (or Cauchy, 1882–1974, Ser. 2, 1:91–169).

———. 1826–1828. *Leçons sur les applications du calcul infinitésimal à la géométrie*. 2 Vols. Paris: Imprimerie royale (or Cauchy, 1882–1974, Ser. 2, 5).

———. 1829. "Sur l'équation à l'aide de laquelle on détermine les inégalités séculaires des mouvements des planètes." *Exercices de mathématiques* 4, 140–160 (or Cauchy, 1882–1974, Ser. 2, 9:174–195).

———. 1844. "Mémoire sur les arrangements que l'on peut former avec des lettres données." *Exercises d'analyse et de physique mathématique* 3, 151–242 (or Cauchy, 1882–1974, Ser. 2, 13:171–282).

———. 1882–1974. *Oeuvres complètes d'Augustin-Louis Cauchy*. 2 Ser. 28 Vols. Paris: Gauthier-Villars.

Cayley, Arthur. 1843. "Chapters in the Analytical Geometry of (*n*) Dimensions." *Cambridge Mathematical Journal* 4, 119–127 (or Cayley, 1889–1898, 1:55–62).

———. 1845a. "On the Theory of Linear Transformations." *Cambridge Mathematical Journal* 4, 193–209 (or Cayley, 1889–1898, 1:80–94).

———. 1845b. "On Jacobi's Elliptic Functions, in Reply to the Revd. Bronwin: and on Quaternions." *Philosophical Magazine* 26, 208–211 (or Cayley, 1889–1898, 1:127).

———. 1846a. "On Linear Transformations." *Cambridge and Dublin Mathematical Journal* 1, 104–122 (or Cayley, 1889–1898, 1:95–112).

———. 1846b. "Sur quelques propriétés des déterminants gauches." *Journal für die reine und angewandte Mathematik* 32, 119–123 (or Cayley, 1889–1898, 1:332–336).

———. 1853. "On a Property of the Caustic by Refraction of the Circle." *Philosophical Magazine* 6, 427–431 (or Cayley, 1889–1898, 2:118–122).

———. 1854a. "On the Theory of Groups, as Depending on the Symbolic Equation $\theta^n = 1$ – Part I." *Philosophical Magazine* 7, 40–47 (or Cayley, 1889–1898, 2:123–130).

———. 1854b. "An Introductory Memoir Upon Quantics." *Philosophical Transactions of the Royal Society of London* 144, 244–258 (or Cayley, 1889–1898, 2:221–234).

———. 1855a. "Remarques sur la notation des fonctions algébriques." *Journal für die reine und angewandte Mathematik* 50, 282–285 (or Cayley, 1889–1898, 2:185–188).

———. 1855b. "Sur la transformation d'une fonction quadratique en elle même par des substitutions linéaires." *Journal für die reine und angewandte Mathematik* 50, 288–299 (or Cayley, 1889–1898, 2:192–201).

———. 1856. "A Second Memoir Upon Quantics." *Philosophical Transactions of the Royal Society of London* 146, 101–126 (or Cayley, 1889–1898, 2:250–281).

———. 1858a. "A Memoir on the Theory of Matrices." *Philosophical Transactions of the Royal Society of London* 148, 17–37 (or Cayley, 1889–1898, 2:475–496).

———. 1858b. "A Memoir on the Automorphic Linear Transformation of a Bipartite Quadric Function." *Philosophical Transactions of the Royal Society of London* 148, 39–46 (or Cayley, 1889–1898, 2:497–505).

———. 1859. "On the Theory of Groups, as Depending on the Symbolic Equation $\theta^n = 1$ – Part III." *Philosophical Magazine* 18, 34–37 (or Cayley, 1889–1898, 4:88–91).

———. 1871. "A Ninth Memoir on Quantics." *Philosophical Transactions of the Royal Society of London* 161, 17–50 (or Cayley, 1889–1898, 7:334–353).

———. 1878a. "On the Theory of Groups." *Proceedings of the London Mathematical Society* 9, 126–133 (or Cayley, 1889–1898, 10: 324–330).

————. 1878b. "The Theory of Groups." *American Journal of Mathematics* 1, 50–52 (or Cayley, 1889–1898, 10: 401–403).

————. 1889–1898. *The Collected Mathematical Papers of Arthur Cayley*. Ed. Arthur Cayley and A. R. Forsyth. 14 Vols. Cambridge: Cambridge University Press.

Chace, Arnold Buffum. 1979. *The Rhind Mathematical Papyrus*. Reprinted Reston, VA: National Council of Teachers of Mathematics.

Chemla, Karine and Guo Shuchun. 2004. *Les neuf chapîtres: Le Classique mathématique de la Chine ancienne et ses commentaires. Édition critique bilingue traduite, présentée et annotée*. Paris: Dunod.

Cifoletti, Giovanna. 1992. "Mathematics and Rhetoric: Jacques Peletier, Guillaume Gosselin and the French Algebraic Tradition." PhD Dissertation. Princeton University.

————. 1996. "The Creation of the History of Algebra in the Sixteenth Century." In Goldstein, Gray, and Ritter, Ed. 1996, 121–142.

————. 2004. "The Algebraic Art of Discourse: Algebraic Dispositio, Invention and Imitation in Sixteenth-Century France." In *History of Science/History of Text*. Ed. Karine Chemla. New York: Springer, 123–135.

Clapham, Chistopher and Nicholson, James. 2005. *The Concise Oxford Dictionary of Mathematics*. 3rd Ed. Oxford: Oxford University Press.

Clebsch, Alfred. 1861a. "Ueber eine Transformation der homogenen Functionen dritter Ordnung mit vier Veränderlichen." *Journal für die reine und angewandte Mathematik* 58, 109–126.

————. 1861b. "Ueber symbolische Darstellung algebraischer Formen." *Journal für die reine und angewandte Mathematik* 59, 1–62.

————. 1872. *Theorie der binären algebraischen Formen*. Leipzig: B. G. Teubner.

Clifford, William Kingdon. 1878. "Applications of Grassmann's Extensive Algebra." *American Journal of Mathematics* 1, 350–358.

Colebrooke, H. T. 1817. *Algebra with Arithmetic and Mensuration from the Sanskrit of Brahmegupta and Bhāscara*. London: John Murray.

Colson, John. 1707. "Aequationum cubicarum et biquadraticarum, tun analytica, tum geometrica et mechanica, resolutio universalis," *Philosophical Transactions of the Royal Society* 25, 2353–2368.

Cooke, Roger. 2008. *Classical Algebra: Its Nature, Origins and Uses*. New York: Wiley-Interscience.

Corry, Leo. 1996/2004. *Modern Algebra and the Rise of Mathematical Structures*. 1st Ed./2nd Ed. Basel: Birkhäuser.

————. 2004. *David Hilbert and the Axiomatization of Physics (1898–1918): From Grundlagen der Geometrie to Grundlagen der Physik*. Dordrecht: Kluwer Academic.

Cramer, Gabriel. 1750. *Introduction à l'analyse des lignes courbes algébriques*. Geneva: Les Frères Cramer & Cl. Philibert.

Crilly, Tony. 1986. "The Rise of Cayley's Invariant Theory (1841–1862)." *Historia Mathematica* 13, 241–254.

———. 1988. "The Decline of Cayley's Invariant Theory (1863–1895)." *Historia Mathematica* 15, 332–347.

———. 2006. *Arthur Cayley: Mathematician Laureate of the Victorian Age*. Baltimore: Johns Hopkins University Press.

Crowe, Michael J. 1967. *A History of Vector Analysis: The Evolution of the Idea of a Vectorial System*. Notre Dame: University of Notre Dame Press.

Cullen, Christopher. 2004. *The Suan shu shu: Writings on Reckoning*. Needham Research Institute Working Papers. Cambridge: Needham Research Institute.

———. 2009. "People and Numbers in Early Imperial China." In Robson and Stedall, Ed. 2009, 591–618.

Cuomo, Serafina. 2001. *Ancient Mathematics*. London and New York: Routledge.

Curtis, Charles W. 1999. *Pioneers of Representation Theory: Frobenius, Burnside, Schur, and Brauer*. AMS/LMS Series in the History of Mathematics. Vol. 15. Providence: American Mathematical Society and London: London Mathematical Society.

Dalmas, André. 1956. *Évariste Galois: Révolutionnaire et géomètre*. Paris: Fasquelle (1981. 2nd Ed. Paris: Le Nouveau commerce).

Dardi of Pisa. 1344/2001. *Aliabraa argibra*. Ed. Rafaella Franci. Quaderni del Centro Studi della Matematica Medioevale. Siena: Servizio Editoriale dell'Università di Siena.

Darlington, Oscar. 1947. "Gerbert, the Teacher." *American Historical Review* 52, 456–476.

Datta, Bibhutibhusan and Singh, Avadhesh Nasayan. 1935–1938. *History of Hindu Mathematics*, 2 Vols. Bombay: Asia. (Reprinted in 1961 and 2004.)

Dauben, Joseph. 2007. "Chinese Mathematics." In Katz, Ed. 2007, 187–384.

———. 2008. "Suan shu shu: A Book on Numbers and Computations." *Archive for History of Exact Sciences* 62, 91–178.

De Moivre, Abraham. 1707. "Aequationum quarundam potestatis tertiæ, quintæ, septimæ, nonæ, et superiorum, ad infinitum usque pergendo, in terminis finitis, ad instar regularum procubivis quae vocantur Cardani, resolutio analytica." *Philosophical Transactions of the Royal Society* 25, 2368–2371.

De Morgan, Augustus. 1842. "On the Foundations of Algebra. No. 1 and No. 2." *Transactions of the Cambridge Philosophical Society* 7, 173–187 and 287–300, respectively.

———. 1847. *Formal Logic*. London: Taylor and Walton.

Dedekind, Richard. 1877/1996. *Sur la théorie des nombres entiers algébriques*. Paris: Gauthier-Villars; Trans. 1996. *Theory of Algebraic Integers*. Trans. John Stillwell. Cambridge: Cambridge University Press.

———. 1885. "Zur Theorie der aus *n* Haupteinheiten gebildeten komplexen Grössen." *Nachrichten von der Königlichen Gesellschaft der Wissenschaften und der Georg-Augusts-Universität zu Göttingen*, 141–159.

Dedekind, Richard and Weber, Heinrich. 1882/2012. "Theorie der algebraische Functionen einer Veränderlichen." *Journal für die reine und angewandte Mathematik* 92, 181–290; Trans. 2012. *Theory of Algebraic Functions of One Variable.* Trans. John Stillwell. Providence: American Mathematical Society and London: London Mathematical Society.

Derbyshire, John. 2006, *Unknown Quantity: A Real and Imaginary History of Algebra.* Washington: Joseph Henry Press.

Descartes, René. 1637/1954. *The Geometry of René Descartes.* Trans. David Eugene Smith and Marcia L. Latham. New York: Dover.

———. 1637/1968. *Discourse on Method and the Meditations.* Trans. F. E. Sutcliffe. New York: Penguin Books.

Dick, Auguste. 1981. *Emmy Noether 1882–1935.* Trans. H. I. Blocher. Boston: Birkhäuser.

Dickson, Leonard Eugene. 1901. *Linear Groups: With an Exposition of the Galois Field Theory.* Leipzig: B. G. Teubner.

Dieudonné, Jean. 1985. *History of Algebraic Geometry: An Outline of the History and Development of Algebraic Geometry.* Trans. Judith D. Sally. Monterey, CA: Wadsworth Advanced Books.

Dirichlet, Peter Lejeune. 1871. *Vorlesungen über Zahlentheorie.* Ed. Richard Dedekind. 2nd Ed. Braunschweig: Friedrich Vieweg und Sohn.

Dodgson, Charles. 1867. *Elementary Treatise on Determinants with Their Application to Simultaneous Linear Equations and Algebraical Geometry.* London: Macmillan.

Dold-Samplonius, Yvonne; Dauben, Joseph; Folkerts, Menso; and Van Dalen, Benno, Ed. 2002. *From China to Paris: 2000 Years Transmission of Mathematical Ideas.* Stuttgart: Frans Steiner.

Dorier, Jean-Luc. 1995. "A General Outline of the Genesis of Vector Space Theory." *Historia Mathematica* 22, 227–261.

Drake, Stillman. 1957. *Discoveries and Opinions of Galileo.* Garden City, NY: Doubleday Anchor Books.

Dunham, William. 1999. *Euler: The Master of Us All.* Washington: Mathematical Association of America.

Dupuy, Paul. 1896. "La vie d'Évariste Galois." *Annales scientifiques de l'École Normale Supérieure.* 3rd Ser. 13, 197–266.

Dyck, Walter von. 1882. "Gruppentheoretische Studien." *Mathematische Annalen* 20, 1–44.

———. 1883. "Gruppentheoretisch Studien II. Über die Zusammensetzung einer Gruppe discreter Operationen, über ihre Primitivität und Transitivität." *Mathematische Annalen* 22, 70–108.

Dzielska, Maria. 1995. *Hypatia of Alexandria.* Trans. F. Lyra. Cambridge, MA: Harvard University Press.

Edwards, Harold. 1977. *Fermat's Last Theorem: A Genetic Introduction to Algebraic Number Theory.* New York: Springer.

————. 1980. "The Genesis of Ideal Theory." *Archive for History of Exact Sciences* 32, 321–378.

————. 1984. *Galois Theory*. New York: Springer.

Ehrhardt, Caroline. 2010. "A Social History of the 'Galois Affair' at the Paris Academy of Sciences (1831)." *Science in Context* 23, 91–119.

————. 2011. *Évariste Galois, la fabrication d'une icône mathématique*. Paris: École des Hautes Études en Sciences Sociales.

Eisenstein, Gotthold. 1844a. "Théorèmes sur les formes cubiques et solution d'une équation du quatrième degré à quatre indéterminées." *Journal für die reine und angewandte Mathematik* 27, 75–79 (or Eisenstein, 1975, 1:1–5).

————. 1844b. "Allgemeine Auflösung der Gleichungen von den ersten vier Graden." *Journal für die reine und angewandte Mathematik* 27, 81–83 (or Eisenstein, 1975, 1:7–9).

————. 1844c. "Untersuchungen über die cubischen Formen mit zwei Variabeln." *Journal für die reine und angewandte Mathematik* 27, 89–104 (or Eisenstein, 1975, 1:10–25).

————. 1844d. "Lois de réciprocités." *Journal für die reine und angewandte Mathematik* 28, 53–67 (or Eisenstein, 1975, 1:126–140).

————. 1844e. "Allgemeine Untersuchungen über die Formen dritten Grades mit drei Variabeln, welche der Kreistheilung ihre Entstehung verdanken." *Journal für die reine und angewandte Mathematik* 28, 289–374 (or Eisenstein, 1975, 1:167–286).

————. 1850. "Über die Irreducibilität und einige andere Eigenschaften der Gleichung, von welcher die Thielung der ganzen Lemniscate abhängt." *Journal für die reine und angewandte Mathematik* 39, 160–179 and 224–287 (or Eisenstein, 1975, 2:536–619).

————. 1975. *Mathematische Werke*. 2 Vols. New York: Chelsea.

Elliott, Edwin Bailey. 1895. *An Introduction to the Algebra of Quantics*. Oxford: Oxford University Press.

Enros, Philip C. 1983. "The Analytical Society (1812–1813): Precursor to the Revival of Cambridge Mathematics." *Historia Mathematica* 10, 24–47.

Euclid. 2002. *Elements*. Trans. Thomas L. Heath. Ed. Dana Densmore. Santa Fe: Green Lion Press.

Euler, Leonhard. 1738. "De formis radicum aequationum cuiusque ordinis coniectatio." *Commentarii Academicae scientiarum petropolitanae* 6, 216–231 (or Euler, 1911–, Ser. 1, 6:1–19 or http://eulerarchive.org, entry E30.)

————. 1748/1988. *Introductio in analysin infinitorum*. 2 Vols. Lausanne: M. M. Bousquet; Trans. 1988. *Introduction to Analysis of the Infinite*. Trans. John Blanton. New York: Springer.

————. 1750. "Sur une contradiction apparente dans la doctrine des lignes courbes." *Mémoires de l'Académie des sciences de Berlin* 4, 219–233 (or Euler, 1911–, Ser. 1, 26:33–45 or http://eulerarchive.org, entry E147).

———. 1751. "Recherches sur les racines imaginaires des équations," *Mémoires de l'Académie des sciences de Berlin* 5, 222–288 (or Euler, 1911–, Ser. 1, 6:78–150 or http://eulerarchive.org, entry E170).

———. 1764. "De resolutione aequationum cuiusvis gradus." *Novi commentarii Academiae scientiarum petropolitanae,* 9, 70–98 (or Euler, 1911–, Ser. 1, 6:170–196 or http://eulerarchive.org, entry E282).

———. 1765a. "Recherches sur la connoissance méchanique des corps." *Mémoires de l'Académie des sciences de Berlin* 14, 131–153 (or Euler, 1911–, Ser. 2, 8:178–199 or http://eulerarchive.org, entry E291).

———. 1765b. "Du mouvement de rotation des corps solides autour d'un axe variable." *Mémoires de l'Académie des sciences de Berlin* 14, 154–193 (or Euler, 1911–, Ser. 2, 8:200–235 or http://eulerarchive.org, entry E292).

———. 1770/1984. *Vollständige Anleitung zur Algebra.* 2 Vols. St. Petersburg: Kays. Akademie der Wissenschaft; Reprinted 1984. *Elements of Algebra.* [Trans. Rev. John Hewlett. London: Longman, Orme, 1840]. New York: Springer.

———. 1911–. *Opera omnia.* 76(+) Vols. Basel: Birkhäuser.

Fauvel, John and Gray, Jeremy. 1987. *The History of Mathematics: A Reader.* London: Macmillan Education.

Fearnley-Sander, Desmond. 1979. "Grassmann and the Creation of Linear Algebra." *American Mathematical Monthly* 86, 809–817.

Fermat, Pierre. 1679. *Varia opera mathematica D. Petri de Fermat Senatoris Tolosani.* Ed. Clément-Samuel de Fermat. Toulouse: Joannis Pech.

Fibonacci, Leonardo. 1202/2002. *Fibonacci's Liber abaci: A Translation into Modern English of Leonardo Pisano's Book of Calculation.* Trans. Laurence Sigler. New York: Springer.

———. 1225/1987. *The Book of Squares: An Annotated Translation into Modern English.* Trans. Laurence Sigler. Boston: Academic Press.

Fierz, Markus. 1983. *Girolamo Cardano 1501–1576: Physician, Natural Philosopher, Mathematician, Astrologer, and Interpreter of Dreams.* Boston: Birkhäuser.

Fisch, Menachem. 1994. " 'The Emergency Which Has Arrived': The Problematic History of Nineteenth-Century British Algebra—A Programmatic Outline." *British Journal for the History of Science* 27, 247–276.

———. 1999. "The Making of Peacock's *Treatise on Algebra*: A Case of Creative Indecision." *Archive for History of Exact Sciences* 54, 137–179.

Flament, Dominique. 1992. "La 'lineale Ausdehnungslehre' (1844) de Hermann Günther Grassmann." In Boi, Flament, and Salanskis, Ed. 1992, 205–221.

Flegg, Graham; Hay, Cynthia; and Moss, Barbara, Ed. 1985. *Nicolas Chuquet, Renaissance Mathematician: A Study with Extensive Translation of Chuquet's Mathematical Manuscript Completed in 1484.* Dordrecht/Boston/Lancaster: D. Reidel.

Flood, Raymond; Rice, Adrian; and Wilson, Robin, Ed. 2011. *Mathematics in Victorian Britain.* Oxford: Oxford University Press.

Fox, Robert, Ed. 2000. *Thomas Harriot: An Elizabethan Man of Science*. Aldershot, UK: Ashgate.

Fraenkel, Adolf. 1914. "Ueber die Teiler der Null und die Zerlegungen von Ringen." *Journal für die reine und angewandte Mathematik* 145, 139–176.

Franci, Raffaella. 2002. "Trends in Fourteenth-Century Italian Algebra." *Oriens-Occidens* 4, 81–105.

Franci, Raffaella and Toti Rigatelli, Laura. 1983. "Maestro Benedetto e la storia dell'algebra." *Historia Mathematica* 10, 297–317.

———. 1985. "Towards a History of Algebra from Leonardo of Pisa to Luca Pacioli." *Janus* 72, 1–82.

Frend, William. 1796. *The Principles of Algebra*. London: J. Davis for G. G. and J. Robinson.

Friberg, Jöran. 2007. *A Remarkable Collection of Babylonian Mathematical Texts*. New York: Springer.

Frobenius, Georg. 1877. "Über das Pfaffsche Problem." *Journal für die reine und angewandte Mathematik* 82, 230–315 (or Frobenius, 1968, 1:249–334).

———. 1878. "Ueber lineare Substitutionen und bilineare Formen." *Journal für die reine und angewandte Mathematik* 84, 1–63 (or Frobenius, 1968, 1:343–405).

———. 1879. "Über homogene totale Differentialgleichungen." *Journal für die reine und angewandte Mathematik* 86, 1–19 (or Frobenius, 1968, 1:435–453).

———. 1887. "Neuer Beweis des Sylowschen Satzes." *Journal für die reine und angewandte Mathematik* 100, 179–181 (or Frobenius, 1968, 2:301–303).

———. 1968. *Gesammelte Abhandlungen*. Ed. Jean-Pierre Serre. 3 Vols. Berlin: Springer.

Funkhouser, H. Gray. 1930. "A Short Account of the History of Symmetric Functions of Roots of Equations." *American Mathematical Monthly* 37, 357–365.

Fuss, Paul H. 1843. *Correspondence mathématique et physique de quelques célèbres géomètres du XVIIIème siècle*. St. Petersburg: L'Imprimerie de l'Académie impériale des sciences.

Galileo, Galilei. 1638/1974. *Two New Sciences*. Trans. Stillman Drake. Madison: University of Wisconsin Press.

Galois, Évariste. 1831. "Mémoire sur les conditions de résolubilité des équations par radicaux." In Galois, 1962, pp. 43ff. First published in Galois, 1846.

———. 1846. "Oeuvres mathématiques." *Journal de mathématiques pures et appliquées* 11, 381–444.

———. 1962. *Ecrits et mémoires mathématiques d'Evariste Galois*. Ed. Robert Bourgne and Jean-Pierre Azra. Paris: Gauthier-Villars.

Gaukroger, Stephen. 1995. *Descartes: An Intellectual Biography*. Oxford: Clarendon Press.

Gauss, Carl Friedrich. 1801/1966. *Disquisitiones Arithmeticae*. Trans. Arthur A. Clarke, S.J. New Haven: Yale University Press (or Gauss, 1870–1929, Vol. 1).

———. 1818. "Neue Beweise und Erweiterungen des Fundamentalsatzes in der Lehre von den quadratisches Resten." In Gauss, 1889, pp. 496–510 (originally published in Latin in *Commentationes Societatis regias scientiarum gottingensis recentiores* 4, 3–20 or Gauss, 1870–1929, 2:47–64).

———. 1828. "Theorie der biquadratischen Reste: Erste Abhandlung." In Gauss, 1889, 511–533 (originally published in Latin in *Commentationes Societatis regias scientiarum gottingensis recentiores* 6 (1828), 27–56 or Gauss, 1897–1929, 2:65–94).

———. 1832. "Theorie der biquadratischen Reste: Zweite Abhandlung." In Gauss, 1889, 496–510 (originally published in Latin in *Commentationes Societatis regias scientiarum gottingensis recentiores* 7 (1832), 89–148 or Gauss, 1870–1929, 2:95–148).

———. 1870–1929. *Carl Friedrich Gauss Werke*. Ed. Konigliche Gesellschaft der Wissenschaften zu Göttingen. 12 Vols. Göttingen: Dieterichsche Universitätsdruckerei.

———. 1889. *Untersuchungen über höhere Arithmetik*. Trans. H. Maser. Berlin: Julius Springer.

Gavagna, Veronica. 2012. "L'*Ars magna arithmeticæ* nel corpus matematico di Cardano." In Rommevaux, Spiesser, and Massa Esteve, Ed. 2012, 237–268.

Gibbs, Josiah Willard. 1901. *Vector Analysis*. New York: Charles Scribner's Sons.

Gilio of Siena. 1384/1983. *Questioni d'algebra*. Ed. Raffaella Franci. Quaderni del Centro Studi della Matematica Medioevale. Siena: Servizio Editoriale dell'Università di Siena.

Gillings, Richard J. 1972. *Mathematics in the Time of the Pharaohs*. Cambridge, MA: MIT Press.

Gillispie, Charles C., Ed. 1970–1990. *Dictionary of Scientific Biography*. 16 Vols. and 2 Supps. New York: Charles Scribner's Sons.

Ginsburg, Jekuthiel. 1934. "A Hitherto Unpublished Letter by Benjamin Peirce." *Scripta Mathematica* 2, 278–282.

Goldstein, Catherine. 2011. "Charles Hermite's Stroll through the Galois Fields." *Revue d'histoire des mathématiques* 17, 211–270.

Goldstein, Catherine; Gray, Jeremy; and Ritter, Jim, Ed. 1996. *L'Europe mathématique: Histoire, mythes, identités/Mathematical Europe: History, Myth, Identity*. Paris: Éditions de la Maison des sciences de l'homme.

Gordan, Paul. 1868. "Beweiss, dass jede Covariante und Invariante einer binären Form eine ganze Function mit numerischen Coefficienten einer endlichen Anzahl solcher Formen ist." *Journal für die reine und angewandte Mathematik* 69, 323–354.

Grassmann, Hermann. 1844. *Die lineale Ausdehnungslehre, ein neuer Zweig der Mathematik dargestellt und durch Anwendungen auf die übrigen Zweige der Mathematik, wie auch auf die Statik, Mechanik, die Lehre vom Magnetismus und die Kristallonomie erläutert.* Leipzig: O. Wigand.

———. 1862. *Die Ausdehnungslehre: Vollständig und in strenger Form bearbeitet.* Berlin: von T. C. F. Enslin.

———. 1894–1911. *Gesammelte mathematische und physikalische Werke.* Ed. Friedrich Engel et al. 3 Vols. in 5 Pts. Leipzig: B. G. Teubner.

Grattan-Guinness, Ivor. 1997. "Benjamin Peirce's 'Linear Associative Algebra' (1870): New Light on Its Preparation and Publication." *Annals of Science* 54, 597–606.

———. 2004a. "History or Heritage? An Important Distinction in Mathematics and for Mathematics Education." *American Mathematical Monthly* 111, 1–12.

———. 2004b. "The Mathematics of the Past: Distinguishing Its History from Our Heritage." *Historia Mathematica* 31, 163–185.

———. 2011. "Victorian Logic." In Flood, Rice, and Wilson, Ed. 2011, 359–374.

———. 2012. "On the Role of the École Polytechnique, 1794–1914, with Especial Reference to Mathematics." In *A Master of Science History.* Ed. Jed Z. Buchwald. *Archimedes* 30, 217–234.

Graves, Robert Perceval. 1882–1889/1975. *The Life of Sir William Rowan Hamilton.* 3 Vols. Dublin: Hodges, Figgis, and Co. Reprinted 1975. New York: Arno Press.

Gray, Jeremy J. and Parshall, Karen Hunger. 2007. *Episodes in the History of Modern Algebra (1800–1950),* AMS/LMS Series in the History of Mathematics. Vol. 32. Providence: American Mathematical Society and London: London Mathematical Society.

Gregory, Duncan. 1837–1839. "On the Integration of Simultaneous Differential Equations." *Cambridge Mathematical Journal* 1, 190–199.

Grootendorst, Albertus and Bakker, Miente. 2000. *Jan de Witt's Elementa curvarum linearum, Liber primus.* New York: Springer.

Grootendorst, Albertus; Bakker, Miente; Aarts, Jan; and Erné, Reinie. 2010. *Jan de Witt's Elementa curvarum linearum, Liber secundus.* New York: Springer.

Hahn, Hans. 1922. "Über Folgen linearer Operationen." *Monatshefte für Mathematik und Physik* 32, 3–88.

Hamilton, William Rowan. 1831. "Theory of Conjugate Functions, or Algebraic Couples; With a Preliminary and Elementary Essay on Algebra as the Science of Pure Time." *Transactions of the Royal Irish Academy* 17, 293–422.

———. 1844. "On Quaternions; Or a New System of Imaginaries in Algebra." *Philosophical Magazine* 25, 10–13.

———. 1846. "On Quaternions; Or a New System of Imaginaries in Algebra." *Philosophical Magazine* 29, 26–31.

———. 1853. *Lectures on Quaternions*. Dublin: Hodges and Smith.

Hankel, Hermann. 1867. *Theorie der complexen Zahlensysteme insbesondere der gemeinen imaginären Zahlen und der Hamilton'schen Quaternionen nebst ihrer geometrischen Darstellung*. Leipzig: Leopold Voss.

Hankins, Thomas L. 1980. *Sir William Rowan Hamilton*. Baltimore: Johns Hopkins University Press.

Happel, Dieter. 1980. "Klassifikationstheorie endlich-dimensionaler Algebren in der Zeit von 1880 bis 1920." *L'Enseignement mathématique* 26, 91–102.

Harriot, Thomas. 2003. *The Greate Invention of Algebra: Thomas Harriot's Treatise on Equations*, Ed. and Trans. Jacqueline Stedall. Oxford: Oxford University Press.

Hart, Roger. 2011. *The Chinese Roots of Linear Algebra*. Baltimore: Johns Hopkins University Press.

Hawkins, Thomas. 1972. "Hypercomplex Numbers, Lie Groups, and the Creation of Group Representation Theory." *Archive for History of Exact Sciences* 8, 243–287.

———. 1975a. "Cauchy and the Spectral Theory of Matrices." *Historia Mathematica* 2, 1–29.

———. 1975b. "The Theory of Matrices in the 19th Century." In *Proceedings of the International Congress of Mathematicians: Vancouver, 1974*. 2 Vols. N.p.: Canadian Mathematical Congress, 2:561–570.

———. 1977a. "Another Look at Cayley and the Theory of Matrices." *Archives internationales d'histoire des sciences* 26, 82–112.

———. 1977b. "Weierstrass and the Theory of Matrices." *Archive for History of Exact Sciences* 17, 119–163.

———. 2000. *Emergence of the Theory of Lie Groups: An Essay in the History of Mathematics, 1869–1926*. New York: Springer.

———. 2005. "Frobenius, Cartan, and the Problem of Pfaff." *Archive for History of Exact Sciences* 59, 381–436.

Hayashi, Takao and Takanori Kusuba. 1998. "Twenty-One Algebraic Normal Forms of Citrabhānu." *Historia Mathematica* 25, 1–21.

Heath, Thomas. 1921. *A History of Greek Mathematics*. 2 Vols. Oxford: Clarendon Press.

———. 1964. *Diophantus of Alexandria: A Study in the History of Greek Algebra*. New York: Dover.

Heaviside, Oliver. 1894. *Electromagnetic Theory*. London: "The Electrician" Printing and Publishing Company.

Hedman, Bruce A. 1999. "An Earlier Date for 'Cramer's Rule'." *Historia Mathematica* 26, 365–368.

Heeffer, Albrecht. 2010a. "Algebraic Partitioning Problems from Luca Pacioli's Perugia Manuscript (Vat. Lat. 3129)." *SCIAMVS: Sources and Commentaries in Exact Sciences* 11, 3–51.

———. 2010b. "From the Second Unknown to the Symbolic Equation." In *Philosophical Aspects of Symbolic Reasoning in Early Modern Mathematics*.

Ed. Albrecht Heeffer and Maarten Van Dyck. Milton Keynes, UK: College Publications, 57–101.

———. 2012. "The Rule of Quantity by Chuquet and de la Roche and Its Influence on German Cossic Algebra." In Rommevaux, Spiesser, and Massa Esteve, Ed. 2012, 127–147.

Hermite, Charles. 1854a. "Remarques sur un mémoire de M. Cayley relatif aux déterminants gauches." *Cambridge and Dublin Mathematical Journal* 9, 63–67.

———. 1854b. "Théorie des fonctions homogènes à deux indéterminées." *Cambridge and Dublin Mathematical Journal* 9, 172–217.

———. 1858. "Sur la résolution de l'équation du cinquième degré." *Comptes rendus hebdomadaires des séances de l'Académie des sciences* 46, 508–515.

Hesse, Otto. 1844a. "Über die Elimination der Variabeln aus drei algebraischen Gleichungen von zweiten Grade mit zwei Variabeln." *Journal für die reine und angewandte Mathematik* 28, 68–96.

———. 1844b. "Über die Wendepuncte der Curven dritter Ordnung." *Journal für die reine und angewandte Mathematik* 28, 97–107.

Hilbert, David. 1888–1889. "Zur Theorie der algebraischen Gebilde." *Nachrichten der Gesellschaft der Wissenschaften zu Göttingen, Mathematisch-Physicalische Abteilung*, 450–457, 25–34, and 423–430.

———. 1890. "Über die Theorie der algebraischen Formen." *Mathematische Annalen* 36, 473–534.

———. 1893. "Ueber die vollen Invariantensysteme." *Mathematische Annalen* 42, 313–373.

———. 1897. "Die Theorie der algebraischen Zahlkörper." *Jahresbericht der deutschen Mathematiker-Vereinigung* 4, 175–535.

———. 1902. "Mathematical Problems." Trans. Mary Francis Winston. *Bulletin of the American Mathematical Society* 8, 437–479.

Hogendijk, Jan. 1984. "Greek and Arabic Constructions of the Regular Heptagon." *Archive for History of Exact Sciences* 30, 197–330.

Hölder, Otto. 1889. "Zurückführung einer beliebigen algebraischen Gleichung auf eine Kette von Gleichungen." *Mathematische Annalen* 34, 26–56.

———. 1893. "Die Gruppen der Ordnungen p^3, pq^2, pqr, p^4." *Mathematische Annalen* 43, 301–412.

Høyrup, Jens. 1987. "The Formation of 'Islamic Mathematics': Sources and Conditions." *Science in Context* 1, 281–329.

———. 1988. "Jordanus de Nemore, 13th-Century Mathematical Innovator: An Essay on Intellectual Context, Achievement, and Failure." *Archive for History of Exact Sciences* 38, 307–363.

———. 1996. "The Four Sides and the Area: Oblique Light on the Prehistory of Algebra." In *Vita Mathematica: Historical Research and Integration with Teaching*. Ed. Ronald Calinger. Washington: Mathematical Association of America, 45–65.

———. 2002. *Lengths, Widths, Surfaces: A Portrait of Old Babylonian Algebra and Its Kin*. New York: Springer.

———. 2007. *Jacopo da Firenze's Tractatus Algorismi and Early Italian Abbacus Culture*. Science Networks Historical Studies. Vol. 34. Basel/Boston/Berlin: Birkhäuser.

Hsia, Florence. 2009. *Sojourners in a Strange Land: Jesuits and Their Scientific Missions in Late Imperial China*. Chicago: University of Chicago Press.

Huff, Toby. 1993. *The Rise of Early Modern Science: Islam, China, and the West*. Cambridge: Cambridge University Press.

Hughes, Barnabas. 1981. *Jordanus de Nemore: De numeris datis, A Critical Edition and Translation*. Berkeley: University of California Press.

———. 1982. "Medieval Latin Translations of Al-Khwarizmi's *Al-Jabr*." *Manuscripta* 26, 31–37.

Hutton, Charles. 1795-96. *A Mathematical and Philosophical Dictionary*. London: J. Davis, for J. Johnson, in St. Paul's Church-Yard.

Imhausen, Annette. 2007. "Egyptian Mathematics." In Katz, Ed. 2007, 7–56.

Jami, Catherine. 2012. *The Emperor's New Mathematics: Western Learning and Imperial Authority During the Kangxi Reign (1662–1722)*. Oxford: Oxford University Press.

Jayawardene, S. A. 1963. "Unpublished Documents Relating to Rafael Bombelli in the Archives of Bologna." *Isis* 54, 391–395.

———. 1965. "Rafael Bombelli, Engineer-Architect: Some Unpublished Documents of the Apostolic Camera." *Isis* 56, 298–306.

———. 1973. "The Influence of Practical Arithmetics on the Algebra of Rafael Bombelli." *Isis* 64, 510–523.

Jordan, Camille. 1870. *Traité des substitutions et des équations algébriques*. Paris: Gauthier-Villars.

Kantor, Isai and Solodovnikov, Aleksandr. 1989. *Hypercomplex Numbers: An Elementary Introduction to Algebras*. Trans. Abe Shenitzer. New York: Springer.

Karpinski, Louis. 1930. *Robert of Chester's Latin Translation of the Algebra of al-Khowarizmi*. Ann Arbor: University of Michigan Press.

Kasir, Daoud S. 1931. *The Algebra of Omar Khayyam*. New York: Columbia Teachers College.

Katz, Victor J., Ed. 2007. *The Mathematics of Egypt, Mesopotamia, China, India, and Islam: A Sourcebook*. Princeton: Princeton University Press.

Katz, Victor J. 2009. *A History of Mathematics: An Introduction*. 3rd Ed. Boston: Addison-Wesley.

Kaunzner, Wolfgang. 1996. "Christoff Rudolff, ein bedeutender Cossist in Wien." In *Rechenmeister und Cossisten der frühen Neuzeit*. Ed. Rainer Gebhardt and Helmuth Albrecht. Annaberg-Buchholz: Schriften des Adam-Ries-Bundes, 113–138.

Keller, Agathe. 2006. *Expounding the Mathematical Seed*. Vol. 1. Basel: Birkhäuser.

Klein, Felix. 1893. "A Comparative Review of Recent Researches in Geometry." (Translation of the original German paper of 1872, by Mellon W. Haskell.) *Bulletin of the New York Mathematical Society* 2, 215–249.

Klein, Jacob. 1968. *Greek Mathematical Thought and the Origin of Algebra*. Cambridge, MA: MIT Press.

Kleiner, Israel. 1998. "From Numbers to Rings: The Early History of Ring Theory." *Elemente der Mathematik* 53, 18–35.

———. 2007. *A History of Abstract Algebra*. Boston: Birkhäuser.

Kline, Morris. 1972. *Mathematical Thought from Ancient to Modern Times*. New York: Oxford University Press.

Kouteynikoff, Odile. 2012. "Règle de fausse position ou d'hypothèse dans l'œuvre de Guillaume Gosselin, algébriste de la Renaissance française." In Rommevaux, Spiesser, and Massa Esteve, Ed. 2012, 151–183.

Koppelman, Elaine. 1971–1972. "The Calculus of Operations and the Rise of Abstract Algebra." *Archive for History of Exact Sciences* 8, 155–242.

Kronecker, Leopold. 1870. "Auseinandersetzung einiger Eigenschaften der Klassenanzahl idealer complexer Zahlen." *Monatsberichte der Berliner Akademie*, 881–889.

Kummer, Eduard. 1844. "De numeris complexis, qui radicibus unitatis et numeris integris realibus constant, Gratulationschrift der Universität Breslau zur Jubelfeier der Universität Königsberg." Reprinted 1847. *Journal de mathématiques pures et appliquées* 12, 185–212.

———. 1846. "Zur Theorie der complexen Zahlen." *Monatsberichte Königlichen Preussischen Akademie der Wissenschaften zu Berlin*, 87–96. Reprinted 1847. *Journal für die reine und angewandte Mathematik* 35, 319–326.

———. 1847. "Über die Zerlegung der aus Wurzeln der Einheit gebildeten complexen Zahlen in ihre Primfactoren." *Journal für die reine und angewandte Mathematik* 35, 327–367.

Labarthe, Marie-Hélène. 2012. "L'Argumentation dans le traité d'algèbre de Pedro Nunes: La Part de l'arithmètique et celle de la géométrie." In Rommevaux, Spiesser, and Massa Esteve, Ed. 2012, 185–213.

Lacroix, Sylvestre. 1802/1816. *Traité élémentaire de calcul différentiel et de calcul intégral*. Paris: Duprat; Trans. 1816. *An Elementary Treatise on the Differential and Integral Calculus*. Trans. Charles Babbage, George Peacock, and John Herschel. Cambridge: J. Deighton and Sons.

Lagrange, Joseph-Louis. 1770–1771. "Réflexions sur la théorie algébrique des équations." *Nouveaux mémoires de l'Académie royale des sciences et belles-lettres de Berlin*, 1, 134–215; 2, 138–253 (or Lagrange, 1867–1892, 3:203–421).

———. 1788. *Méchanique analitique*. 2 Vols. Paris: Desaint.

———. 1867–1892. *Oeuvres de Lagrange*. Ed. Joseph Serret. 14 Vols. Paris: Gauthier-Villars.

Lam Lay-Yong and Ang Tian-Se. 1992. *Fleeting Footsteps: Tracing the Conception of Arithmetic and Algebra in Ancient China*. Singapore: World Scientific.

Laubenbacher, Reinhard and Pengelley, David. 2010. " 'Voici ce que j'ai trouvé': Sophie Germain's Grand Plan To Prove Fermat's Last Theorem." *Historia Mathematica* 37, 641–692.

Legendre, Adrien-Marie. 1785. "Recherches d'analyse indéterminée." *Mémoires de mathématique et de physique de l'Académie royale des sciences*, 465–559.

Lehto, Olli. 1998. *Mathematics Without Borders: A History of the International Mathematical Union*. New York: Springer.

Leibniz, Gottfried Wilhelm. 1849. *Leibnizens mathematische Schriften*. Ed. C. I. Gerhardt. 7 Vols. Berlin: A. Asher; Vol. 1. *Briefwechsel*.

Lejbowicz, Max. 2012. "La découverte des traductions latines du Kitab al-jabr wa l-muqābala d'al-Khwārizmī." In Rommevaux, Spiesser, and Massa Esteve, Ed. 2012, 15–32.

Levey, Martin. 1966. *The Algebra of Abū Kāmil, Kitāb fi'l-muqābala, in a Commentary by Mordecai Finzi*. Madison: University of Wisconsin Press.

Lewis, Albert C. 1977. "H. Grassmann's 1844 *Ausdehnungslehre* and Schleiermacher's *Dialektik*." *Annals of Science* 34, 103–162.

Li Yan and Du Shiran. 1987. *Chinese Mathematics: A Concise History*. Trans. John Crossley and Anthony Lun. Oxford: Clarendon Press.

Libbrecht, Ulrich. 1973. *Chinese Mathematics in the Thirteenth Century: The Shu-shu chiu-chang of Ch'in Chiu-shao*. Cambridge, MA: MIT Press.

Lie, Sophus. 1874. "Über Gruppen von Transformationen." *Nachrichten der Königlichen Gesellschaft der Wissenschaften zu Göttingen*, 529–542. Reprinted in Lie, 1924, 5:1–8.

———. 1876. "Theorie der Transformationsgruppen. Erste Abhandlung." *Archive for Mathematik og Naturvidenskab* 1, 19–57. Reprinted in Lie, 1924, 5:9–41.

———. 1924. *Gesammelte Abhandlungen*. Ed. Friedrich Engel and Poul Heegaard. 7 Vols. Oslo: H. Aschehoug and Leipzig: B. G. Teubner.

Lindberg, David C. 1978. "The Transmission of Greek and Arabic Learning to the West." In Lindberg, Ed. 1978, 52–90.

———. Ed. 1978. *Science in the Middle Ages*. Chicago: University of Chicago Press.

———. 1992. *The Beginnings of Western Science: The European Scientific Tradition in Philosophical, Religious, and Institutional Context, 600 B.C. to A.D. 1450*. Chicago: University of Chicago Press.

Lloyd, Geoffrey. 1996. *Adversaries and Authorities: Investigations into Ancient Greek and Chinese Science*. Cambridge: Cambridge University Press.

Loget, François. 2012. "L'Algèbre en France au XVIe siècle: Individus et réseaux." In Rommevaux, Spiesser, and Massa Esteve, Ed. 2012, 69–101.

Maclaurin, Colin. 1748. *A Treatise of Algebra, in Three Parts*. London: A. Millar.

Mahāvīra. 1912. *Gaṇitā-sara-saṅgraha*. Trans. M. Rangācārya. Madras: Government Press. Parts are reproduced in Plofker, 2007.

Mahoney, Michael. 1971. "The Beginnings of Algebraic Thought in the Seventeenth Century." Online translation of "Die Anfänge der algebraischen Denkweise im 17. Jahrhundert." *RETE: Strukturgeschichte der Naturwissenschaften* 1, 15–31 or http://www.princeton.edu/~hos/Mahoney/articles/beginnings/beginnings.htm.

———. 1973. *The Mathematical Career of Pierre de Fermat 1601–1665*. Princeton: Princeton University Press.

———. 1978. "Mathematics." In Lindberg, Ed. 1978, 145–178.

Martini, Laura. 2002. "An Episode in the Evolution of a Mathematical Community: The Case of Cesare Arzelà at Bologna." In Parshall and Rice, Ed. 2002, 165–178.

Maseres, Francis. 1758. *Dissertation on the Use of the Negative Sign in Algebra: Containing a Demonstration of the Rules Usually Given Concerning It; And Shewing How Quadratic and Cubic Equations May be Explained, without the Consideration of Negative Roots.* London: Samuel Richardson.

Massa Esteve, Maria Rosa. 2012. "Spanish 'Arte Major' in the Sixteenth Century." In Rommevaux, Spiesser, and Massa Esteve, Ed. 2012, 103–126.

Melville, Duncan J. 2002. "Weighing Stones in Ancient Mesopotamia." *Historia Mathematica* 29, 1–12.

Möbius, August Ferdinand. 1827. *Der barycentrische Calcul.* Leipzig: J. A. Barth.

Molien, Theodor. 1893. "Ueber Systeme höherer complexer Zahlen." *Mathematische Annalen* 41, 83–156.

Moore, Gregory H. 1995. "The Axiomatization of Linear Algebra: 1875–1940." *Historia Mathematica* 22, 262–303.

Moyon, Marc. 2012. "Algèbre & Pratica geometriæ en Occident médiéval latin: Abū Bakr, Fibonacci et Jean de Murs." In Rommevaux, Spiesser, and Massa Esteve, Ed. 2012, 33–65.

Muir, Thomas. 1906–1923. *The Theory of Determinants in the Historical Order of Development.* 4 Vols. London: Macmillan. Reprinted 1960. New York: Dover.

Muth, Peter. 1899. *Theorie und Anwendung der Elementartheiler.* Leipzig: B. G. Teubner.

Needham, Joseph. 1959. *Science and Civilization in China.* Vol. 3. Cambridge: Cambridge University Press.

Nesselmann, Georg H. F. 1842. *Die Algebra der Griechen.* Berlin: G. Reimer.

Netto, Eugen. 1882. *Die Substitutionentheorie und ihre Anwendung auf die Algebra.* Leipzig: B. G. Teubner.

Netz, Reviel. 2004a. *The Transformation of Mathematics in the Early Mediterranean World: From Problems to Equations.* Cambridge: Cambridge University Press.

———. 2004b. *The Works of Archimedes* Vol. 1. Cambridge: Cambridge University Press.

Neugebauer, Otto. 1951. *The Exact Sciences in Antiquity.* Princeton: Princeton University Press.

Neumann, Olaf. 2006. "The *Disquisitiones Arithmeticae* and the Theory of Equations." In *The Shaping of Arithmetic after C. F. Gauss's Disquisitiones Arithmeticae.* Ed. Catherine Goldstein, Norbert Schappacher, and Joachim Schwermer. Berlin: Springer, 107–127.

———. 2007. "Divisibility Theories in the Early History of Commutative Algebra and the Foundations of Algebraic Geometry." In Gray and Parshall, Ed. 2007, 73–105.

Neumann, Peter M. 2006. "The Concept of Primitivity in Group Theory and the Second Memoir of Galois." *Archive for History of Exact Sciences* 60, 379–429.

———. 2011. *The Mathematical Writings of Évariste Galois.* Zürich: European Mathematical Society.

Newton, Isaac. 1707. *Arithmetica universalis, sive de compositione et resolutione arithmetica liber.* London: Benj. Tooke.

———. 1728. *Universal Arithmetick; or, a Treatise of Arithmetical Composition and Resolution.* Trans. Mr. Ralphson. London: J. Senex. (Available in *The Sources of Science.* No. 3. New York: Johnson Reprint Corporation.)

Nicholson, Julia J. 1993. "The Development and Understanding of the Concept of a Quotient Group." *Historia Mathematica* 20, 68–88.

Noether, Emmy. 1921. "Idealtheorie in Ringbereichen." *Mathematische Annalen* 83, 24–66.

———. 1929. "Hyperkomplexe Grössen und Darstellungstheorie." *Mathematische Zeitschrift* 30, 641–692.

Nový, Luboš. 1973. *Origins of Modern Algebra.* Prague: Academia.

Nuñes, Pedro. 1567. *Libro de algebra.* Antwerp: House of the Heirs of Arnold Birckman. Accessible at http://www.archive.org/details/librodealgebrae00nunegoog.

Oaks, Jeffrey and Alkhateeb, Haitham. 2005. "*Māl,* Enunciations, and the Prehistory of Arabic Algebra." *Historia Mathematica* 32, 400–425.

Ore, Oystein. 1953. *Cardano: The Gambling Scholar.* Princeton: Princeton University Press.

———. 1974. *Niels Henrik Abel: Mathematician Extraordinary.* New York: Chelsea.

Parshall, Karen Hunger. 1985. "Joseph H. M. Wedderburn and the Structure Theory of Algebras." *Archive for History of Exact Sciences* 32, 223–349.

———. 1988a. "The Art of Algebra from al-Khwārizmī to Viète: A Study in the Natural Selection of Ideas." *History of Science* 26, 129–164.

———. 1988b. "America's First School of Mathematical Research: James Joseph Sylvester at The Johns Hopkins University 1876–1883." *Archive for History of Exact Sciences* 38, 153–196.

———. 1989. "Toward a History of Nineteenth-Century Invariant Theory." In *The History of Modern Mathematics.* Ed. David E. Rowe and John McCleary. 2 Vols. Boston: Academic Press, 1:157–206.

———. 1995. "Mathematics in National Contexts (1875–1900): An International Overview." In *Proceedings of the International Congress of Mathematicians: Zürich.* 2 Vols. Basel/Boston/Berlin: Birkhäuser, 1995, 2:1581–1591.

———. 1998. *James Joseph Sylvester: Life and Work in Letters.* Oxford: Clarendon Press.

———. 1999. "The Mathematical Legacy of James Joseph Sylvester." *Neuw Archief voor Wiskunde* 17, 247–267.

———. 2006. *James Joseph Sylvester: Jewish Mathematician in a Victorian World.* Baltimore: Johns Hopkins University Press.

———. 2009. "The Internationalization of Mathematics in a World of Nations: 1800–1960." In Robson and Stedall, Ed. 2009, 85–104.

———. 2011. "Victorian Algebra." In Flood, Rice, and Wilson, Ed. 2011, 339–356.

Parshall, Karen Hunger and Rice, Adrian C. Ed. 2002. *Mathematics*

Unbound: The Evolution of an International Mathematical Community, 1800–1945. AMS/LMS Series in the History of Mathematics. Vol. 23. Providence: American Mathematical Society and London: London Mathematical Society.

Parshall, Karen Hunger and Rowe, David E. 1994. *The Emergence of the American Mathematical Research Community (1876–1900): J. J. Sylvester, Felix Klein, and E. H. Moore.* AMS/LMS Series in the History of Mathematics. Vol. 8. Providence: American Mathematical Society and London: London Mathematical Society, 1994.

Parshall, Karen Hunger and Seneta, Eugene. 1997. "Building an International Reputation: The Case of J. J. Sylvester." *American Mathematical Monthly* 104, 210–222.

Partington, James R. 1952. *An Advanced Treatise on Physical Chemistry.* 3 Vols. London: Longmans, Green.

Peacock, George. 1820. *A Collection of Examples of the Applications of the Differential and Integral Calculus.* Cambridge: J. Deighton and Sons.

———. 1830. *A Treatise on Algebra.* Cambridge: J. and J. J. Deighton.

———. 1834. "Report on the Recent Progress and Present State of Certain Branches of Analysis." In *Report on the Third Meeting of the British Association for the Advancement of Science Held at Cambridge in 1833.* London: Murray, pp. 185–352.

Peano, Giuseppe. 1888. *Calcolo geometrico secondo l'Ausdehnungslehre di H. Grassmann, preceduto dalle operazioni della logica deducttiva.* Turin: Bocca.

Peirce, Benjamin. 1881. "Linear Associative Algebra with Notes and Addenda by C. S. Peirce, Son of the Author." *American Journal of Mathematics* 4, 97–229.

Petsche, Hans-Joachim; Lewis, Albert C.; Liesen, Jörg; and Russ, Steve, Ed. 2011. *Hermann Graßmann: From Past to Future: Graßmann's Work in Context.* Basel: Birkhäuser.

Plofker, Kim. 2007. "Mathematics in India." In Katz, Ed. 2007, 385–515.

———. 2009. *Mathematics in India.* Princeton: Princeton University Press.

Pollard, Justin and Reid, Howard. 2006. *The Rise and Fall of Alexandria: Birthplace of the Modern Mind.* New York: Viking Press.

Poincaré, Henri. 1884. "Sur les nombres complexes." *Comptes rendus hebdomadaires des scéances de l'Académie des sciences* 99, 740–742.

Puig, Luis. 2011. "Researching the History of Algebraic Ideas from an Educational Point of View." In *Recent Developments on Introducing a Historical Dimension in Mathematics Education.* Ed. Victor J. Katz and Constantinos Tzanakis. Washington: Mathematical Association of America, 29–43.

Pycior, Helena. 1979. "Benjamin Peirce's 'Linear Associative Algebra.'" *Isis* 70, 537–551.

———. 1982. "Early Criticism of the Symbolical Approach to Algebra." *Historia Mathematica* 9, 392–412.

———. 1983. "Augustus De Morgan's Algebraic Work: The Three Stages." *Isis* 74, 211–226.

Rashed, Roshdi. 1986. *Sharaf al-Dīn al-Tūsī, oeuvres mathématiques: Algèbre et géométrie au XIIe siècle*. Paris: Les Belles Lettres.

———. 1994a. *The Development of Arabic Mathematics: Between Arithmetic and Algebra*. Dordrecht: Kluwer Academic.

———. 1994b. "Fibonacci et les mathématiques arabes." *Micrologus: Natura, scienze e società medievali* 2, 145–160.

Rashed, Roshdi and Vahabzadeh, Bijan. 2000. *Omar Khayyam the Mathematician*. New York: Bibliotheca Persica Press.

Recorde, Robert. 1557. *The Whetstone of Witte*. London: John Kyngston.

Reich, Karin. 1968. "Diophant, Cardano, Bombelli, Viète: Ein Vergleich ihrer Aufgaben." In *Rechenpfennige: Aufsätze zur Wissenschaftsgeschichte Kurt Vogel zum 80. Geburtstag*. Munich: Forschungsinstitut des Deutschen Museums für die Geschichte der Naturwissenschaften und der Technik.

Reid, Constance. 1972. *Hilbert*. New York: Springer.

Rivoli, M. T. and Simi, Annalisa. 1998. "Il Calcolo delle radici quadrate e cubiche in Italia da Fibonacci a Bombelli." *Archive for History of Exact Sciences* 52, 161–193.

Robson, Eleanor. 2007. "Mesopotamian Mathematics." In Katz, Ed. 2007, 57–186.

———. 2008. *Mathematics in Ancient Iraq: A Social History*. Princeton: Princeton University Press.

Robson, Eleanor and Stedall, Jacqueline, Ed. 2009. *The Oxford Handbook of the History of Mathematics*. Oxford: Oxford University Press.

Roman, Steven. 2005. *Advanced Linear Algebra*. 2nd Ed. New York: Springer.

Rommevaux, Sabine. 2012. "Qu'est-ce que l'algèbre pour Christoph Clavius?." In Rommevaux, Spiesser, and Massa Esteve, Ed. 2012, 293–309.

Rommevaux, Sabine; Spiesser, Maryvonne; and Massa Esteve, Maria Rosa, Ed. 2012. *Pluralité de l'algèbre à la Renaissance*. Paris: Honoré Champion Éditeur.

Rose, Paul Lawrence. 1975. *The Italian Renaissance of Mathematics: Studies of Humanists and Mathematicians from Petrarch to Galileo*. Geneva: Librairie Droz.

Rosen, Frederic. 1831. *The Algebra of Muhammed ben Musa*. London: Oriental Translation Fund; reprinted 1986. Hildesheim: Olms.

Rowe, David E. 1996. "On the Reception of Grassmann's Work in Germany during the 1870s." In Schubring, Ed. 1996, 131–145.

Rudolff, Christoff. 1525. *Coss*. Strassburg: Johannes Jung.

Ruffini, Paolo. 1799. *Teoria generale delle equazioni, in cui si dimostra impossibile la soluzione algebrica delle equazioni generali di grado superiore al quarto*. 2 Vols. Bologna: Stamperia di S. Tommaso d'Aquino.

———. 1802a. "Della soluzione delle equazioni algebriche determinate particolari di grado superiore al quarto." *Memorie di matematica e di fisica della Società italiana delle scienze* 9, 444–526.

———. 1802b. "Riflessioni intorno alla rettificazione, ed alla quadratura del circolo." *Memorie di matematica e di fisica della Società italiana delle scienze* 9, 527–557.

———. 1803. "Della insolubilità delle equazioni algebriche generali di grado superiore al quarto." *Memorie di matematica e di fisica della Società italiana delle scienze* 10, 410–470.

Saliba, George. 2007. *Islamic Science and the Making of the European Renaissance.* Cambridge, MA: MIT Press.

Sandifer, C. Edward. 2007. *How Euler Did It.* Washington: Mathematical Association of America.

Sarma, K. V. 2008. *Ganita-Yukti-Bhāṣā (Rationales in Mathematical Astronomy) of Jyeṣṭhadeva.* Vol. 1. New Delhi: Hindustan Book Agency.

Sarma, Sreeramula Rajeswara. 2002. "Rule of Three and Its Variations in India." In Dold-Samplonius et al., Ed. 2002, 133–156.

Sarton, George. 1938. "The Scientific Literature Transmitted through the Incunabula." *Osiris* 5, 41–123 and 125–245.

Sayili, A. 1962. *Logical Necessities in Mixed Equations by 'Abd al-Hamid ibn Turk and the Algebra of His Time.* Ankara: Türk Tarih Kurumu Basimevi.

Schappacher, Norbert. 2007. "A Historical Sketch of B. L. van der Waerden's Work in Algebraic Geometry: 1926–1946." In Gray and Parshall, Ed., 2007, 245–283.

Scheffers, Georg. 1889a. "Zur Theorie der aus n Haupteinheiten gebildten complexen Grössen." *Berichte über die Verhandlungen der königlich sächsischen Gesellschaft der Wissenschaften zu Leipzig. Mathematisch-physische Klasse* 41, 290–307.

———. 1889b. "Ueber die Berechnung von Zahlensysteme." *Berichte über die Verhandlungen der königlich sächsischen Gesellschaft der Wissenschaften zu Leipzig. Mathematisch-physische Klasse* 41, 400–457.

———. 1891. "Zurückführung complexer Zahlensysteme auf typische Formen." *Mathematisische Annalen* 39, 294–340.

———. 1893. "Ueber die Reducibilität complexer Zahlensysteme." *Mathematische Annalen* 41, 601–604.

Scholz, Erhart, Ed. 1990. *Geschichte der Algebra: Eine Einführung.* Mannheim: Bibliographisches Institut Wissenschaftsverlag.

Schubring, Gert, Ed. 1996. *Hermann Günther Grassmann (1809–1977): Visionary Mathematician, Scientist and Neohumanist Scholar.* Dordrecht: Kluwer Academic.

Séguier, Jean-Armand de. 1904. *Éléments de la théorie des groupes abstraits.* Paris: Gauthier-Villars.

Serafati, Michel. 2005. *La Révolution symbolique: La Constitution de l'écriture symbolique mathématique.* Paris: Éditions Petra.

Serret, Joseph. 1866. *Cours d'algèbre supérieure.* 3rd Ed. Paris: Gauthier-Villars.

Sesiano, Jacques. 1977a. "Les Méthodes d'analyse indéterminée chez Abū Kāmil." *Centaurus* 21, 89–105.

———. 1977b. "Le Traitement des équations indéterminées dans le *Badī' fi'l-Ḥisāb* d'Abū Bakr al-Karajī." *Archive for History of Exact Sciences* 17, 297–379.

———. 1982. *Books IV to VII of Diophantus' Arithmetica in the Arabic Translation Attributed to Qusṭā ibn Lūqā*. New York: Springer.

———. 1985. "The Appearance of Negative Solutions in Mediaeval Mathematics." *Archive for History of Exact Sciences* 32, 105–150.

———. 1999. *Une Introduction à l'histoire de l'algèbre: Résolution des équations des mésopotamiens à la Renaissance*. Lausanne: Presses polytechniques et universitaires romandes.

———. 2009. *An Introduction to the History of Algebra: Solving Equations from Mesopotamian Times to the Renaissance*. Trans. (of Sesiano, 1999) Anna Pierrchumbert. Providence: American Mathematical Society.

Shen Kangshen; Crossley, John N.; and Lun, Anthony W.-C. 1999. *The Nine Chapters on the Mathematical Art*. Oxford: Oxford University Press.

Shukla, K. S., 1954. "Ācārya Jayadeva, the Mathematician." *Gaṇita* 5, 1–20.

Sinaceur, Hourya. 1991. *Corps et modèles: Essai sur l'histoire de l'algèbre réelle*. Paris: J. Vrin.

Slembek, Silke. 2007. "On the Arithmetization of Algebraic Geometry." In Gray and Parshall, Ed. 2007, 285–300.

Smith, David Eugene. 1959. *A Source Book in Mathematics*. New York: Dover.

Sørenson, Henrik Kragh. 1999. *Niels Henrik Abel and the Theory of Equations*. Online at http://www.henrikkragh.dk/pdf/part199911g.pdf.

Speziali, Pierre. 1973. "Luca Pacioli et son oeuvre." In *Sciences de la Renaissance: VIIIe Congrès international de Tour*. Paris: Librairie philosophique J. Vrin, 93–106.

Spiesser, Maryvonne. 2006. "L'Algèbre de Nicolas Chuquet dans le context français de l'arithmétique commerciale." *Revue d'histoire des mathématique* 12, 7–33.

———. 2012. "Pedro Nunes: Points de vue sur l'algèbre." In Rommevaux, Spiesser, and Massa Esteve, Ed. 2012, 269–291.

Srinivasan, Bhama and Sally, Judith, Ed. 1983. *Emmy Noether at Bryn Mawr*. New York: Springer.

Srinivasiengar, C. N. 1967. *The History of Ancient Indian Mathematics*. Calcutta: World Press Private.

Stedall, Jacqueline. 2011. *From Cardano's Great Art to Lagrange's Reflections: Filling a Gap in the History of Algebra*. Zürich: European Mathematical Society.

———. 2012. "Narratives of Algebra in Early Printed European Texts." In Rommevaux, Spiesser, and Massa Esteve, Ed. 2012, 217–235.

Steinitz, Ernst. 1910. "Algebraische Theorie der Körper." *Journal für die reine und angewandte Mathematik* 137, 167–302.

Stifel, Michael, 1544. *Arithmetica integra*. Nuremberg: Johann Petreius.

———. 1553. *Die Coss Christoffs Rudolffs*. Königsberg: Alexander Behm.

Stillwell, John. 2012. "Translator's Introduction." In *Theory of Algebraic Functions of One Variable*. By Richard Dedekind and Heinrich Weber. Trans. John Stillwell. Providence: American Mathematical Society and London: London Mathematical Society, 1–37.

Struik, Dirk. 1986. *A Source Book in Mathematics, 1200–1800*. Princeton: Princeton University Press.

Stubhaug, Arild. 2000. *Niels Henrik Abel and His Times: Called Too Soon by Flames Afar*. Trans. Richard H. Daly. Berlin: Springer.

Sylow, Ludwig. 1872. "Théorèmes sur les groupes de substitutions." *Mathematische Annalen* 5, 584–594.

Sylvester, James Joseph. 1850. "Additions to the Articles, 'On a New Class of Theorems,' and 'On Pascal's Theorem'." *Philosophical Magazine* 37, 363–370 (or Sylvester, 1904–1912, 1:145–151).

———. 1851a. "On the Relation between the Minor Determinants of Linearly Equivalent Quadratic Functions." *Philosophical Magazine* 1, 295–305 (or Sylvester, 1904–1912, 1:241–250).

———. 1851b. "Sketch of a Memoir on Elimination, Transformation, and Canonical Forms." *Cambridge and Dublin Mathematical Journal* 6, 186–200 (or Sylvester, 1904–1912, 1:184–197).

———. 1852a. "A Demonstration of the Theorem That Every Homogeneous Quadratic Polynomial Is Reducible by Real Orthogonal Substitutions to the Form of a Sum of Positive and Negative Squares." *Philosophical Magazine*, 4, 138–142 (or Sylvester, 1904–1912, 1:378–381).

———. 1852b. "On the Principles of the Calculus of Forms." *Cambridge and Dublin Mathematical Journal* 7, 52–97 and 179–217 (or Sylvester, 1904–1912, 1:284–363).

———. 1853. "On a Theory of the Syzygetic Relations of Two Rational Integral Functions, Comprising an Application to the Theory of Sturm's Functions, and That of the Greatest Algebraical Common Measure." *Philosophical Transactions of the Royal Society of London*, 143, 407–548 (or Sylvester, 1904–1912, 1:429–586).

———. 1879. "Tables of Generating Functions and Ground-Forms for the Binary Quantics of the First Ten Orders." *American Journal of Mathematics* 2, 223–251 (or Sylvester, 1904–1912, 3:489–508).

———. 1882. "A Word on Nonions." *Johns Hopkins University Circulars* 1, 241–242 (or Sylvester, 1904–1912, 3:647–650).

———. 1883. "Sur les quantités formant un groupe de nonions analogues aux quaternions de Hamilton." *Comptes rendus des séances de l'Académie des sciences* 97, 1336–1340 (or Sylvester, 1904–1912, 4:118–121).

———. 1884. "Lectures on the Principles of the Universal Algebra." *American Journal of Mathematics* 6, 270–286 (or Sylvester, 1904–1912, 4:208–224).

———. 1904–1912. *The Collected Mathematical Papers of James Joseph Sylvester*. Ed. Henry F. Baker. 4 Vols. Cambridge: Cambridge University Press. Reprinted 1973. New York: Chelsea.

Taber, Henry. 1889. "On the Theory of Matrices." *American Journal of Mathematics* 12, 337–396.

Taisbak, Christian Marinus. 2003. *Euclid's Data, or The Importance of Being Given*. Copenhagen: Museum Tusculanum Press.

Taylor, R. Emmett. 1942. *No Royal Road: Luca Pacioli and His Times*. Chapel Hill: University of North Carolina Press.

Tignol, Jean-Pierre. 2001. *Galois' Theory of Algebraic Equations*. Singapore: World Scientific.

Tobies, Renate. 1996. "The Reception of Grassmann's Mathematical Achievements by A. Clebsch and His School." In Schubring, Ed. 1996, 117–130.

Thomaidis, Yannis. 2005. "A Framework for Defining the Generality of Diophantos's Method in 'Arithmetica'." *Archive for History of Exact Sciences* 59, 591–640.

Unguru, Sabetai. 1975. "On the Need to Rewrite the History of Greek Mathematics." *Archive for History of Exact Sciences* 15, 67–114.

United States Department of Education. 2008. *The Final Report of the National Mathematics Advisory Panel*. Washington, DC.

Vandermonde, Alexandre-Théophile. 1771. "Mémoire sur la résolution des équations." *Mémoires de l'Académie royale des sciences à Paris*, 365–416.

Van der Waerden, Bartel L. 1930–1931. *Moderne Algebra*. 2 Vols. Berlin: Julius Springer.

———. 1985. *A History of Algebra from al-Khwārizmī to Emmy Noether*. Berlin: Springer.

Van Egmond, Warren. 1978. "The Earliest Vernacular Treatment of Algebra: The *Libro di ragioni* of Paolo Gerardi (1328)." *Physis* 20, 155–189.

———. 1983. "The Algebra of Master Dardi of Pisa." *Historia Mathematica* 10 (1983), 399–421.

Viète, François. 1591/1968. *Introduction to the Analytic Art*. Trans. J. Winfree Smith. In Klein, 1968, pp. 313–353.

———. 1983. *The Analytic Art: Nine Studies in Algebra, Geometry, Trigonometry from the Opus Restitutæ Mathematicæ Analyseos, seu Algebrâ Novâ*. Trans. T. Richard Witmer. Kent, OH: Kent State University Press.

Vogel, Kurt. 1971. "Fibonacci, Leonardo." In Gillispie, Ed. 1970–1990, 4:604–613.

Volkov, Alexei. 2012. "Recent Publications on the Early History of Chinese Mathematics (Essay Review)." *Educação Matemàtica Pesquisa, São Paulo* 14, 348–362. (Online at revistas.pucsp.br/index.php/emp/article/download/12759/9350.)

Wagner, Roy. 2010. "The Nature of Numbers in and around Bombelli's *L'Algebra*." *Archive for History of Exact Sciences* 64, 485–523.

Wantzel, Pierre. 1845. "De l'impossibilité de résoudre toutes les équations algébriques avec des radicaux." *Journal de mathématiques pures et appliquées* 4, 57–65.

Waring, Edward. 1762. *Miscellanea analytica de aequationibus algebraicis, et curvarum proprietatibus*. Cambridge: Typis Academicus excudebat J. Bentham.

———. 1770. *Meditationes algebraicae*. Cambridge: Typis Academicis excudebat J. Archdeacon. English translation by Dennis Weeks: 1991. *Meditationes algebraicae: An English Translation of the Work of Edward Waring*. Providence, RI: American Mathematical Society.

Waterhouse, William C. 1979–1980. "The Early Proofs of Sylow's Theorem." *Archive for History of Exact Sciences* 21, 279–290.

Weber, Heinrich. 1882. "Beweis des Satzs, dass jede eigentlich primitive quadratische Form Unendlich viele Primzahlen fähig ist." *Mathematische Annalen* 20, 301–309.

———. 1893. "Die allgemeinen Grundlagen des Galois'schen Gleichungstheorie." *Mathematische Annalen* 43, 521–549.

———. 1895. *Lehrbuch der Algebra*. Vol. 1. Braunschweig: F. Vieweg und Sohn.

———. 1896. *Lehrbuch der Algebra*. Vol. 2. Braunschweig: F. Vieweg und Sohn.

Wedderburn, Joseph Henry Maclagan. 1907. "On Hypercomplex Numbers." *Proceedings of the London Mathematical Society* 6, 77–118.

Weierstrass, Karl. 1858. "Über ein die homogenen Functionen zweiten Grades betreffendes Theorem, nebst Anwendung desselben auf die Theorie der kleinen Schwingungen." *Monatsberichte der Königlichen Preussischen Akademie der Wissenschaften zu Berlin* 3, 207–220 (or Weierstrass, 1894–1927, 1:233–246).

———. 1868. "Zur Theorie der quadratischen und bilineare Formen." *Monatsbericht der Königlichen Preussischen Akademie der Wissenschaften zu Berlin* 3, 310–338 (or Weierstrass, 1894–1927, 2:19–44).

———. 1884. "Zur Theorie der aus n Haupteinheiten gebildeten complexen Grössen." *Nachrichten von der Königlichen Gesellschaft der Wissenschaften und der Georg-Augusts-Universität zu Göttingen*, 395–414 (or Weierstrass, 1894–1927, 2:311–332).

———. 1894–1927. *Mathematische Werke*. 7 Vols. Berlin: Mayer & Müller. Reprinted 1967. Hildesheim: Georg Olms Verlagsbuchhandlung and New York: Johnson Reprint Corporation.

Westfall, Richard S. 1980. *Never at Rest: A Biography of Isaac Newton*. Cambridge: Cambridge University Press.

Weil, André. 1984. *Number Theory: An Approach through History from Hammurapi to Legendre*. Boston/Basel/Stuttgart: Birkhäuser.

Weyl, Hermann. 1918. *Raum, Zeit, Materie: Vorlesungen über allgemeine Relativitätstheorie*. Berlin: Springer.

———. 1944. "David Hilbert, 1862–1943." *Obituary Notices of Fellows of the Royal Society* 4, 547–553.

Wilson, Robin. 2011. "Combinatorics: A Very Victorian Recreation." In Flood, Rice, and Wilson, Ed. 2011, 377–395.

Woepcke, Franz. 1853. *Extrait du Fakhrī, traité d'algèbre par Aboū Bekr Mohammed ben Alhaçan al-Karkhī*. Paris: L'Imprimerie impériale.

Wolfson, Paul. 2008. "George Boole and the Origins of Invariant Theory." *Historia Mathematica* 35, 37–46.

Wussing, Hans. 1984. *The Genesis of the Abstract Group Concept: A Contribution to the History of the Origin of Abstract Group Theory*. Trans. Abe Shenitzer. Cambridge, MA: MIT Press.

Yaglom, Isaak M. 1988. *Felix Klein and Sophus Lie: Evolution of the Idea of Symmetry in the Nineteenth Century*. Boston: Birkhäuser.

Index

Milton Keynes UK
Ingram Content Group UK Ltd.
UKHW021414080924
447947UK00004B/61

9 780691 149059